ELEMENTS OF TROPICAL ECOLOGY

Elements of Tropical Ecology

With Reference to the African, Asian, Pacific and New World Tropics

J. YANNEY EWUSIE

Professor at the Department of Botany
University of Nairobi, Kenya

HEINEMANN

Heinemann Educational Books Ltd
22 Bedford Square, London WC1B 3HH

IBADAN NAIROBI
EDINBURGH MELBOURNE AUCKLAND
SINGAPORE HONG KONG KUALA LUMPUR
NEW DELHI KINGSTON PORT OF SPAIN

Heinemann Educational Books Inc.
4 Front Street
Exeter
New Hampshire 03833
USA

ISBN 0 435 93700 6

First published 1980

Photoset in Malta by Interprint Limited
Printed in Great Britain
by Spottiswoode Ballantyne Ltd,
Colchester and London

Contents

List of Tables

List of Figures

Preface

Ecology has for many years been casually treated in most tropical school and university courses, usually being rushed through with only a handful of lectures and a few field excursions. Of late, however, there has been an awakening to the need for a sound knowledge of ecology in view of its repercussions on national economy, and there is now a sudden realization of the dearth of specialists in this field in many countries. In the temperate and advanced countries the literature in ecology has been reasonably well-developed but in tropical and under-developed countries this is not so. As a result, the recent realization of the importance of ecology has put a great strain on teachers who cannot get the material with which to teach the subject to the present and future generations.

It is true that ecology, like many other subjects which were first developed in temperate countries, has established a number of principles. Some of these principles do not seem to apply completely to the tropical situation and, even where they do, the examples described in the textbooks written for readers in temperate countries are within neither the experience nor the imagination of most students and teachers in tropical countries. In view of the characteristic features of the tropical environment, it should be possible for the tropical dweller to be able to relate examples from his own local experience to ecological data from other tropical countries. The occurrence of the annual severe winter of the temperate climate, as compared with the almost continuously warm weather of the tropical climate, affects differently the behaviour of plants and animals and this underlines the futility of applying, without due caution, temperate ecological experience to the tropical scene.

It is in order to provide the student in a tropical country with some general picture which will be near enough to his own experience and imagination that this work has been commissioned. It is hoped that the examples quoted from typical tropical places will serve as a guide to tropical readers to enable them to find suitable substitutes in their own local environment even though this may not always be easy. With these difficulties in mind I have laid emphasis on principles as much as possible. My task has been that of putting together some known data from various sources. I do not claim that much of the information is necessarily new. It is hoped that students in the secondary schools including the sixth forms, the training colleges and the first year of university will find the material useful.

It may be noted that botanical aspects of the subject would seem to have received more prominence than zoological ones. While much of this emphasis reflects the author's specialization, it is a fact that it is often the plants and plant products which provide for the presence and activities of animals, including man.

The book is presented in two parts. Part 1 consists of eight chapters aimed at discussing the basic principles, including those applicable to both terrestrial and aquatic situations. Part 2 deals with the structure of the broad climax communities or formations such as the tropical forests, savannas and deserts. Since the broad formations contain specialized minor communities, these have been considered here even though references have been made to them in Part 1 by way of illustrating certain basic principles. The effect of fire on the structure of tropical vegetation is so important that, again, even though this idea runs through certain sections of Part 1, it has been given a special treatment in Part 2 of the book to elaborate on its many effects. Finally, the activities of man in the destruction of vegetation and in the pollution of the environment logically end Part 2.

If this work succeeds in stimulating students and teachers in tropical countries to draw attention to some more interesting data of which I am not aware, to recognize and understand some of the problems with ecological implications which exist within their environments and to utilize ecological principles in trying to solve such problems and thus in the long run help to prolong the good health of their environment for mankind, then the effort put into this work will not have been in vain.

I wish to end by expressing my sincere thanks to my many colleagues for their help in various ways in making the writing of this work possible. In particular I should like to place on record the considerable advice I received from Dr C. D. Adams of the Department of Biological Sciences, University of the West Indies, St Augustine, Trinidad.

J. Yanney Ewusie

Acknowledgements

The author and publisher would like to thank the following for permission to reproduce photographs in this book.

Cadbury Ltd for Figures 20, 72 from *Cocoa Growers' Bulletin*, no. 18, Feb. 1972.

Cambridge University Press for Figures 53, 54 from *The Tropical Forest* by P. W. Richards, 1952.

Dr Forest Shrive for Figure 67 from *The Plant Kingdom* by W. H. Brown, Ginn, 1935.

Ginn for Figures 36, 56, 57, 59 from *The Plant Kingdom* by W. H. Brown, 1935.

Harper and Row for Figure 11 from *Plant Communities: A Textbook of Plant Synecology* by Rexford Daubenmire, 1968.

Inga and Olov Hedberg for Figures 48, 55 from *Conservation of Vegetation in Africa South of the Sahara*. Proceedings of the Sixth Plenary of AEFTAT, Uppsala, 1968.

Professor D. A. Herbert for Figure 66 from *The Plant Kingdom* by W. H. Brown, Ginn, 1935.

Journal of Ecology, 45(3) for Figure 60 from 'Ecology studies of tropical coast-lines. 1. The Gold Coast, West Africa' by A. S. Boughey, 1957.

Professor Lawson for Figures 13, 37, 42, 49 from *Plant Life in West Africa*, Oxford University Press, 1966.

University of Ghana for Figures 11, 64, 66, 68, 69, 70, 71 from *Agriculture and Land Use in Ghana* by Brian J. Wills (ed.), 1962.

Full details of the sources for the line-drawings follows.

Brown, W. H. (1935), *The Plant Kingdom* (London: Ginn).

Ewusie, J. Y. (1973), *Tropical Biology Drawings* (London: George Harrap).

Ewusie, J. Y. (1974), *Tropical Biology* (London: George Harrap).

Lawson, G. W. (1955), 'Rocky shore zonation in the British Cameroons', *Journal of the West African Scientific Association*, 1 (2), 78–88.

Lowe-McConnell, R. H. (1969), *Speciation in Tropical Environments* (London: Academic Press).

Maxwell, D. A. (1971), *An Introduction to Ecology for Tropical Schools* (Ibadan: Onibonoje Press).

Odum, E. P. (1959), *Fundamentals of Ecology* (Eastbourne: Holt-Saunders).

Owen, D. F. (1966), *Animal Ecology in Tropical Africa* (London: Oliver and Boyd).

Polunin, N. (1964), *An Introduction to Plant Geography* (London: Longman).

Richards, P. W. (1952), *The Tropical Rain Forest* (Cambridge: Cambridge University Press).

Wickstead, J. H. (1965), *An Introduction to the Study of Tropical Plankton*, Hutchinson Tropical Monographs (London: Hutchinson).

Wills, J. Brian (1962), *Agriculture and Land Use in Ghana* (London: Oxford University Press).

Part One Broad Ecological Tropical Features and Methods of Study

Chapter 1 An Introduction to Tropical Ecology

The Nature of Ecology

By ecology is meant the study of the reciprocal interactions of individual organisms, between and within populations of the same species, or between communities of different populations, and the many different abiotic factors which constitute the effective environment in which the organisms, populations or communities live. The effective environment includes the tangle of interactions between the living organisms themselves. The study of ecology enables us to understand the community as a whole by throwing light on such questions as how particular organisms behave in nature, how they live in particular places and what is responsible for their numbers. In order to ascertain these facts it may be necessary to conduct experiments in the field, in the laboratory or in both.

A basic assumption of ecology is that most modern plant and animal species have evolved modifications to their ancestral structures and physiological functions in such ways that enable them to live in each kind of environment. The development of a particular set of adaptive characters may make the organism less able than its ancestors were to live in a wide range of environments, but species which have evolved similar adaptations are often found growing together. For example, in a well-defined habitat like a mangrove swamp the populations of plants and animals which are found there are influenced by the fluctuating salinity of the water, the restrictive nature of the substratum, as well as the rather complex inter-relationships which exist between the other adapted plants and animals present.

Many branches of science, and sometimes other disciplines outside science, are bound to be employed in ecology. These include taxonomy, physiology, statistics, pedology, genetics, evolution and sociology. The student of ecology should be prepared to call on his knowledge of any of these disciplines at any time and to realize that ecology is more than a single subject. Above all it provides a way of thinking about all the other branches of biology.

Ecological studies may deal with any of the three levels of the organization of the organism: the individual, the population and communities. It deals also with the way in which the organisms at each of these levels of organization function in harmony with the non-living physical environment.

The Role of Ecology

Apart from being a good academic subject, like other subjects which exercise the mind, ecology has many useful applications aimed at maintaining a healthier and more productive biosphere for the life of man and other living organisms. Not least among the benefits of ecological study are the principles it provides for the judicious use of natural resources, often referred to as conservation. The importance of this has been more widely realized in recent years. The planning of the conservation policy of a country requires the proper integration of knowledge in a number of subjects each of which contributes some elements to applied ecology. Among these fields are agriculture, forestry, wild life and range management, irrigation, the construction of dams, pond management, the control of river pollution and marine fisheries. Knowledge of ecology is needed in all these fields and if those entrusted with particular engineering projects do not have such knowledge they would need to seek the advice of the ecologist.

Lack of this realization has led some of the advanced countries to execute large projects without adequate advice from the ecologist, resulting in serious ecological problems such as environmental pollution and the spread of disease vectors. It is now being realized that it is only through the well-planned management of biological resources that a level of productivity sufficient to feed the ever-growing world population can be maintained in the biosphere.

Levels of Ecological Studies

On the basis of structure, there are three levels at which ecology may be studied, as follows.

1. The Individual

The individual organism, be it a plant or an animal, is genetically a uniform entity. Together with its limited environment the individual forms an ecological unit. The study of the ecology of the individual is concerned with how a particular plant or animal interacts with its micro-environment. The factors of the environment of the individual affect its physiology, such as the supply of energy and raw materials. The way it adapts to the physical and biological factors of the environment is important. The study of the ecology of the individual organism yields useful information for building up a complete picture of the population of organisms of the same species. The study of the ecology of the individual, which may be carried out in the laboratory, zoo, greenhouse or garden, as well as in the field, can be rather time-consuming. Often not enough time is available for the student to do much work at this level of his ecological studies but at the research level intensive investigations of individuals may be vital to better understanding of ecological phenomena.

2. The Population

Groups of individuals of the same species, or of an interbreeding group, form a population. Since the species as a group is kept together by interbreeding, with its exchange of hereditary factors or genes, the species is said to have a common gene pool. The species may form relatively isolated groups comprising local populations. The same species may have more than one local population, each of which is adapted to its local environmental conditions. The small differences in local adaptation between populations are the basis for natural selection and hence evolutionary change.

3. The Community

The community in the ecological sense refers to an assemblage of populations of different species which occupy a given area.

A community is not necessarily a large tract of diverse vegetation with its equally varied animal species, or a mangrove swamp, or a lagoon. A community can in fact be of any size, even as small as a laboratory jar of water containing bacteria, fungi or protozoans. Even the soil by itself supports a community. Plant communities in the tropics are usually complex and cannot easily be named after one or two dominant species as is usually the case in the temperate regions.

One can, of course, distinguish and study separately if one wishes, distinct entities such as plant communities (phytocoenoses), animal communities (zoocoenoses) or communities of microbes (microbiocoenoses), etc. A collection of communities like these in a given place forms a **biome.**

Let us now consider some differences which may exist in the composition of the different entities being studied in ecology. Plants, for example, do not move about as animals do. This means that plants respond to environmental conditions by growth and structural changes, while animals respond by motion and behaviour changes. While the important plant species in a community can be quantitatively studied with precision, the animal portion of the same community may be less easy to observe, with the result that often only some discrete populations of animals are selected for more intensive study. It is not strange to find that the study of population ecology is more advanced with terrestrial animals, while community ecology is developed to a greater extent with terrestrial plants. Community studies are also better developed with both plants and animals of the aquatic habitat. It is not surprising either that the study of plant communities by itself can give a fairly complete picture provided that the effects of the animals on the plants are taken into account. This is because plants do not depend directly on animals for their food whilst the reverse is usually true.

The Ecosystem

The community and the non-living, or physical, environment function together as an ecological system referred to as the ecosystem. The ecosystem consists of the complex of the community of organisms together with the environment that interacts with it. The ecosystem as a concept is often described as the basic unit of ecology and it embraces every level of organization.

An ecosystem is never entirely stable but exists in a state of delicate balance. The way in which it operates is well illustrated by the cycles of essential elements like carbon and nitrogen which pass between the living and non-living states within the ecosystem. These elements are taken from the soil or atmosphere by the growing plant to form the needed compounds, but through food chains they are incorporated into animals which, by their death and decay, eventually release the original elements to the soil or air. Energy flow through the ecosystem is an even more significant feature. At the global level we can regard the entire biosphere as a giant ecosystem, the biosphere being the portion of the earth in terms of air, radiant energy, soil and the earth's crust which supports living organisms.

It is obvious that a full description of the structure and function of any ecosystem is a rather ambitious task, so that complete understanding seems to be an ideal which is still being pursued. Naturally, the logical approach is to study different aspects of the ecosystem separately. We can study the energy and mineral factors of the particular environment, then the plants and animals separately at first, before trying to synthesize the information from these sources. We can make studies at different ecological levels, as described above, that is of the individual, the population and the community.

The Tropical Environment

The tropics, as covered in this work, consist of the equatorial region and the regions which extend northwards and southwards approximately to the Tropics of Cancer (23½° North) and Capricorn (23½° South) respectively. But the tropics are not limited to these boundaries as far as plants and animals are concerned, because tropical plants and animals are found beyond these latitudes in some places. This happens where the flow of ocean currents is from the equator towards the poles, and the land masses whose shores they pass become warmer than they would otherwise be. An example is the Gulf Stream which gives the Bermuda Islands a warmer and hence nearly tropical climate. There are also areas outside the tropics where the vegetation is to all intents and purposes tropical rain forest. Examples of these are found in northern Burma, eastern Brazil and eastern Australia (Figure 1). It is worth noting that within tropical latitudes we often find, at higher altitudes, a contrary situation where a montane flora, consisting of plants otherwise typical of temperate latitudes, exists.

Between latitudes 23½° North and 23½° South, which form the principal limits of the tropics, the sun is overhead or at its maximum height in the sky twice a year, whereas it is so only once outside the tropics. The angle of incidence of the sun's rays is also steeper than in the temperate latitudes (Figure 2) so that light is more evenly spread on the ground in the tropics and the direction of the slope of the land has much less importance than in temperate latitudes. The result is fairly high temperatures in virtually all tropical areas throughout the year. This situation is conducive to constant metabolism. The annual range of temperature is far less variable and therefore less seasonal than in the temperate regions. The range of day lengths and night lengths within the tropics is also less than in temperate latitudes. Although some temperate plants are found on tropical mountains they do not experience the same extremes of long summer and short winter days that they would experience further from the equator.

1 (over) The world's distribution of tropical and sub-tropical rain forests
SOURCE: R H Lowe-McConnell (1969)

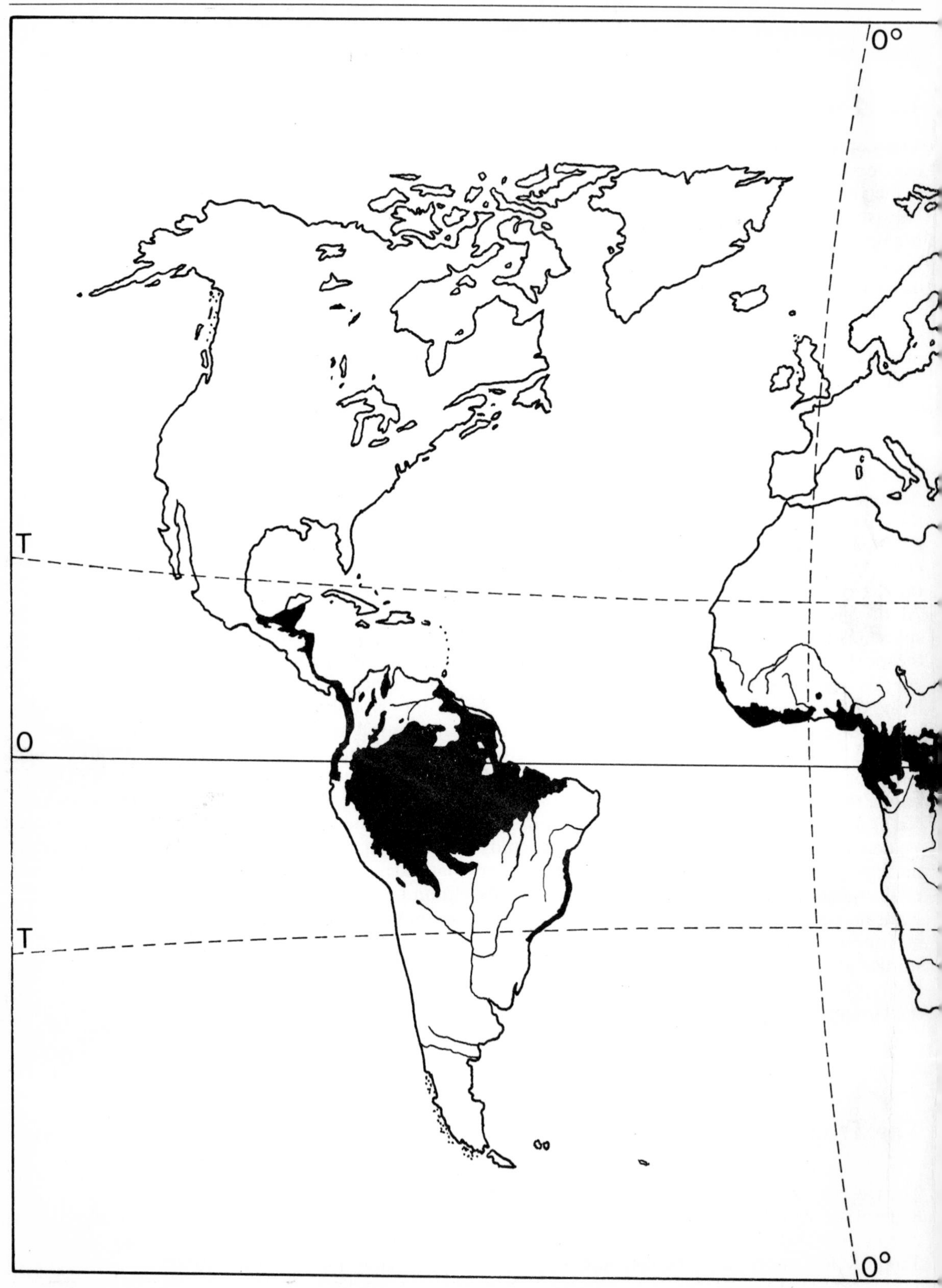
0°
T
O
T
0°

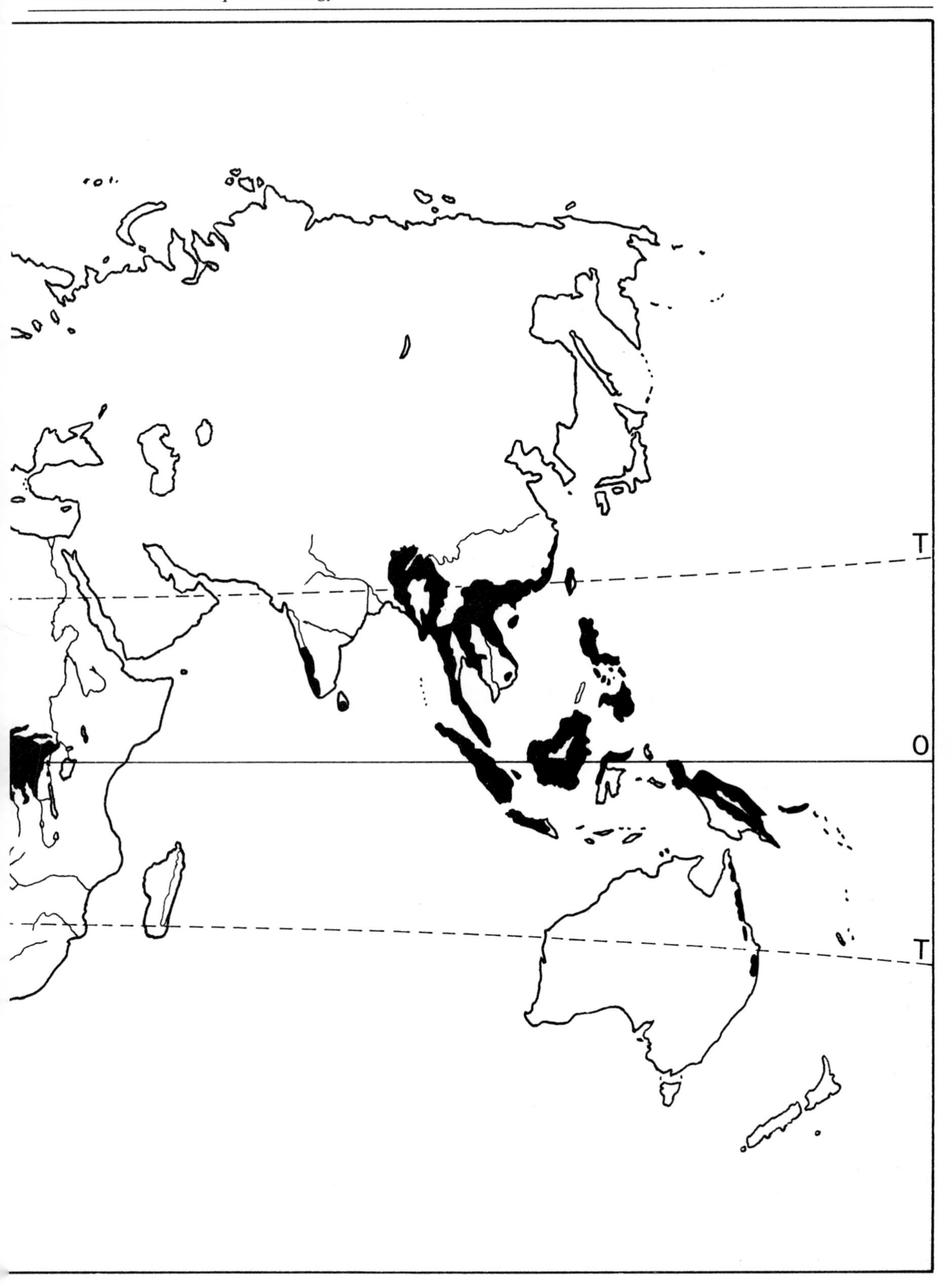
T
O
T

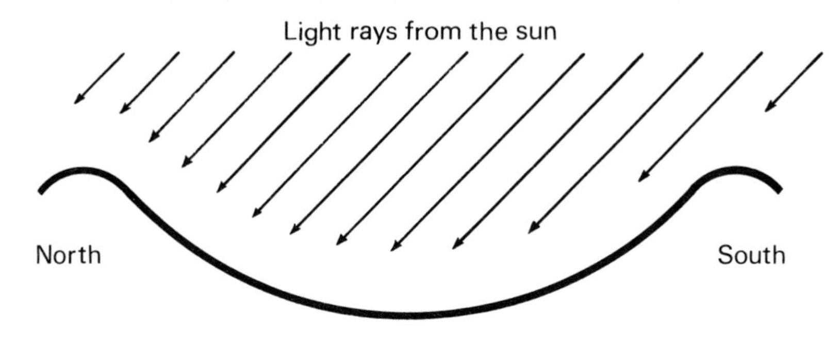

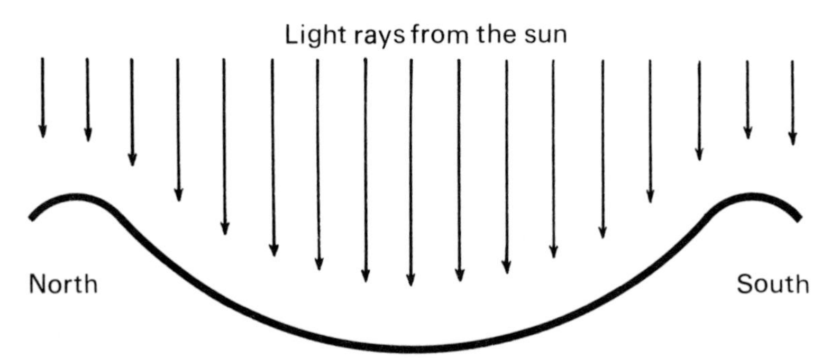

2 Diagram to show the angle of incidence of the sun's rays in temperate and tropical latitudes
SOURCE: G W Lawson (1966)

Some Ecological Features of Tropical Areas

1. Flora and Fauna Generally Luxuriant Throughout the Year

In the tropics, except for the deserts, much of the environment maintains a fairly high temperature throughout the year and the temperature in the lowlands never falls as low as in the temperate climates in winter. There are no extreme seasonal changes and most of the flora and fauna generally remain luxuriant throughout the year. The plants may be subjected to reduced activity at the dry periods of the year but they are not adapted to the prolonged low-temperature dormancy that is found in temperate plants. This condition is typified by the tropical rain forest of the equatorial belt. Also, because in the temperate regions, growth nearly ceases in the winter the trees show marked annual growth rings, but in the tropics, as may be expected, these rings do not always have the same relation to the number of years or age of the plant. Flowering and fruiting of different species occur throughout the year in the tropics and with these a wider variety of animals also seems to have evolved in accordance with enhanced opportunities for feeding and their associated roles as pollen vectors and seed dispersers.

Lowe-McConnell (1969*a*) states that unlike temperate birds, tropical birds of one species or another nest nearly the whole year and no bird breeding is interrupted to avoid an off-season. Thus birds, other animals and plants indicate by their responses that the tropical climate is much less seasonal than the temperate climate.

2. Rainfall as a Limiting Factor on the Biota

Rainfall seems to be the major climatic factor in the tropics, just as temperature is in the temperate areas of the world. The amount and distribution of rainfall are important in determining the principal vegetation types. A minimum annual rainfall of 1300 mm, evenly distributed throughout the year, will support tropical rain forest. Rainfall below about 1100 mm with seasonal incidence gives rise to savanna vegetation, while 250 mm or less, falling irregularly, is typical of tropical deserts. Again in the tropics a number of species are seen to be adapted to the rains either for flowering or for the 'flushing' of new leaves.

3. Some Areas Have More than One Period of Rainfall Associated with More than One Period of Flowering and Fruiting

Except on the equator proper, and the near-desert and desert areas which have one peak period of rainfall a year, much of the intervening latitudes experience two peak periods with corresponding intervening dry periods. In these areas the two periods of rainfall are reflected in the frequency of flowering and fruiting which is twice a year in a number of tree species (Ewusie, 1969). Other plants flower even thrice a year while a number of species may flower at any time in the year. This contrasts sharply with the temperate climates where most plants flower and fruit only once a year with special reference to the summer periods. This multiple flowering accounts partly for the greater annual productivity of tropical vegetation when compared with temperate vegetation in which most of the species find the winter critical.

4. High Rate of Decay and Decomposition

The relatively high and constant temperature of the tropical areas is often accompanied by high humidity as is typically found in the tropical rain forest. It is only in the drier savannas and the tropical deserts that humidity is low. The combination of high temperature and high humidity creates an environment conducive to high metabolic activity in organisms like bacteria and fungi which are responsible for the rapid decay of dead plant and animal materials. So the tropical rain forest is marked not only by active growth but also by active decay.

5. Richness in Species

The tropical flora and fauna are typified by the extraordinary numbers of species of most taxa when compared with temperate flora and fauna. Thus there is more uniform relative abundance of species (which means less conspicuous dominance). There are, of course, some areas of apparently uniform vegetation and hence distinct dominance, such as the freshwater swamps of Africa, the Mora Forest of Trinidad or the peculiar rain forest of the Amazon basin, but when compared with temperate vegetation these communities still have a much greater number of tree species. In Europe or North America, the forest may have only one, two or very few species making up as much as 90 per cent of the trees. Where only one species dominates, the community is referred to as a consociation and receives its name from that species; for example, pine forest (*Pinus sylvestris*) in Scotland. If, as seems often to be the case, two species dominate together, the community is referred to as an association; for example, the beech – maple forest of North America. In contrast, tropical forest rarely has so few species dominating it. In Trinidad the richness of species of trees which can grow to large sizes results in the evergreen seasonal forest there having, in four typical examples, as many as 23, 25, 31 and 33 species respectively reaching the emergent layer (open top stratum). The species are not equally represented numerically, but all individuals are structurally of equivalent status. The idea of the association is very difficult to apply to forests of such complexity and diverse composition, although ecologists from temperate countries have attempted to do so. This has often resulted in some sort of compromise being reached between floristic and abundance estimates which may change rapidly from place to place. The cause of the tropical diversity in flora and fauna has been the subject of several hypotheses.

6. A High Proportion of Woody Plants

In the tropics the vegetation seems to contain a high proportion of woody plants, the exceptions being, of course, the drier grasslands and the hot deserts. In the forests in particular, apart from trees being predominant, most of the climbers (lianes) and a number of the epiphytes are woody. Families of plants which are represented by woody species in the tropics are found to be largely herbaceous in the temperate climates. This is illustrated in a comparison of life forms of plants in Chapter 3. The woodiness of the tropical vegetation has resulted in the prominence of epiphytes, and some arboreal animals in the tropical biome.

7. Richness in Habitats and Ecological Diversity

The tropical environment is typified by the wide range of plant and animal habitats and micro-habitats. For example, while a temperate woodland often consists of one layer of trees with shrubs and a ground flora, a moist tropical forest has not only three storeys of trees but also subordinate communities (synusiae), comprising vines and epiphytes which increase further the micro-habitats in the tree storeys, as well as shrubs and a ground flora.

It is not only in lowland forest micro-environments that habitat diversity occurs. There are extensive habitats where unusually low temperatures occur within the tropical belt. As mentioned earlier with respect to the mountains reaching about 4000 m in the tropical latitudes, low temperatures are accompanied by the presence of plant species which are otherwise typical of temperate climates. Together with some common tropical plants these species form a montane flora which has features of temperate vegetation. The higher humidity of the tropical rain forests contrasts with the relative aridity of the temperate habitats.

In summary it may be said that the tropical environment and its flora and fauna are typified by different levels of variety and probably greater annual productivity than are found elsewhere. There is a greater variety of species and micro-environments. The effect of topography is seen in the appearance of typical temperate plants on tropical mountains. Warm areas of high humidity with luxuriant growth and rapid decay contrast with those of extreme dryness or cold where activity, rather than being prolonged and continuous, may be sudden and short-lived.

Chapter 2 Tropical Populations

Extent of Species Diversity

Before considering the characteristics of populations it is worth emphasising the extent of species diversity of tropical populations. Diversity means the condition of being different or of having various differences in forms or qualities. Diversity of species in the tropics can be seen at two levels, namely, large numbers of species of similar life form and the presence of many species of widely different life forms not found in other parts of the world.

On the question of diversity of number of species, it is observed that in most temperate regions there are usually about 50 species of trees and shrubs per hectare in a woodland. Even in eastern North America, where relatively more species are often found, about 100 or 150 species are found per hectare. In tropical lowland evergreen forests, 750 or more species may be found per hectare, with the greatest numbers occurring in the forests of tropical Asia and South America. There is little doubt that the tropical forest supports a much greater number of species than the tropical savanna. Richards (1964) and others have shown that individually, tropical species are relatively rarer in their communities than temperate species, but this is not always so because in some situations tropical species show local dominance. Holdridge (1967) has shown that decreasing diversity in terms of number of species, tree heights, basal area and stem numbers sometimes found in tropical vegetation may be related to increase in altitude and reduced rainfall.

Where there is a great diversity of plant species there is often a large number of animal species. This is because each species of animal may depend in one way or another on a select group of species of plants for food and other requirements. For insects which feed on trees, it can be seen that the number of insect species in the community is more closely correlated with the number of plant genera (even though not with the number of plant species) present. The layering of tree species in tropical forests by itself appears to contribute to an increased animal population. This is found even in the rare forests of single species dominance. It is only freshwater invertebrates and diatoms that do not seem to show an increase in diversity of species in the tropics over their temperate counterparts (Patrick *et al.*, 1966).

In the case of fishes, the richest freshwater fish faunas have been found in the Amazon, equatorial Africa and tropical Asia. With even the incomplete data that are now available as many as 1383 species belonging to some 46 families are recorded from tropical Brazil, as opposed to 339 species in subtropical Argentina. It is remarkable that in one haul of a 45 metre seine, 70 to 90 species of fish were obtained from the Essequibo (Eigenmann, 1912). Two areas in Zaïre (formerly the Congo) in Central Africa are reported to have about 408 species of fish belonging to some 24 families. In tropical Asia 546 species representing 49 families have been reported from Thailand. When it is realized that the whole of Europe has only 192 species of fish belonging to 25 families and that the Great Lakes of North America have about 172 species representing 29 families, then the richness of the tropics in fish populations becomes quite evident.

The second type of diversity, namely the unusual species, is often shown by the ways in which species that feed on different foods show adaptation of the relevant parts of the body. This may be illustrated by tropical birds of prey which differ so much in form and structure, each revealing a different adaptation to cope with the food or prey.

A number of suggestions have been put forward to explain the high species diversity of the tropical environment, but before coming to discuss these, it is necessary to explain certain ecological terms, namely the habitat, the micro-habitat and the ecological niche.

The Habitat, the Micro-habitat and the Ecological Niche

The Habitat

The environment consists of the physical and biological factors of the surroundings of an organism, a population or a community. The sum total of the conditions of the environment at a particular locality may be referred to as the habitat. However, the term is sometimes used also in a more general sense to denote areas with similar physical and chemical composition and which sustain life of a certain kind. It is only in this broad sense that we can recognize, say, the marine, the freshwater, the terrestrial and the arboreal habitats. We sometimes speak also of a habitat factor of a given locality when referring to the environmental factors of that locality.

While we find that some plants grow over a wider variety of environments than others, most plants tolerate a certain rather narrow range of environmental conditions. Thus, for most plants and animals, if the conditions of their habitat happen to change to become conditions which they cannot tolerate, they may die or fail to grow healthily. Many tropical forest herbs which do not grow so well under the forest canopy are found to grow more vigorously in places where there is more light, as in clearings, roadsides or along river banks in grassland areas, showing that under the forest canopy, low light intensity is a limiting factor to the growth of these species. Some species may even fail to reproduce or may reproduce rather poorly; the seeds produced may not be able to germinate at all or they may show a very reduced percentage of germination.

It may be recalled that in the definition of ecology given at the beginning of this book stress was placed on the factors which constitute the effective environment. This is because some factors of the environment are more important in particular habitats or at particular times than in other habitats or at other times in the same habitat. Salinity and waterlogged soil conditions, for example, are more effective controlling factors than, say, temperature in the mangrove vegetation. Also, a large deciduous tree such as *Antiaris africana* or *Ceiba pentandra* may mean shade as well as shelter against heavy rains to plants growing below it during the period of the year when it is in full leaf. When the leaves are shed at another time of the year, these restrictive factors for the plants growing below it are removed and more sunlight reaches the ground.

In considering the habitat, it should always be recognized that the soil is one of the important factors. Environmental factors are further discussed in Chapter 4 and soils in Chapter 8.

The Micro-habitat (Micro-environment)

In a community which has the same general habitat, it may happen that some of the organisms are in fact living under conditions which differ markedly from the general one, such as a crab bore-hole in a coastal scrub or a termite mound in either a grassland or a deciduous forest. The local environmental factors are evidently different from those of the rest of the wider habitat. Within termite mounds, for example, humidity is much higher and temperature less variable than in the surrounding habitat. These small areas are often referred to as micro-environments, micro-climates or micro-habitats.

The importance of appropriate micro-climates for some organisms is illustrated by the behaviour of two species of mosquitoes in the rain forest of Trinidad. Here it is found that the mosquito species *Anopheles homonculus* requires a more humid atmosphere than *Anopheles billater*. The former always remains nearer the ground than the latter since there is more moisture lower down. As the humidity decreases with the rising of the sun in the morning to about midday both mosquito species move higher in the forest, but maintain their relative positions. After midday, when the humidity increases again, they move downwards. In this way these mosquito species try to remain within a fairly constant micro-environment.

The Ecological Niche

Unlike the habitat or micro-habitat an ecological niche does not refer to a specific place. It is to be considered rather as the sum total of the many effective environmental factors available to or used by an individual organism or a population during its life-span. We can also use the term to cover the requirements of the species at a particular point in time. To illustrate, even though a plant remains in the same place during its life, the effective combination of factors of the environment of the seedling may be quite different from that of the full-grown plant of the same species, and often when the plant flowers the environmental factors may also be different from the pre-reproductive phase. During flowering the species may depend on particular agencies for pollination and on different factors for fruiting and seed dispersal. Dry weather, which may be necessary for the dispersal of seeds, may be utterly unsuited to the growth of seedlings or the development of fresh leaves in the case of most perennials. It is the combination and sequence of these effective factors as needed by a species at different points in time in its life-history which constitute its ecological niche. A general example may be shown by the sequence of conditions necessary for the establishment of a fern prothallus, the fertilization of the archegonium, the growth of the sporophyte and the shedding of the spores.

Although for plants a niche is experienced only in one place, for animals it may be experienced in different places by the same species as a result of migration. For example, the larvae of an insect like the mosquito live in water and being primary consumers feed on small diatoms, while the adult mosquito flies in the air and sucks blood from man or animals. A tadpole lives as a primary consumer by feeding on plants like *Pistia* while the adult frog lives as a secondary consumer by feeding on insects and other small animals.

An ecological niche at a particular point in time in the life of a species is regarded by some authors simply as the status of a species in its community. This is taken literally as reflecting the influence which the species exerts on other biological and physical factors of the habitat. In addition, it is reflected in what influence these factors in turn exert on the species.

In a given community, or a given habitat, different species have their own ecological niches, so that a community would contain as many niches as there are species comprising it. This means that only one species is theoretically associated with one ecological niche. If a habitat is not occupied by any species, as may be found on a fresh volcanic island, it may be said to have no ecological niches. For example, in the same pond of water different species of water-bugs may play different roles: *Notonecta* or the backswimmer feeds on other insects of its size, while *Corixa* feeds on dead and decaying remains of other organisms. A given plant community may have all its niches occupied when it may be supposed to have attained its climax development. In theory in such a community, a new species could only invade or rise to prominence by displacing an already existing one.

Hypotheses for the High Species Diversity of Tropical Communities

Here are some of the hypotheses that have been put forward to account for the high diversity of species in tropical communities:

1. Tropical species can better tolerate overlapping ecological niches, the latter being considered to be a feature of the variety of micro-environments of the complex tropical habitat, particularly in forests. In this way, it is argued, similar species have less exclusive requirements. This in effect means that competitors could, as it were, be packed closer together (McArthur and Levins, 1967). We should, however, take into account the possibility of the same niche being occupied by different species at different times of the year and at different stages in the development of the community. As an example one may consider the scatter in time of flowering and fruiting periods of species in the tropics (see Chapter 5).

   Although theoretically no two species may occupy the same niche, yet there are situations where it is observed that species populations may utilize the same territory. For example, in some of the East African game reserves it is common to find three predator species, the lion, the leopard and the cheetah, co-existing within the same environmental complex but feeding on different animals. The lion may feed on buffaloes, the leopard on baboons and monkeys, and the cheetah on young waterbuck and other smaller animals like antelopes. This phenomenon by which the same 'territory' is utilized by co-existing species is sometimes referred to as niche diversification. The same phenomenon also operates through the utilization of the same resource by different species populations at different seasons or at different times in the day. The dung of elephants and buffaloes, for example, is utilized by the tropical scarab beetles of Africa in the rainy season and by termites in the dry period. Diurnal squirrels feed on the same food in the day as is used by nocturnal rats.

Lowe-McConnell (1969*b*) has shown by her studies on tropical freshwater fishes in Africa and in South America that there is often much overlap in the food eaten by co-habiting species in lakes as well as in rivers. Her findings showed that two or more species of fish share the same niche if factors other than food keep the numbers below the level of competition for food. She conceded, however, that it is unlikely that two species would share the niche at all times of the year or throughout their lives.

2. Another hypothesis tries to account for the greater diversity of tropical species in terms of availability of resources, especially of food. It is argued here that since the rate of production is high in the tropics there are always food resources adequate for new species which are thus enabled to enter the community (Odum, Cantlon and Kornicker, 1960). It is further suggested that there may be a feedback, as more species cause more stability, so that up to a point more species would be enabled to enter (Cornell and Orias, 1964).

   It is important to demonstrate the more productive nature of the lowland tropical environment compared to that of the temperate regions. This may be done by showing that evapotranspiration is distinctly greater in the tropical lowlands (Holdridge, 1967) than in the lowlands of temperate regions (Rozenweig, 1968). (It has been shown that actual evapotranspiration, that is, total evaporation plus transpiration, is given as precipitation less percolation and run-off.) It appears that the tropical lowland is more productive than the temperate zone, where the winter is rather unfavourable to the growth of many organisms.
3. Yet another hypothesis suggests that the high incidence of predation and parasitism in tropical environments tends to limit the abundance of any given species and thus makes it difficult for many species to increase their densities. In that way, more species can fit into the habitat. This hypothesis would seem to account more for the low numbers of particular species than for the abundance of species in the community.
4. Finally there is the hypothesis which suggests that the tropics have offered more opportunities for speciation in the past, but that these species have been slow in adapting to climates that include a winter (Fischer, 1960). Although this last hypothesis appears to be at variance with the others which predict a saturation with species, it would appear that the hypotheses enumerated above are not entirely mutually exclusive. For example, it is possible to find in a habitat some taxa whose presence may be explained on historic grounds while others could be justified on ecological considerations. Again, it is possible that some taxa in a habitat may continue to increase in diversity while others may have already reached species saturation.

In her efforts to integrate the above hypotheses and situations Lowe-McConnell (1969*a*) puts forward a new hypothesis. It states that species interactions in the form of competition and predation are important and influence species existence and abundance. According to this hypothesis the tropical environment is not to be likened to a box that will hold only so many eggs but rather should be viewed more like a balloon which resists further inflation proportionately to its present contents and which can at the same time always hold a little bit more if necessary. Furthermore, the contents of the balloon always try to escape to regions of lower pressure. Assuming a global equilibrium in species diversity, then one would say that it is achieved by a balance between speciation and immigration on the one hand and extinction and emigration on the other. If the tropical inflow of species were greater, or its outflow were less, then it would equilibrate with more species.

In conclusion it is worth referring also to MacArthur (1965) who maintained that the number of species within a habitat can be expected to increase with productivity, structural complexity of the habitat, lack of seasonality in resources, degree of specialization and reduced number of individuals within a species. He observed that there is no shortage of these potential causes of species diversity in the tropics. He therefore suggested that any combination of these rather than only one of them would operate in any situation.

Some Features of Populations

Having considered the high species diversity of the tropics, we may now consider the basic features of populations. As has been evident from the above, a population is defined as a group of organisms of the same species occupying a particular space. Even though the term can embrace varieties, ecotypes or other groups which may have ecological unity, a population has certain unique characteristics which are distinct from, and others that are additional to, the general characteristics of the individuals which make the group. Among the characteristics which the population shares with an individual are the fact that it has a life-history as seen by the fact that it grows, differentiates and maintains itself and that it has a definite organization and structure which can be described in the same terms as the individual.

The group characteristics of the population include features like natality or reproductive rate, mortality or death rate, sex composition or reproductive system, age structure, distribution and social structure. These will now be considered in detail.

Natality and Reproductive Rates

In ecology the term natality is used to refer to the inherent ability of a population to increase. Thus natality rate is the same as birth rate in normal language and refers to the rate of production of new individuals of any organism. Natality rate is, therefore, always positive or at least zero, but never negative. For every population there is theoretically a maximum number of new individuals that it is capable of producing under ideal conditions when there are practically no limiting factors. This is the maximum natality of the population. Since in nature limiting factors are never absent, this maximum natality is probably never achieved. What is attained in practice is referred to as the realized natality or the ecological natality or simply the natality of the organism in a particular habitat. This varies with the size and composition of the population and the physical environmental factors.

In practice, maximum natality can be approximated by the average reproductive rate of a population when placed in a favourable environment or when the major factors that are limiting to the growth of the population are temporarily inoperative for brief periods or seasons. It may also be determined from the highest average number of offspring achieved by a group of the species in a series of experiments. It is statistically invalid to determine it on the basis of the performance of any single individual of the species. The value of maximum natality lies in the fact that it provides a yardstick for comparison with realized natality, and as such it is necessary always to state the conditions under which the estimate is made.

Natality is normally expressed as a rate determined by dividing the number of new individuals produced by the time within which they are produced. It may also be expressed as the number of new individuals per unit of time per unit of population. It is thus clear that the kind of birth rate that is being measured, which depends on the available experiment, ought to be clearly stated, and this is better achieved by using mathematical notations.

Thus

1. $\Delta Nn =$ natality, here defined as the production of new individuals in the population,
2. $\frac{\Delta Nn}{\Delta t} =$ natality rate per unit time,
3. $\frac{\Delta Nn}{N\Delta t} =$ natality rate per unit time per individual,

where N represents the total population, and Nn the new individuals produced. It is important to note that in particular situations it may not be correct to calculate N on the assumption that N represents the total population because it may only represent the reproductive part of the population, as is the case with populations of higher organisms where only the female part of it may be reproductive. Dasman (1964) reports, for example, that in Rhodesia species of impala had an annual production of 50 yearlings per 100 adult females.

It is necessary to show how natality differs from growth rate. As mentioned above, natality rate is always either positive or zero but never negative. This is because, as a parameter, it represents only the net increase, if any, in the population. It does not take into account factors like mortality, emigration or immigration as apply in the case of growth rate. Thus, unlike natality, growth rate may not only be positive or zero but may indeed be negative.

It is, of course, important to ensure that in comparing natalities of different species populations, particularly those with complicated life histories, comparable stages in the life history are used.

Mortality and Death Rate

In ecology the term mortality is used to refer to the inherent ability of a population to decrease through death. It is thus the antithesis of natality and so has specific definitions as has natality. Thus mortality rate $\Delta Mn/t$ may be expressed as the number of individuals dying in a given time, while

$\Delta Mn/N\Delta t$ is the mortality rate per unit time per individual of the population. Theoretically there is a minimum mortality for every population which represents the loss under ideal or non-limiting conditions. This means that even under the best of conditions parts of the population will die.

In most populations in nature the average longevity is far less than the maximum life span. As a result, the actual or specific mortality rates are often greater than the minimum. The average longevity can be determined in the field or under various ecological environments of a population and used as a basis for comparing different habitats.

Ecological or realized mortality is the average loss of individuals of a population under a given set of environmental conditions. Generally, however, the term specific mortality is used to express mortality as a percentage of the initial population dying within a given time.

One way of obtaining the full picture of the changes in mortality with age is to estimate specific mortalities at different stages of the life history of the population. This complete picture is presented in the form of a life-table which shows the numbers of individuals out of a given population (1000 or any other convenient number) which survive after regular intervals of time, the numbers that die during successive time intervals, the mortality rate during successive intervals (in terms of initial population) and the life expectancy at the end of each time interval.

Data from the life-table are often used in producing survivorship curves that are very instructive. For example, if the numbers of individuals out of a given population which survive are plotted on the *y*-axis against the time interval of the *x*-axis, the resulting curve is called a survivorship curve. By using a semi-logarithmic plot with the time interval expressed as a percentage of the mean length of life, it is possible to compare the survivorship curves of species populations of different life spans. Various studies along these lines have shown three basic survivorship curves as shown in Figure 3. One type is found where the average physiological longevity is realized; here all individuals will live out their life span and die almost at the same time. This situation is shown by a highly convex and sharply-angled curve. This is exemplified in Figure 3 by the experiments on fruit flies starved on emerging from the pupa. Another type is that of a straight diagonal line as shown for *Hydra* on Figure 3, which results where the specific mortality rate at all ages is constant. The third case is a concave curve and this results from a high mortality during the young stages, as found in the oyster. It should be mentioned that population density can affect the eventual shape of the survivorship curve in that high density, or over-crowding, reduces the life expectancy of the population. It is evident that survivorship curves are useful also in providing information on population dynamics. For example, they make it possible to determine the most vulnerable age periods of a population during which a change in mortality can have marked effects on the population.

Although both mortality and natality are important in determining population size and trend, various studies seem to suggest that mortality is more affected than natality by various environmental changes.

Reproduction and Mortality Rates in the Tropics

Reproductive rates depend on the number of progeny produced at a time, the length of the breeding season and the age at which breeding takes place. In general, tropical passerine birds lay smaller clutches of eggs as compared with those of similar species in the temperate climates. While tropical passerines often have clutches of two, similar temperate species have five or six. This is explained on the basis of a greater but temporary availability to the young birds of food during the long daylight hours of the summer in the temperate climates, as opposed to the relatively lower and uniform availability of food at all times in the tropics. The absence of long hours of daylight for feeding in the tropics, which is supposed to affect the clutch size, is found not only in the lowlands but also in the highlands, even though the species diversity on tropical highlands is similar to that of temperate climates.

Numbers of animals, in general, are not controlled so much by reproductive rates as by death rates. In the tropics it is found that the bird nest mortality rates are higher than in the temperate climates. An example of this is the red-winged blackbird (*Agelaus phoeniceus*) which experiences higher nest mortality in the tropical Costa Rica than in the temperate Washington State.

The reproductive rate realized by a population is also much influenced by the relation of the species to its environment. Laws (1966) has shown that the difference between the density of the elephant population on the south (1 per km^2) and the north banks (2 per km^2) of the Victoria Nile in the

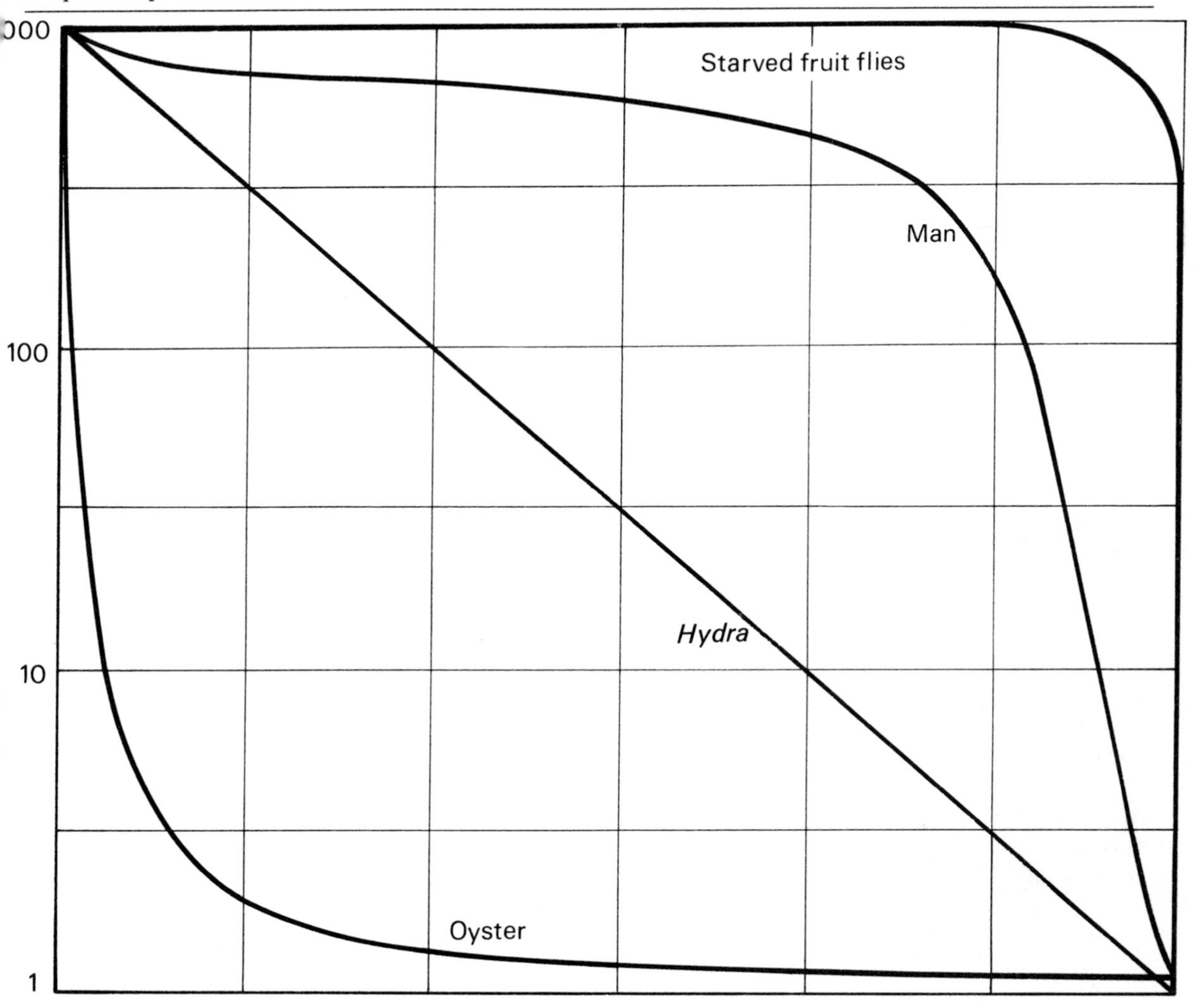

3 Types of survivorship curves plotted on the basis of survivors per thousand log scale (vertical coordinate) and age in relative units of mean life span (horizontal co-ordinate)
SOURCE: E P Odum (1959)

Murchison Falls National Park (Uganda) is due to the state of the vegetation of the two areas. On the south bank the vegetation is much destroyed whereas that on the north bank is less so. The age of first breeding is later on the south bank and the young do not survive as well. As a result the reproductive rate on the south bank has dropped by 30 per cent, but by only 10 per cent on the north bank.

Age Structure

Nearly every population has three ecological ages, namely pre-reproductive, reproductive and post-reproductive. These refer to the relative duration of each of these stages in the life spans of populations. The actual periods vary considerably in different species. Many plants and animals have a rather long pre-reproductive period and some animals, especially insects, have extremely long pre-reproductive periods. Wild animals characteristically show very rapid growth rates and early maturity compared with domestic stock of similar weight in similar areas (Talbort *et al.*, 1965). The duration of the ecological ages has a bearing on the age distribution of a population. It should also be noted that some tropical species have no post-reproductive stages, since the individuals die shortly after reproduction. Examples are the sisal plant (*Agave sisalana*) and the chameleon (*Chameleo bitaeniatus*).

Reproduction is often restricted to the middle and older age groups, especially in higher animals and plants. The ratio of the three age groups in a given population at any particular time is an index of the reproductive status of the population at that time and it enables the ecologist to make projections about the future development of the population. Roughly three types of age structures can be seen. These are (a) a rapidly expanding population which usually contains a large proportion of young individuals; (b) a stationary population with a more even distribution of age classes; and (c) a declining population with a large proportion of old individuals.

Data on the age structure of a population are often presented in the form of an **age pyramid**. This is done by showing the number of individuals or the percentage in the different age classes in terms of the relative widths of successive horizontal bars. Figure 4 illustrates three theoretical forms of the pyramid representing the three types of age structure mentioned above: (a) is a pyramid with a broad base indicating a high percentage of young individuals; (b) is a bell-shaped polygon for a moderate proportion of young to old individuals; and (c) is an urn-shaped figure for a low percentage of young individuals, characteristic of a senile or declining population. These types of age structure may be exemplified respectively by a fresh growth of grass after burning, an intermediate stage after about five months, and a final stage in the dry season before the next burning.

In spite of these three age structures, it is found that every population has a characteristic age structure towards which actual age distribution progresses, and, once this is reached, only temporary changes arise through changes in natality or mortality so that between these the population reverts to its stable age structure, unless the upset is too drastic. The idea of a stable age distribution provides a base for evaluating the actual age structure of a population in nature. This is in line with the concept of population as a biological unit with definite biological constants and definite limits to variations that may occur around these constants.

In temperate climates, because of the distinct annual cycle of extremes of temperature between summer and winter, a number of organisms have features that help to determine the age of the individuals. Thus fish, for example, show growth rings on the scales or other hard parts and trees show annual rings. In the tropics, however, these may not always be distinct because of the absence of temperature extremes. In tropical trees where seasonal rings are distinct they appear to be related not only to bursts of growth but also to other phenomena in the plant, such as the number of times in a year it sheds its leaves or produces flowers. In mammals, various structures are used in determining age including tooth wear, tooth replacement, the layering of cementum in teeth according to season and the growth and wear of horns (Mosby, 1963 and Mitchell, 1965).

Biotic Potential

It is worth considering now how natality, mortality and age structure can be considered together to provide a picture of how the population as a whole is growing, as well as its best possible performance with which we can compare its normal performance. The concept of the biotic potential is useful provided that (a) the average natality and

4 Age pyramids depicting the percentage of population in the different age classes a) High percentage of young individuals b) Moderate percentage of young individuals c) Low percentage of young individuals
SOURCE: E P Odum (1959)

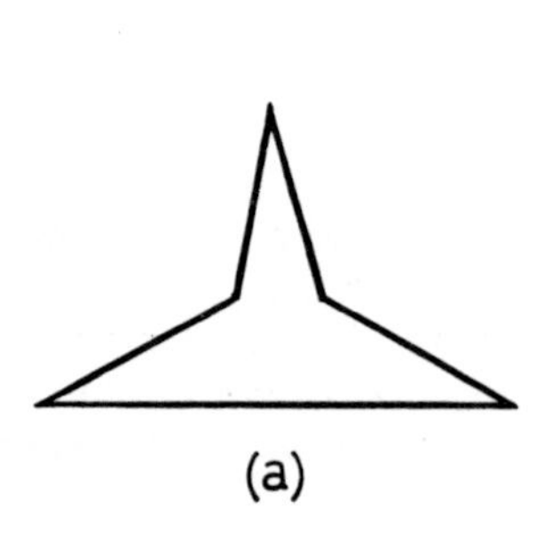

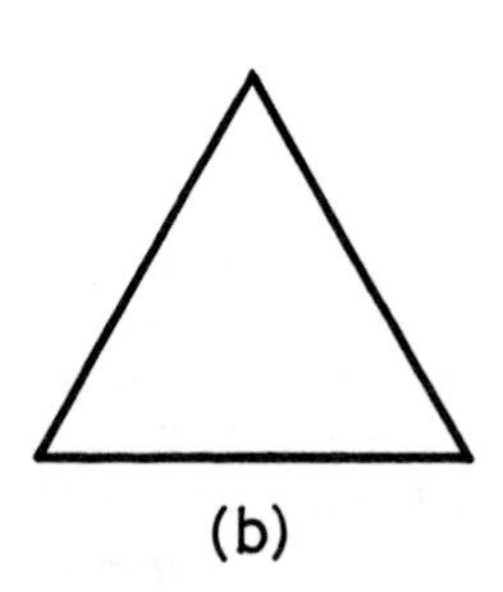

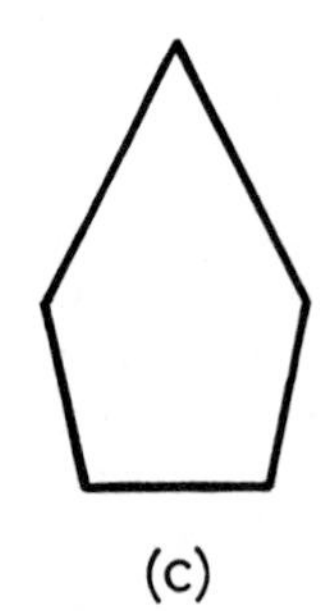

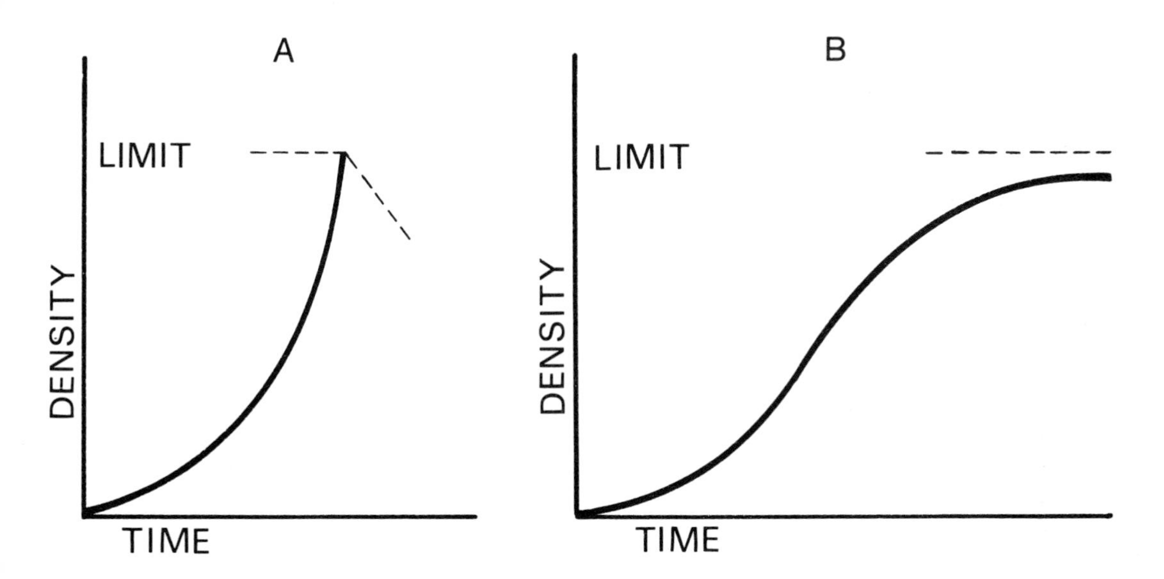

5 Two basic patterns of population growth form, showing the J-shaped (exponential) A, and the S-shaped (sigmoidal) B forms
SOURCE: E P Odum (1959)

mortality remain the same for any specific age group when environmental conditions are optimum and (b) that a population with constant age schedules of natality and mortality in an unlimited environment assumes a fixed age distribution.

From what has been said so far, it becomes evident that the overall population growth rate under non-limiting environmental conditions depends on the age composition and the specific growth rates due to reproduction by the appropriate age groups. When a stationary and stable age distribution obtains in a population, the specific growth rate is referred to as the intrinsic rate of natural increase. The maximum value of this is called the biotic potential and the rate of increase which occurs in nature or under laboratory conditions is often taken as a measure of the environmental resistance. Environmental resistance is the total of the limiting environmental factors which prevent the biotic potential from being reached.

Population Growth Form and the Concept of Carrying Capacity

Populations have characteristic patterns of increase which are called population growth forms. Two basic patterns are the J-shaped growth form and the S-shaped or sigmoidal growth form, which may be combined or modified in various ways depending on the population and its environment. As shown in Figure 5(a), the J-shaped form is obtained when density increases rapidly in an exponential or compound interest fashion and then stops abruptly. Environmental resistance becomes effective more or less abruptly. This type may be represented by the model

$$\frac{N}{t} = rN, \text{ with a definite limit on } N$$

Here N represents the amount of increase in density over a given period of time, t represents the time interval, and r represents the rate of increase in density. In the sigmoidal type, density increase is slow at first, becomes rapid as in the J-shaped type, but then slows down gradually as the environmental resistance increases, until a more or less stable equilibrium is reached and maintained. This form may be represented by the logistic model

$$\frac{N}{t} = rN(K - N)$$

The upper level or equilibrium beyond which no major increase can occur, as determined by the constant K, is referred to as the **carrying capacity**. Carrying capacity is a measure of the quantity of any species that a particular environment can support.

The sigmoidal or S-shaped growth form, Figure 5(b), is quite common and is found in various organisms both small and large. It may be described as the result of greater action of detrimental factors (environmental resistance) as the density of the population increases, in contrast to the J-shaped form where the environmental resistance is delayed until near the end of the increase. Thus the sigmoidal type may be said to be the one in which detrimental factors are linearly proportional to the density.

Data on growth of field populations are few and incomplete. This is largely because determination of numbers is often difficult and many natural populations are in the mature stage when they are being counted. The best situation is found when the population enters, or is introduced into, a new and unoccupied environment.

Population Density and Biomass

The concept of the carrying capacity cannot be of much use to us unless we have data on the numbers and densities of different species in different habitats. Density is the average number of the individuals of a species in a given unit area. Actual examples from the tropics are not readily available owing to the labour involved in routine counting and sampling. As already pointed out, although most tropical species are relatively low in numbers in nature, some of them are in fact extremely abundant locally. In plant communities a metre quadrat is often used for sampling in a herbaceous flora, a three-metre square one for shrubs and a 30 metre square one for trees. The use of quadrats (and transects) for sampling can be effective for small animals that do not move about much such as land snails and soil invertebrates. Fast moving and flying animals such as insects, birds and bats are quite difficult to assess.

Capture, marking, release and recapture are a technique commonly used for animals, but a large proportion must be marked. It is also necessary to understand the behaviour of the animals if errors are to be avoided. Some data on the population size and density of large ungulates and primates have been provided by Bourlière (1963) in the Tano Nimri Forest Reserve in Ghana (Table 1). Bourlière also provides the average weight of each species and the biomass.

Biomass is a useful measure of the weight per unit area and it is obtained by multiplying the aver-

Table 1. Numbers, population density and biomass of ungulates in the Tano Nimri Forest Reserve, Ghana, in 1954 (from Bourlière, 1963)

SPECIES	AVERAGE ADULT WEIGHT (KG)	NUMBER PER 250 KM²	DENSITY (NUMBER/KM²)	BIOMASS (KG/KM²)
duiker, *Philantomba maxwelli*	8	79	0.31	2.48
duiker, *Cephalophus dorsalis*	20	38	0.15	3
antelope, *Neotragus pygmaeus*	4	7	0.03	0.12
black colobus, *Colobus polykomos*	10	916	3.6	36
red colobus, *Colobus badius*	8	621	2.4	19
diana monkey, *Cercopithecus diana*	5	144	0.57	2.8
mona monkey, *Cerecocebus torquatus*	5	127	0.50	2.5
mangabey, *Cercocebus torquatus*	8	83	0.33	2.6
olive colobus, *Colobus verus*	4	5	0.02	0.08
chimpanzee, *Pan troglodytes*	40	22	0.09	3.6

age weight of the animals by the density. The total biomass obtained by Bourlière from the ungulates and primates in the Ghana Reserve was only 72 kg/km². On the other hand, a similar estimate by Bourlière from the Rwindi – Rutshuru plains of eastern Zaïre (formerly the Congo) gave an exceptionally high total biomass of 23 556 kg/km² (Table 2).

In general large mammals are much scarcer in the forest than in the savanna, largely because the large ungulates feed on grass which is scarcer in the forests than in the savanna. Most forest animals are relatively small compared with savanna species. Indeed it seems that generally tropical savannas support a much greater mammal biomass than any comparable vegetation in the temperate climate, even though there are exceptions like the savannas of South America.

Data on densities of species in tropical soils are rare. It is estimated that there are over 50 000 arthropods per square metre of the top fifteen centimetres in parts of tropical Africa, but about twice that number of species of ants and arthropods in soils in England.

Annual Cycle and Occasional Oscillations of Population Densities

Everyone is familiar with the conspicuous increases that occur in the population density of some species at certain times of the year and their virtual disappearance at other times of the year. Examples of this include the flying phases of termites, an abundance of fruit flies, mosquitoes or a particular kind of bird or a bloom of phytoplankton on a freshwater pond. Some of these fluctuations may be due to some intrinsic factors within the organism itself, such as predation, disease or an inherent type of growth form; but they seem more often to be caused by certain changes in factors of the external environment, and such factors are described as extrinsic. In the tropics much explosion in the population densities of species takes place in response to a change in water regime. Very few organisms maintain a uniform density throughout the year. Bates (1945) found that relative abundance of only one out of seven tropical species of mosquitoes failed to show seasonal variations (Figure 6).

Occasional population explosions unconnected with annual seasons are of special ecological interest. In the tropical world the migratory locust or grasshopper shows such an occasional explosion. In most years they are non-migratory and do not feed on crops. Under such conditions they attract little attention; but, once in a while, there is a sudden explosion in population density when they develop longer wings and emigrate to cultivated fields, consuming everything in their way. It is shown that the mosaic of vegetation and bare ground produced by shifting cultivation and overgrazing by cattle is conducive to the outbreak

Table 2. Numbers, population density and biomass of some large herbivorous mammals on the Rwindi – Rutshuru plains, eastern Zaïre. The figures are the means of six counts made in 1959 (from Bourlière, 1963)

SPECIES	AVERAGE ADULT WEIGHT (KG)	NUMBER PER 600 KM²	DENSITY (NUMBER/ KM²)	BIOMASS (KG/KM²)
elephant, *Loxodenta africana*	3000	1026	1.7	5100
hippopotamus, *Hippopotamus amphibius*	1400	4800	8.0	11 200
buffalo, *Syncerus caffer*	500	7402	12.3	6150
topi, *Damaliscus korrigum*	130	1199	2.0	260
waterbuck, *Kobus defassa*	150	760	1.2	195
Uganda kob, *Adenota kob*	70	4976	8.3	581
warthog, *Phacochoerus aethiopicus*	70	603	1.0	70

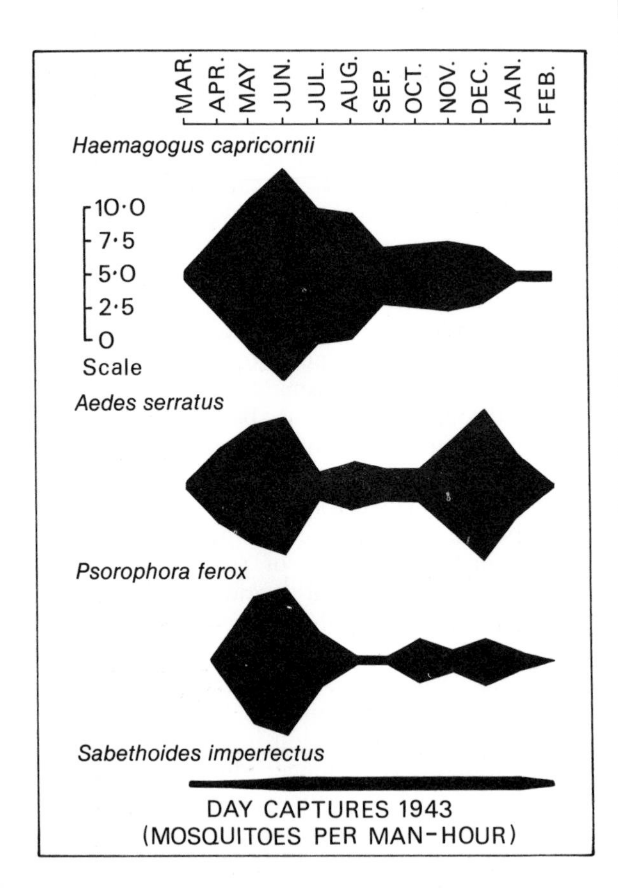

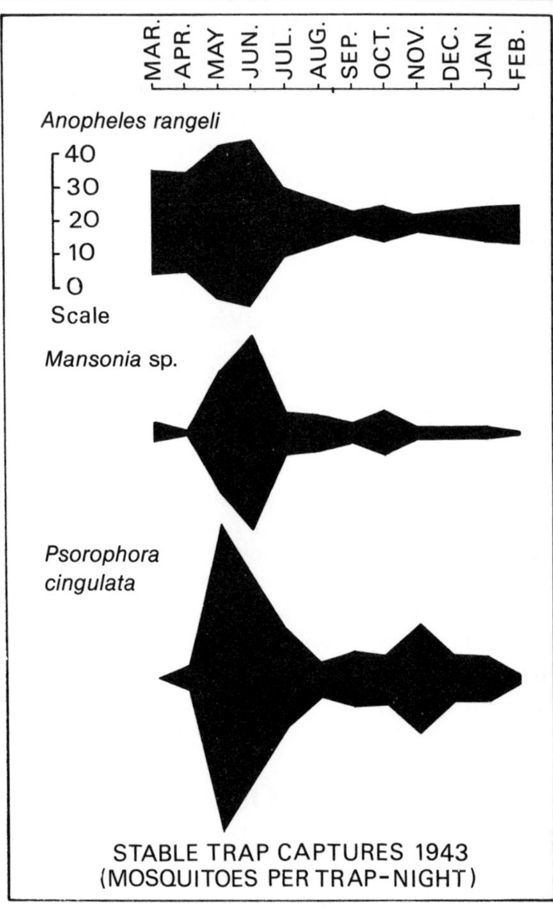

6 Seasonal changes in the abundance of seven species of mosquitoes in the tropical environment of eastern Columbia
SOURCE: E P Odum (1959)

of locust swarms since they lay their eggs on the bare ground. In Africa the climaxes of invasions took place shortly after the First World War, in the early 1930s, during the Second World War and during the early 1950s. Thus it has been suggested that the locust cycle may be roughly correlated with the eleven year cycle of highest and lowest intensities of sunspots, temperature, evaporation and of levels of lake waters.

Factors Regulating Population Size

In considering fluctuations in population densities in time, it is necessary to examine some of the factors which control their size. These are considered under two categories, namely density-independent and density-dependent factors or events. Density-independent factors refer to events that result in sudden environmental change that affects all members of the population equally, irrespective of their local densities. For example a flood in an area could kill virtually all insect populations irrespective of their local densities, except for those that may have some special adaptations for tolerating an increased water regime. Also a bush fire would normally kill almost all centipedes or termites in the area irrespective of their densities. Density-independent factors thus tend to produce large fluctuations in density as well as occasional extermination of the population. It is, however, argued that if an environmental factor is to regulate the population density of a species it should be able to destroy a greater proportion of it when the density is high than when it is low. In this regard, then, density-independent factors are not seen as regulating factors by some ecologists.

Density-dependent factors, on the other hand, are environmental factors that operate to different extents depending on the population densities. In general food supply is density dependent. The more animals available the less food there is for each individual, and the greater the effect of food scarcity. Again, the removal of predators from an area would result in an increase in the density of the animals preyed upon but the extent of the increase would depend on the density of the original population. Other examples are competition, disease and effects of migration. Density-dependent events tend to keep a population relatively stable depending on the environment. In tropical weather when rather high densities of larger animals are confined to a limited space, the temperature of the space rises rather quickly owing to the combined heat of respiration from the animals, and this in turn leads to increased mortality which brings down the population density. It does not, however, often lead to extermination of the population as is the case with density-independent factors. It is conceivable that climatic factors control the more normal fluctuations in density, while density-dependent factors operate mainly when maximum and minimum densities are approached.

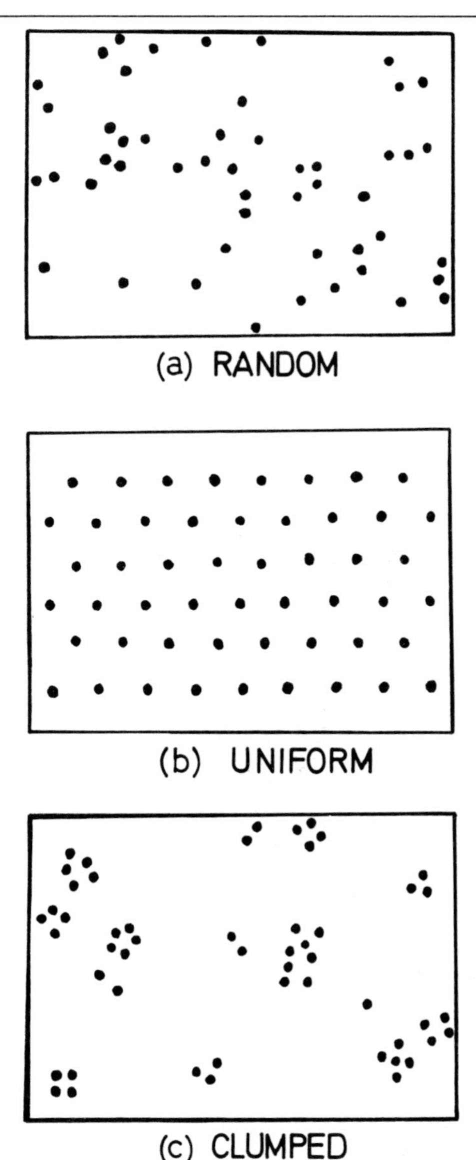

7 Patterns of distribution of individuals in a population
SOURCE: E P Odum (1959)

Population Structure: Dispersion, Aggregation and Territoriality

Dispersion

The distribution of individuals of a population may be in one of three patterns, namely random, uniform (being more regular than random) and clumped (being irregular and non-random) as illustrated in Figure 7. It is not common to find random distribution in nature. It is only where the environment is very uniform or where many small factors act together on the population that this may occur. Uniform distribution on the other hand does occur where competition between individuals is severe or where some positive antagonism exists. In general clumping of varying degrees in the commonest pattern encountered when we study the distribution of individuals in nature. It is only where individuals of a population form groups as found in vegetative clones in plants that the distribution of the clones as units would tend to be random.

Random distribution follows the so-called 'normal' curve on which standard statistical methods are based. Thus, one method of determining the distribution of the individuals of a species is to compare the actual frequency of occurrence of different sized groups or individuals obtained in a series of samples with a Poisson series that gives the frequency with which groups of 0, 1, 2, 3, etc. individuals will be encountered together if the distribution is random. In general, if the occurrence of small-sized groups including blanks and large-sized groups is more frequent and the occurrence of middle-sized groups less frequent than expected, then the distribution is said to be clumped. Statistical tests are used to determine whether the observed deviation from the Poisson curve is significant or not. It is important to use an appropriate sample unit size depending on the nature of the vegetation in order to obtain valid results. A study of one population of the herbivorous land snail *Limicolaria martensiana* in Uganda showed that the number of snails in each metre quadrat varied from 9 to 202. It was clear that their dispersion was not random. The snails were found to become progressively more frequent in line with the gradient of wetness and availability of food. Another population of this snail was found to be associated with one of its food plants, *Bryophyllum pinnatum.*

Similarly, a study of the fish eagle, *Cuncuma vocifer*, a large and conspicuous bird in tropical Africa, showed that its distribution on the shores of Lake Albert in Uganda was not random or even but was concentrated in places where food was available and abundant and related to the availability of perches, the depth of parts of the lake and the presence of fishing villages around which the fish were most available. Also colonies of social birds such as weavers and colonies of termites are spaced out in such a way as to ensure effective utilization of food and other resources.

Aggregation of Individuals

It is known that different degrees of clumping are found in nature as a result of aggregation of individuals depending on the specific nature of the habitat, the weather and other physical factors, the type of reproductive pattern characteristic of the species and the degree of sociability. Aggregation may result in an increase in competition between individuals for nutrients or space but this is often more than counterbalanced by increased survival potential of the group. When organisms are in groups they often experience a lower mortality rate during unfavourable periods or during attacks by other organisms than do isolated individuals. This is because the surface area exposed to the environment is less in proportion to the mass and also the group may be able to modify the microclimate favourably.

Group survival value is an important result of aggregation. A group of plants, not necessarily of the same species, may be able to withstand the action of the wind or be able to reduce water loss more effectively than isolated individuals. In green plants the deleterious effects of competition for light and nutrients often outweigh the advantage of the survival value referred to above. It is in animals that the survival value of aggregation is more pronounced and noticeable. Experiments have shown, for example, that groups of fish could withstand a given dose of poison introduced into the water much better than isolated individuals. It has also been demonstrated that isolated individuals are more resistant when placed in water not previously used by fish. This effect is traced to the secretion of mucus and other secretions produced to counteract the poisons.

Territoriality

Individuals, pairs or family groups of vertebrates and the higher invertebrates often restrict their activities to a definite area. If this area is actively defended it is called a territory. Animals which show territoriality often have complicated reproductive behaviour patterns involving nest building, egg laying and care as well as protection of the young. Thus in many bird and fish populations territories are defended during the breeding season only. Territoriality seems to reduce competition, conserve energy during critical periods and in some cases prevent overcrowding and exhaustion of food supply. Ecologically, territoriality also leads to the spacing out of the organisms. It also increases the chances of some males obtaining a female as a mate.

In view of the lack of pronounced seasonality in the tropics, territorial behaviour is not so common since many tropical species remain in the same area all their lives. Territoriality is therefore more common in temperate species which live in one area and migrate to other areas for breeding purposes. Among the well-observed examples of territoriality in the tropics is the *Tilapia* fish species of East Africa in which the males congregate in the breeding grounds in the lakes. Here each male establishes a territory which it defends against other males of its species (Lowe-McConnell, 1956). Some species of weaver birds breed in colonies in trees as a way of securing protection from their enemies which consist of large predators like eagles. Buechner (1961) described the kob *Adenota*, one of the common antelopes in East Africa, as showing the territorial habit all the year round, thus enabling the females to enter the defended area at any time of the year to mate.

Vicarious Species and Ecological Segregation

Ecological segregation seems to be a phenomenon that might work to reduce competition between closely related species. The separation may be due to genetically determined differences in ecological requirements. However, vicarious species tend to grade into each other. The whole phenomenon is not sufficiently understood. For example, in the West African vegetation *Nauclea pobeguinii* is found in the forest while *N. latifolia* is more commonly found in the savanna. Similarly, *Lophira alata* generally occurs in the forest and *L. lanceolata* in the savanna. When the forest species was transplanted to the savanna it retained its forest features, and so it appears that there is a genetic basis for the segregation.

Examples are also known in tropical animals. One such case is found in two species of *Schistosoma* which are endoparasites causing bilharzia in man. One of these, *Schistosoma haematobium*, uses aquatic snails, which are abundant in drier areas of Africa, as its intermediate host. This species is more commonly found in areas where the drought becomes serious periodically. On the other hand *S. mansoni* uses, for its intermediate hosts, snails which are associated with wetter parts of Africa near the equator. Thus the two species of *Schistosoma* tend to be separated or segregated on ecological grounds into drier and wetter habitat populations respectively. Here water requirement differences seem to underlie the segregation of the species.

It is assumed that in the course of evolution, and by a process referred to as adaptive radiation, animals have become modified each to a particular habitat to which it will have recourse in difficult times. When food resources are plentiful many species may feed on a common food source, but when the food becomes scarce they come to depend on different restricted micro-habitats. However, the factors which determine the dependence on different habitats may be quite complex and varied, so that ecological segregation is achieved not only by differences in food but also by such differences as body size, ability to do without water and migratory movements. This is illustrated by the work of Lamphrey (1963) in the Tarangire Game Reserve of Tanzania where he demonstrated as many as five different ways by which ecological segregation is achieved in this Reserve.

They are:

1. The occupation of different vegetation types or habitats (wildebeest and buffalo).
2. The selection of different types of food (grazers, mixed feeders, browsers).
3. The use of different feeding levels in the vegetation (giraffe, rhino and duiker).
4. The occupation of different areas at the same season (zebra and wildebeest).
5. The occupation of the same area at different seasons (Grant's gazelle and wildebeest).

Discontinuous Distribution on a Continental Scale

There are a number of cases where the same species is found at two or more places, the distances between which are greater than that which the dispersal capacity of their propagules would normally be able to bridge. The species here would not be found in the intervening areas. This constitutes a discontinuous or disjunct distribution. They do not grade into each other as may be found with vicarious species just discussed above.

An illustration may be afforded by the occurrence of the giant lobelias at great heights on both the Cameroon Mountains of West Africa and the East Africa mountains which are separated by nearly 2500 km. This has been explained on the basis that there was a certain time in the past when climatic conditions were such that this and other species at these heights were not so limited but were found in the intervening lowland areas, and survived only where they could migrate to much higher elevations. This has been justified along with other evidence by the discovery on the East African mountains that old snowlines once existed lower down than at present. The view that during the glaciations the tropics were almost certainly cooled by about 5–7°C seems to be supported by this evidence. However, it is difficult to accept that a fall in temperature of a few degrees was the only factor that would make that possible. As warmer conditions returned, the species concerned died off in the intervening lowland areas, leaving only those on the higher elevations on the mountains of West and East Africa as we find today.

Not all disjunct distributions, however, can be explained by changes in past climate. Some affinities between the floras of tropical Africa and America are undoubtedly the result of long distance dispersal, in a combination of chance events which has brought seed great distances into a new but suitable environment. For example, the cactus family is naturally confined to America, except for one epiphytic species, *Rhipsalis baccifera*, which is common in tropical forests throughout the world. This plant has a berry-like fruit with small seeds embedded in sticky pulp. Seeds could have been carried across the Atlantic Ocean by migratory birds. Likewise, the only member of the family *Bromeliaceae* occurring naturally outside the American tropics is *Pitcairnia feliciana* in Guinea. It is a distinct species, unknown elsewhere, but its ancestors could have been carried to West Africa in a similar fashion.

Many useful and ornamental trees have been introduced by man from one part of the tropics to another during the past 500 years or so, but some others have disjunct distributions which are almost certainly due to natural causes. The forests of tropical West Africa and tropical America and Asia comprise trees belonging mostly to the same families; in West Africa and America there are many plants belonging to the same genera and there are even some native species which occur naturally in forests on both sides of the Atlantic. *Andira inermis*, called angelin in tropical America and dog almond in Nigeria, is one of these and *Lonchocarpus sericeus*, the Senegal Lilac, is another. These distributions are currently explicable through our new and growing knowledge of plate tectonics, the movements of the earth's crust through geological time. The affinities of widely distant floras are of ecological importance because they enable us to make comparisons with much greater confidence when we know that the same species grow there naturally.

Chapter 3 The Development and Nature of Tropical Communities

Formation and Dynamics

A study of ecology is not complete without a careful analysis of the principles which govern how plant communities are developed and how they grow to reach a particular status. In other words it is important to understand the processes by which order comes to a plant community to make it more than a mere collection of plants. Three broad factors appear to play significant roles. These are the chance availability of colonizing or invading material, selection from the material available in the environment and the modification of the environment by the plants. These will now be considered in some detail.

1. Chance Availability of Colonizing or Invading Material

The chance availability of colonizing or invading plant material such as seeds, fruits and spores and the order in time in which the various colonizers arrive on the scene are very important factors in the development of a plant community, because the species making up the community at any particular time depend on what materials happen to have been brought to that locality.

An area that is available for colonization will not be reached at one and the same time by the seeds and spores of all the colonizing plants. The first to arrive cannot exploit all the possibilities of the habitat. It is a matter of chance as to which species arrive first. They will be followed by other species over an indefinite period of time. It is noticed that the number of species increases rapidly at first and then more slowly. Assuming that there is a fair variation in the successive colonizing material then the opportunities provided by the habitat niches would become filled and this would account for the falling off of the rate of increase in the number of species. Thus the positions in place and time taken by particular colonizing species in a habitat are due to chance.

Within the same area, one part of it may be more favourably situated to receive more colonizing species than another part, and this can result in one part being richer in species. The direction of the wind or of a stream may be the cause of such a difference in the richness in species of parts of a habitat. Also, the normal range of dispersal of the seeds of a tree in relation to the position of that tree may be such that one part of an otherwise uniform habitat would have more of the seeds than another part. Even so, the exact spot on which a particular seed falls is largely a matter of chance, and from that stage the suitability of that spot for germination of the seed would determine whether or not the seedling would grow there.

This can be illustrated by the observations of Hasselo and Swarbrick (1960) on the recolonization of the lava flows after the volcanic eruption of Cameroon Mountain in West Africa in 1959. These authors found that, during the first year after the eruption, the first colonizers were brought by high winds in the form of seeds and spores. Those that ultimately became lodged in cracks and crevices under and between stones had the greatest chance of survival and development. The other main method of colonization was effected by birds which flew over and perched on the lava flow, depositing their faeces. Within the first six months after the eruption and during a period of strong winds after the main rainy period they noticed algal cover of about 20 per cent, together with some mosses and ferns. There were also seedlings and seeds of some 18 species of flowering plants; but a few months later, when the weather was dry and the wind velocity was very low, only 4 species of the flowering plant seedlings and seeds survived, the other 14 seedlings and seeds having perished. If a species enters a fresh area much too early when the habitat is not yet suitable for it, it is most likely to perish. Sometimes a species which may be better suited for the spot may be rather late in appearing in the area simply because its invading material may happen not to have been carried to the area in time.

2. Selection from Material Available in the Environment

After different colonizing seeds and seedlings have started life on a habitat, only those whose range of tolerance corresponds to that of the environment would survive. The environment, however, may be such that for some seeds no germination would take place while for others the seedlings would not survive. This stage is quite critical because in general the range of tolerance at the seedling stage is much narrower than that of the mature plant. Of course, different environments exert different degrees of selection. In extreme cases like rock surfaces or sand dunes only a small number of specialized species would be suited.

Even after the environment has made a selection from the material available by chance, a stable community could not have been formed yet. If unexploited niches still exist the community is described as 'open'. At this stage we have only a collection of plants which we might say have passed an examination set by the environment. This is because no settled inter-dependence has been established amongst the members of the community. Active erosion, leaching, silting up or inundation may be going on which might impose further demands on the continued survival of some of the species. Before a proper balance is established and a true community emerges, the environment would have been modified by the plants themselves. This is what is regarded as the third factor in colonization, as discussed below.

3. Modification of the Environment by the Plants

From the time that the first colonizers arrive on the bare habitat and start to grow they begin to modify the environment. The degree of influence gathers momentum in later stages in the development. The ways by which plants modify their environment are summed up by the process of succession which will be considered under soil formation, dominance and layering, and cyclic change and pattern. These will now be considered briefly.

Succession

1. Soil formation

Unless the colonization takes place on soil which is already formed, soil formation first results from the presence of the colonizing plants on say a bare rock. (The processes of soil formation are more fully described in Chapter 8.) The first few colonizers, referred to as **pioneers**, consist of rather specialized hardy species such as lichens. They are able to live in such conditions. Their remains begin the very slow and gradual process of building an organic content in the form of a film of humus. The decomposition of the latter begins to supply mineral salts and also to improve the aeration and the water-holding capacity of the substrate, and finally to improve the crumb structure. Algae and sometimes liverworts and mosses may also join the early colonizers. Some ferns also are pioneers (Figure 8). The substrate undergoes a gradual change in this direction until a soil is formed. In the lowland tropics where bryophytes are uncommon, these do not contribute to many types of succession. (Bryophytes are, however, common in tropical montane forests.) See p. 157.

2. Dominance and Layering

When soil forms, new niches are created and the balance of factors is changed so that other species are able to enter the area. Thus annual flowering plants of a weedy nature are often the next to invade the habitat and, as the soil conditions change further, perennial herbs, including grasses, follow. The dead remains of these species contribute to the fertility of the soil and as soon as this reaches a point at which it can support the growth of taller species, shrubs and trees succeed in growing.

The vegetation changes gradually from an open to a closed community. The taller plants begin to shade out some of the smaller species which may die out, but the shade-loving species among these survive and increase. Thus the taller trees exercise a major controlling effect on the other plant (and animal) species beneath and around them and they are said to be the **dominant** species in the community. A major part of the energy transfer in the community is effected through the dominants. Whereas in temperate climates only a few tree species are dominant in a mature plant community, in the tropics

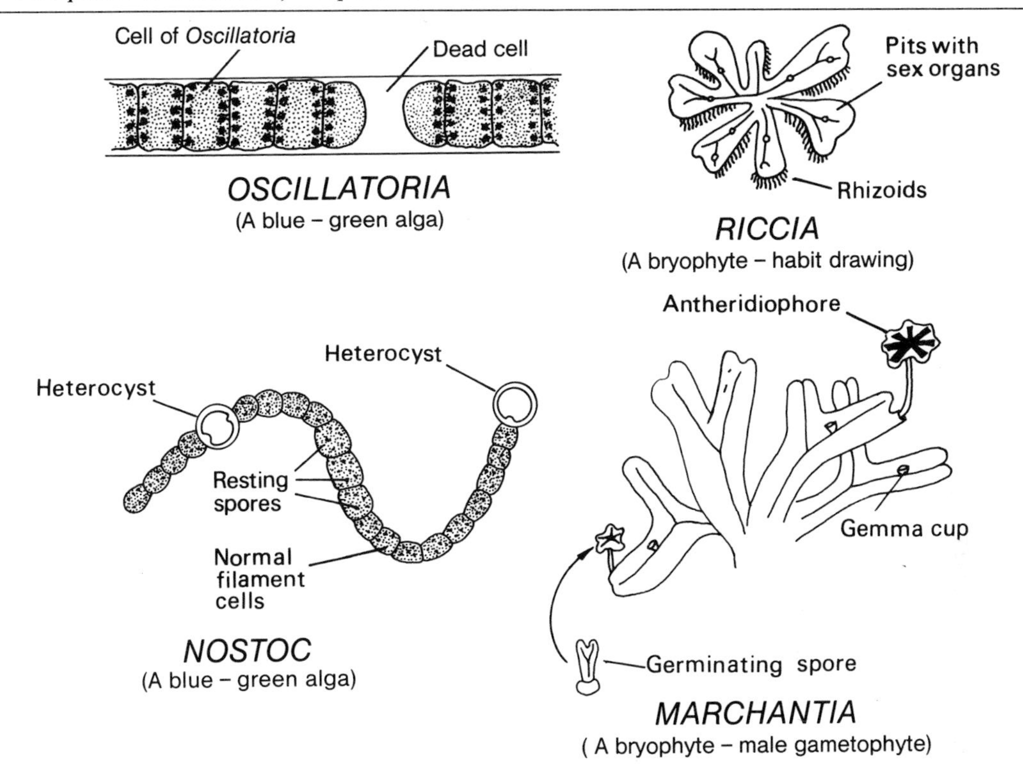

8 Some early colonizers in terrestrial primary succession: *Oscillatoria*, *Nostoc*, *Riccia* and *Marchantia*

quite a large number of species may be involved. This is a reflection of the fact that in tropical successions the number of species increases up to the climax, whereas in a temperate climate the number falls before the climax is reached.

Once species of different heights are present in a community, particularly where trees are included, **layering** or stratification of the vegetation is likely to develop. In layering, the crowns of trees or shrubs appear to form more or less distinct layers running parallel to the surface of the ground. These layers are also referred to sometimes as storeys, and the vegetation is then said to be storeyed. A storeyed vegetation generally has a tree layer, a shrub layer, a herbaceous layer and a ground layer. Layering arises in response to the differing amounts of light which penetrate the canopy of the trees. Much of this is discussed further in this book.

Patterns of layering of course, differ in different communities. The tree canopy itself may be layered as can be illustrated in lowland tropical rain forests which have three such tree layers or storeys. The highest of these tree layers, the 'A' storey, consists of trees about 40–50 m high. The crowns of these trees tend to be wider than they are deep and often touch each other to form a continuous layer. The second or 'B' storey is about 18–30 m high. The third or 'C' storey consists of trees about 4.5–18 m high. The crowns vary in shape but they form a continuous and rather dense layer.

Apart from the tree layers or storeys most tree communities have shrub and ground or herbaceous layers. It may be stressed that fewer storeys exist in other vegetation types. Layering in tropical forests is more fully discussed in Chapter 9.

3. Cyclic Change and Maintenance of Dynamic Equilibrium

When a plant community reaches the stage of having layers or storeys it does not mean that it ceases to change. It is always in a dynamic state and the various parts of it are in a delicate balance with each other. Any change in one of the factors or species is bound to have consequences resulting in other changes. For example, the microclimates created beneath trees of different ages of the same species may show distinct differences.

These micro-climatic differences may be reflected in the nature and density of the ground flora and fauna. Thus if the tree canopy is broken as a result of death of some of the trees or the shedding of leaves at some period of the year, there will be complementary responses.

As the trees go through a step-wise cyclic change from the sapling stage to maturity, old age and death, an opening of the canopy will make way for new saplings, and complementary cyclic changes in the ground vegetation will take place. It is not only the ground vegetation which is affected by the cyclic change in the dominant species. Sometimes the dominant trees may produce a micro-climate in which their own seedlings do not survive, so that the way is paved for the invasion of the habitat by other species whose seedlings can survive the new conditions and which may then take over the dominance of the habitat in the long run.

Main Climaxes

The broad stages of plant succession have been described above. They show that when an area is left undisturbed, the vegetation on it tends to change in a definite direction; broadly from simple communities consisting of low plants to more complex communities of taller plants, as time goes on. This phenomenon is referred to as succession. Succession is a product of the plants themselves, in that the plants present on the area at any particular time modify the environment, consisting of the soil, the plants and the micro-climate above it, in such a way as to make it more suitable for other species than for themselves.

In no region of the world can the natural succession of vegetation be studied with more profit and interest than in the tropics. The vegetation passes through at least the early successional stages with relatively rapid growth, before the eyes of the observer, in a manner that cannot be seen elsewhere. Nevertheless, it has been estimated that for a tropical forest to develop to maturity, even without disturbance, may take hundreds of years.

It should be mentioned here that one could study succession of animal communities too, as for example in rotting wood. A newly-felled log may be found first to be inhabited by wood-boring beetles which burrow into the wood. They are followed by wood-rotting micro-organisms. Later the log becomes populated by cryptozoic animals and the peeling bark provides shelter for wood-lice, centipedes, spiders and small vertebrates. As the wood decomposes to form humus, the numbers of inhabiting species decrease.

The succession of stages in the development of a particular climax community are referred to as **seres**, and the community that is present at each time constitutes a seral community. The almost stable type of community which is ultimately developed is called the **climax community**. The type of climax formed as a result of succession depends on the broad climate of the region. The climax is thus in a state of dynamic equilibrium with the climate and is self-sustaining. It would persist as long as the climate is unchanged and provided also that no retrogressive change, as may occur with deterioration in soil or accumulation of toxic substances, sets in. Examples of climaxes are the tropical rain forest, the woodland savannas and deserts.

Sometimes a seral community, that is one of the pre-climax communities which is ordinarily destined to make way for the next seral community, may become stable for a considerable period of time or even indefinitely. Such a seral community becomes an **arrested climax** and if it remains at one stage below the climax, it is referred to as a **sub-climax**. When the principal factor resulting in the arrest of a seral community appears to be that of the type of the soil, the arrested climax is referred to as an **edaphic climax**, as exemplified by the tropical mangrove swamp. If the arrest is due to the action of man or animals it is called a **biotic climax**, as exemplified by the tropical savanna grasslands. We should finally recognize the existence of the **post climax**, which refers to vegetation that is more advanced than that of the surrounding climax. An example of a post climax is the fringing forests bordering rivers in savannas.

For some time there has been some controversy between the proponents of the monoclimax hypothesis and those of the polyclimax hypothesis. The monoclimax hypothesis holds that a full regional climatic climax is the highest type of vegetation that can ultimately develop throughout the land area of a given climate. The polyclimax hypothesis, on the other hand, contends that in any one region, several different climax communities, which may even differ in the dominant layer, reach maturity and die. The physiography could make the general climax of the region unattainable in a certain part of the area. We have seen that soil, water and other conditions might prevent the development of vegetation to its climax, and also that it is possible to have differential retrogression resulting from leaching, erosion and similar factors. In spite of this, the monoclimax hypothesis of a regional or climatically controlled climax recognizes the convergence of successional trends toward a similar end. In being able to account for local differences, the monoclimax hypothesis has gained greater acceptance.

When succession starts from a spot or an area not previously occupied by vegetation and goes through all the stages without interference from outside it is known as a **primary succession** or prisere and the natural community that is so developed is known as a primary community. Examples are bare rock or water surfaces. When succession begins on an area that has once been occupied by vegetation or seeds, and which has some parts of the remains of the former vegetation, or if a community is brought into being by the disturbances of man (such as fire, cultivation, etc.), the succession is referred to as **secondary succession** and the community a secondary community. Examples are abandoned farmland, abandoned timber concessions or waste ground.

Primary Succession

Among the extreme conditions commonly found at sites for primary colonization are dry surfaces like bare rock or very wet surfaces like standing water. Pioneer communities are broadly of two kinds, those adapted to tolerate shortage of water, i.e. xerophytes, and those adapted to an excess of water, i.e. hydrophytes. Primary successions which start from a dry habitat are referred to as **xerarch successions** while those which start from standing water are known as **hydrarch successions**. Within each of these types, primary successions start with pioneer communities of remarkable similarity irrespective of climate.

Xerarch Succession (Xeroseres)

Assuming that the xerarch succession begins on a granite rock, it is found that, typically, the first colonizers are crustose lichens which glue themselves onto the surface of the rock. They help in hastening the chemical breakdown or weathering of the rock and their own organic remains form the beginning of soil. In due course they are followed by larger foliose lichens. The situation is rather precarious at this stage for, while water is required for the proper growth of these pioneer plants, there is the danger of the rudimentary soil being washed away. The process of soil formation is assisted later when some herbaceous plants arrive on the scene and begin to trap any particles of dust on the rock.

When things become a little more stable, the seeds of some hardy grasses and perennial herbs are able to germinate and grow. They also further the process of soil accumulation. In due course the depth of soil is able to support larger plants, and this process may lead to the formation of a scrubby vegetation. At this stage a shady habitat is created, in which the soil surface is sheltered from exposure and soil moisture does not constitute a very serious problem to the community.

Later bigger trees enter and these form the tree layer which gradually shades out the light-loving grasses and scrub beneath them. The tree layers later shade out their own seedlings which require more light and so they themselves are replaced in time by trees whose seedlings are more shade-tolerant. At this stage, a fairly stable plant community in a dynamic equilibrium is established. This becomes the climax community, which is considered to be an ecosystem in which the maximum use is being made of the resources of the environment.

Succession on Krakatau

The recolonization that took place after the famous volcanic eruption of the island of Krakatau (one of the group of small volcanic islands between Java and Sumatra) in 1883 forms one of the best known illustrations of primary succession of xeroseres in the tropics, in spite of the controversy over whether all the original rain forest vegetation was destroyed. After the eruption the surviving part of the island was covered with pumice-stone and ash to an average depth of 30 m. It is, therefore, evident that the island must have been completely sterilized so that all organisms which were found there after the eruption must have migrated across the sea. The description given below is limited to the inland region (lowland) of the island (up to 400 m).

1. In 1884, a year after the eruption, the island was a mere 'desert' with no plant life.
2. By 1886 the vegetation consisted of a lower layer of blue-green algae and an upper layer consisting chiefly of ferns. The latter consisted of about 26 kinds of vascular plants.
3. By 1897 the vegetation consisted of a dense growth of grasses, some as high as a man. The main species of grasses were *Saccharum spontaneum, Neyraudia madagascariensis* and *Pennisetum macrostachyum.* Associated with the grasses were various dicotyledons, including several climbers.
4. From 1906 to 1919 the same grasses were present, but the associated plants included *Cyperaceae* and some shrubs. There was also a strip of mixed woodland consisting of mixed forest trees of *Ficus* and *Macaranga*.
5. By 1932 the strip of *Ficus – Macaranga* woodland had developed and extended. It had become so shady that there were few ground herbs, although young trees were abundant. Even the main grass vegetation had scattered trees of *Ficus* and *Macaranga* which foreshadowed the development of mixed woodland from the 'savanna'. In some places the shade under the trees was enough to suppress the grasses and to allow the growth of shade species, such as the ground orchid *Nervilia aragoana*. More luxuriant growth of trees and shrubs took place in the ravines, where they replaced the grasses, and humus had even begun to accumulate. The mixed woodland was richer in species than the grass savanna but was much less rich than a primary rain forest.

It is obvious that the recolonization of Krakatau illustrates the successive dominance, firstly of cryptogams (here chiefly algae and ferns), followed by herbaceous flowering plants, and finally trees.

The Krakatau successions are typical of tropical primary successions and are unlike temperate primary successions in that there is a continual increase in the number of species and that there are several dominants in the *Ficus – Macaranga* woodland. The total species of angiosperms and gymnosperms on the whole island throughout the successions were as follows: zero in 1883; 26 in 1886; 64 in 1897; 115 in 1908; 184 in 1920; 214 in 1928 and 271 in 1934. In temperate primary successions the number of species reaches its peak just before the change from an open to a closed community. With the change goes a fall in the number of species. This is now seen to be largely due to the fact that temperate vegetation is not rich in woody species.

Even though primary succession is relatively easy to study in the tropics there are very few examples as well-documented as that of Krakatau.

Modifications of the Xerarch Succession

It should be pointed out that the successional stages and the nature of the climax vegetation formed differ in detail depending on the nature of the surface on which it starts. Thus on sites which are comparatively fertile from the start the building-up of the soil is virtually absent as a factor in dictating the rate of the succession. In such cases the early stages are very much shortened and small annual herbs, including a number of weeds, appear early since they can germinate on the soil and grow quickly. Quick-growing perennials follow but they are later shaded out by slow-growing perennials and the succession continues, as in the case of the rock succession, to the climax.

Other sites for modified xerarch succession include surfaces exposed after a landslide and alluvial deposits. The nature of the climax vegetation depends on the climate and the parent material from which the soil was built up. Thus in the same climate the climax communities on soils of different origin would differ to varying degrees, in structure and in species composition.

Hydrarch Succession (Hydroseres)

In a typical hydrarch succession the stages in the succession are towards the accumulation of silt, raising the soil above the water level and lowering the water table in order to create conditions similar to those of the land habitat. The first plants to colonize a pond or a lake include submerged aquatics, which usually have dissected leaves and are anchored in the mud, such as *Ceratophyllum*. They must be at such depth as would afford them sufficient illumination to metabolize at or above their compensation point. If the water is clear the depth can be quite great but where the water contains a great deal of plankton or other suspended matter which absorbs much of the light, the colonizing submerged water plants would be quite near the surface of the water. If the water is itself not deep, then rooted species like water-lilies with floating leaves will also be present.

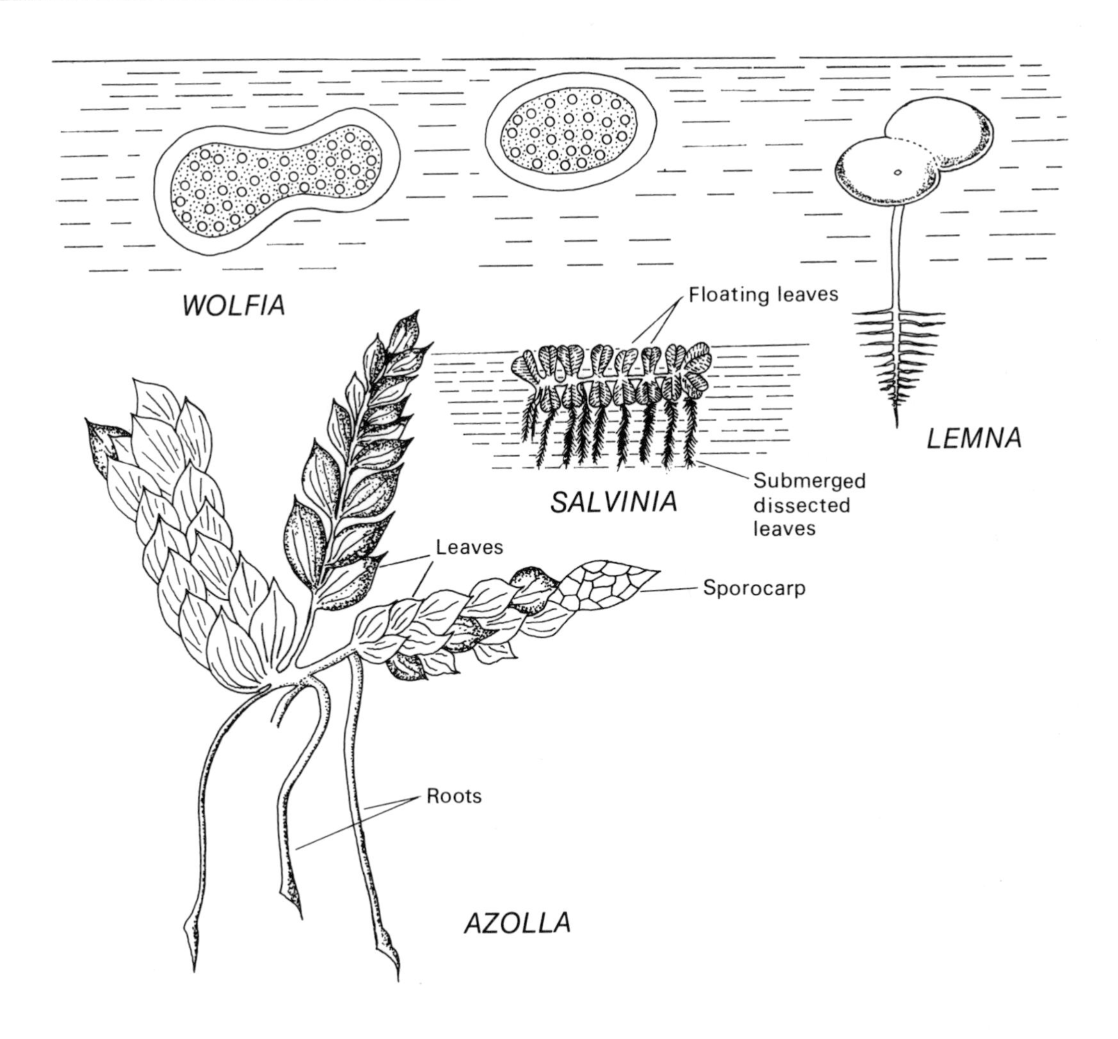

9 Some free-floating early colonizers in aquatic primary succession: *Wolfia*, *Lemna*, *Salvinia* and *Azolla*

The remains of these plants accumulate to raise the level of the mud at the bottom of the pond. Later the floating leaves of species like water-lilies enter and shade out the submerged species. These events make it possible for some free-floating water plants like species of *Lemna* and *Azolla* (Figure 9) to enter the zone of the floating leaves, in the centre, although wind currents often blow them to the margins of the pond. As more humus accumulates, species with aerial leaves, such as *Typha*, increase in numbers and accelerate the accumulation of silt. Thus, in the course of hydrarch succession there is a gradual change from open water to relatively dry conditions, and like the xerarch succession a mesophytic environment may be produced. The nature of the final climax, however, depended on whether the water was entrophic or oligotrophic (see Chapter 10).

Tropical Hydroseres

In tropical lowlands opportunities for studying hydroseres exist in rivers, swamps, lagoons and lakes, where stages in the succession from open water to closed forest abound and can be inferred with reasonable confidence from zonation of the plant communities. Unfortunately, as for the xeroseres, there is not enough well-documented data on hydroseres in the tropics but the information available seems to permit certain basic generalizations.

The course of hydrarch succession, including the species involved, seems to be very similar everywhere in the tropics, and in fact essentially it follows the same course as in the temperate climates with even some species of water plants being common to both the temperate and tropical climates. The succession of tropical hydroseres seems to be as follows:

1. **First Stage**. Free-floating and submerged aquatic plants are the first to enter the open water. The genera and species are often at least closely related if not the same. Tropical hydroseres at this stage differ from temperate ones by the frequent development of a sudd or free-floating community with grasses dominating. Sudd formation rarely occurs in temperate climates.
2. **Second Stage**. Silting results in making the water shallower and rooted floating-leaf vegetation appears, often including water-lilies.
3. **Third Stage**. As silting continues the habitat becomes suitable for a community of emergent aquatics which root below the surface of the water but which have most of their shoots in the air well above the water.

   It is worthy of note that in both tropical and temperate hydroseres, monocotyledons (in the tropics including palms) feature in the intermediate stages. This has been presumed to be due to the exceptional ability of monocotyledons to tolerate poorly-aerated environments.
4. **Final Stages**. The community of emergent aquatics helps to reduce the water by more rapid transpiration and a reed swamp vegetation develops. As the area becomes relatively drier scrub or low forest communities appear. Finally, mature swamp forest vegetation with tall, woody species succeeds the scrub or low forest. The swamp forest has a higher tree density than normal dry forest, but lower herbaceous species density, and characteristic swamp species are present. The canopy is more open and less regular, permitting greater illumination which results in a dense undergrowth, some of whose species are otherwise typical of secondary forest in normal dry conditions. The swamp forest becomes an edaphic climax which may not proceed to a mixed rain forest unless the water table falls.

Succession in the Gatun Lake

Among the few well-documented hydroseres in the tropics is that of the Gatun Lake in the Panama Canal described by Kenoyer (1929). Here five stages were identified as follows:

1. **First Stage**. Floating aquatic sere: among the plants present were *Salvinia auriculata, Ludwigia natans, Utricularia mixta, Pistia stratiotes* and *Eichhornia azurea*.
2. **Second Stage**. Water-lily sere: the species included *Nymphaea ampla*, in addition to the species of the first stage.
3. **Third Stage**. Emergent aquatic sere: here the more common species were *Typha domingensis* and the fern *Acrostichum danaeifolium*. Others were *Crinum erubescens, Hibiscus sororius* and *Sagittaria lancifolia*.
4. **Fourth Stage**. Reed swamp sere: the species in this seral community included *Cyperus giganteus, Scirpus cubensis* and other *Cyperaceae*. Also present were large grasses among which were *Phragmites communis* and *Gynerium sagittatum*. Finally there were ferns and a few dicotyledonous herbs including *Ludwigia octovalvis*.
5. **Fifth Stage**. Marsh scrub sere: the species here included *Dalbergia ecastaphyllum* and *Montrichardia arborescens*, the latter being a tall aroid.

Secondary Succession

As mentioned earlier, secondary succession refers to succession which begins from an area that has been inhabited before by plants, and which has remains of the vegetation or seeds. Examples of these are abandoned farmland in, say, a tropical forest, a burnt forest or savanna grassland, an abandoned timber concession, a heavily grazed grassland vegetation or a dumped heap of surface soil. A diagram depicting the various interrelationships in succession is shown in Figure 10.

Secondary Succession in Tropical Vegetation

The best-studied secondary succession in the tropics comes from the tropical rain forest (see

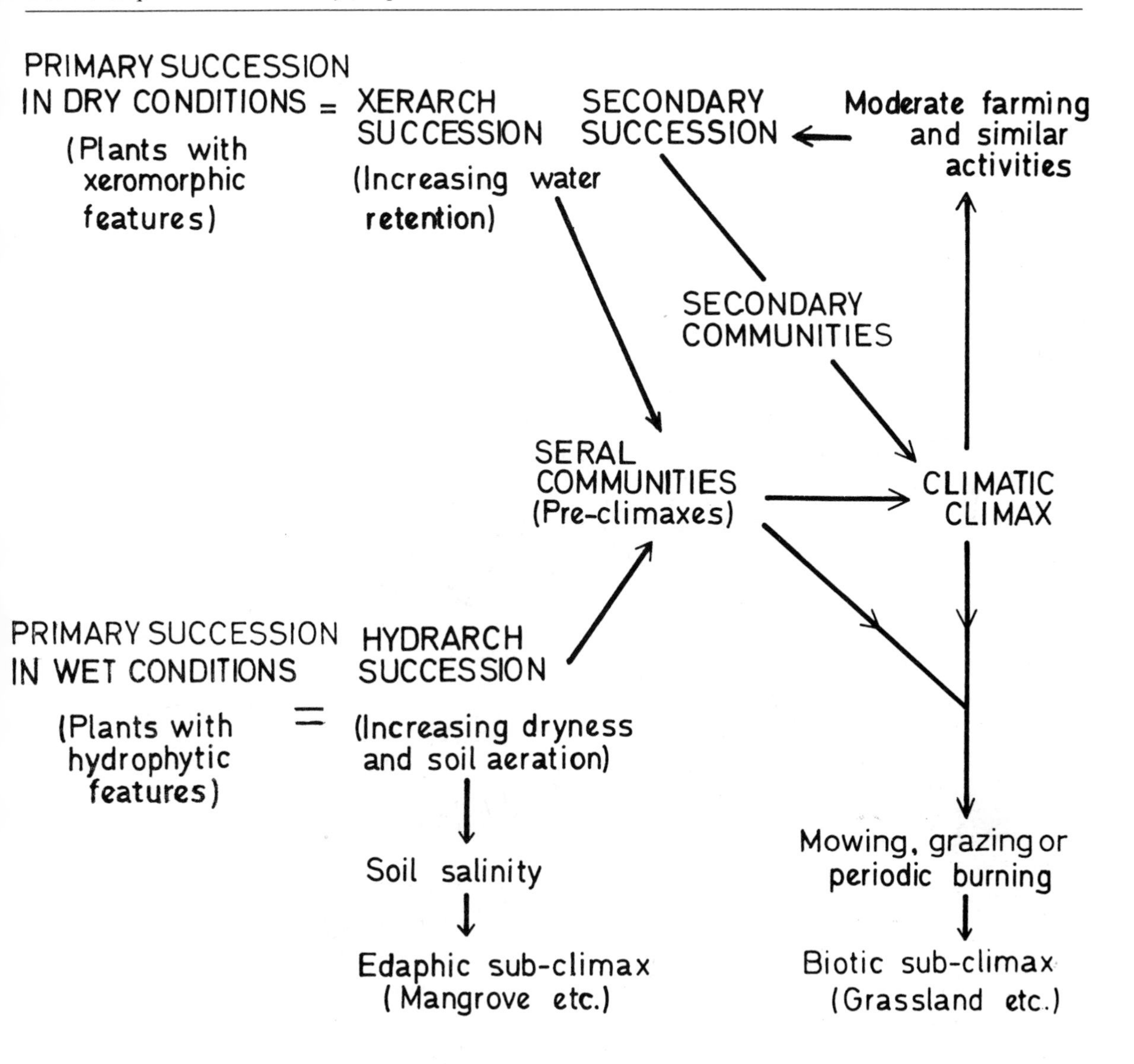

10 Diagram to show the inter-relationships in different types of succession

Chapter 9). The features of secondary succession to be described here will be based on observation following the destruction of this climax vegetation. It is apparent that in general such secondary successions tend to lead to the restoration of the rain forest as a climatic climax, but if fire, grazing or soil deterioration continues in the forest the direction of the succession is altered or deflected, leading to biotic climaxes which are lower in status than the forest.

When a tropical rain forest which has been subjected to a few years' cultivation or timber exploitation is abandoned, secondary succession begins. The changes that take place in the environment and the stages of the regrowth may be as follows:

1. **First Phase**. When the forest cover is removed changes at once take place in light intensity, temperature and humidity. The light intensity is increased from that of the deep shade of the rain forest to that of full daylight. The temperature increases in range and the humidity

is drastically reduced. The array of microclimates of the original forest disappears. Owing to the fact that the ground then becomes exposed to rain and sun, soil deterioration sets in, leading to erosion and rapid loss of humus.

The first phase of the succession is dominated by weeds, including grasses. These are often short-lived and may go through their life cycle in less than a year.

2. **Second Phase**. Shrubs then enter the area and may succeed in dominating the vegetation, but sooner or later trees take over and form a canopy. These trees are mostly short-lived, fast-growing and have mechanisms for wind or animal dispersal of their seeds or fruits. The trees cast more shade, in which their own light-demanding seedlings cannot grow.
3. **Third Phase**. In course of time the more slowly growing, shade-tolerant seedlings of forest species enter. The young secondary forest is often even-aged and dominated by a single species which generally lasts a single generation, after which it is replaced by other species. As time goes on it becomes more mixed in age and species structure and gradually becomes like the climax of the region.

It should be stressed that the course of secondary succession, like primary succession, depends on the characteristics of the soil, such as whether it is a tropical podsol (black) or tropical red soil (see Chapter 8). In deflected successions the course of succession is modified largely through a change in reaction and growth vigour of certain species. The general pattern of secondary succession in the tropical climate is similar to that in the temperate climate, the sequence followed being that of herb, shrub and tree.

The length of time taken for secondary succession in the forest to reach the forest climax will depend on the length of time of disturbance of the forest before the area was left to regenerate. This, of course, is largely a reflection of the degree of deterioration of the soil. Chevalier (1948) estimates in one case in Cambodia that five to six centuries are necessary for the secondary succession to reach almost the original forest.

Secondary Successions in a Tropical African Forest

One of the works on secondary successions in West Africa is that carried out by Ross on the Shasha Forest Reserve in Nigeria, as reported by Richards (1964). The stages described here have so far been found to be quite typical of the West African forests, even though the forest here was of a semi-deciduous type rather than a rain forest. These stages are:

1. **First Stage**. At first, when the destroyed vegetation is left, the flora often consists of only a few surviving trees as well as some of the perennial crop plants, while the ground is otherwise quite bare. A few weeks later various species within the dispersal range invade the area and those whose seedlings are able to establish themselves do so. Soon the area becomes covered with a dense mass of low vegetation. The first invaders may be classified into three types as follows:
   (a) Herbaceous weeds of cultivated land such as *Phyllanthus* and *Solanum* spp. These form a closed stand and dominate for only a short period of time.
   (b) Mainly woody species, which are characteristic of secondary habitats, but which are found also in small openings made in high forest by the felling or death of large trees. Examples are *Musanga cecropioides, Trema guineense, Vernonia conferta, V. frondosa* and *Fugara macrophylla*. These species have more rapid growth than those in group (c).
   (c) Light-demanding high forest species whose seedlings are able to establish themselves under open conditions. Examples are *Erythrophleum ivorense, Khaya ivorensis* and *Lophira procera*.
2. **Second Stage**. The second group of species, that is the woody species of group (b) above, takes over the dominance of the community from the herbaceous weeds and for 15 to 20 years one species or the other in this group is dominant.

   The first dominants are often either *Vernonia conferta* and *V. frondosa* which are woody composites with large leaves, or *Trema guineense*, depending on which genus happens to be fruiting more profusely at the time the recolonization begins. After about three years. the tree *Musanga cecropioides* becomes dominant. *Musanga cecropioides*, probably the most common and characteristic secondary rainforest species in tropical Africa, is able to dominate at this stage by virtue of its extraordinarily rapid growth. *Macaranga barteri*, a tree with stilt roots like the *Musanga cecro-*

pioides, comes close to the height of the latter in forming the tree layer. Below the tree layer is a second storey of shrubs and young trees, among which are species of *Conopharyngia*, *Disco-glypremna, Rinorea* and *Rauwolfia*. After five years this storey attains about 5 m in height. Further down, especially in the early years of the *Musanga* phase, small herbs including species of *Geophila* and some *Commelinaceae*, form a closed carpet. This carpet thins out as increased shade reduces the number of these herbs.

The *Musanga* trees die after 15 to 20 years through senescence and may be blown down by wind. Since *Musanga* is unable to regenerate in the shade, it does not leave its successors. By this time various tree species are present and the height of the secondary forest is about 20 to 25 m.

3. **Third Stage**. Although the further progress of development after the first 20 years was not traced by Ross, it is evident from other studies that it would lead to a decrease in the abundance of the group (b) species and an increase in that of the group (c) species, resulting in the formation of a high forest.

   As more of the light-demanding species die they are not replaced, since their seedlings cannot survive in the increasing shade. As time goes on they are replaced by slow-growing forest trees which are shade tolerant in their younger stages of growth but which will become the dominants in the final climax equilibrium (see Figure 39).

Secondary succession along similar lines has been described by Vermoesen (quoted by Lebrun, 1936) in Zaïre.

Secondary Successions in the Malayan Region

Studies of secondary successions of rain forest sites in the Malayan region show that while the stages are often much the same as in the African region there are a few differences. One main difference is that there is much variation in the floristic composition in the successions as a reflection of the richness of species in the Malayan flora as a whole, in comparison with the African flora. Also, in normal succession in the Malayan region, unlike in Africa, lalang grass (*Imperata cylindrica*) often dominates a stage for some long period. In Africa it is only in deflected successions following over-cultivation that *Imperata* features. Among the principal works on secondary successions in the Malayan region are those of Jochens (1928) and Symington (1933).

Secondary Successions in Tropical America

The pattern is much like that of the Old World except for differences in the actual species. Here the soft-wooded trees of the genus *Cecropia* are a characteristic feature of young secondary forest. *Cecropia* behaves like *Musanga cecropioides* in tropical Africa. Their general physiognomy, shape of leaves, taxonomy, the formation of pure stands and their short life show that they fulfil a similar ecological role. *Imperata cylindrica* does not occur in South America except in Chile, and in general the role of this species is taken by *I. brasiliensis* and other grasses.

Greig-Smith (1952) studied disturbed forest and secondary succession in Trinidad and, on the strength of detailed quantitative surveys, established certain interesting features. By means of determining comparative height class ratios he was able to show that mortality on the secondary sites was less in the smaller height classes and greater in the taller height classes than in less disturbed forests. This means that natural thinning, which takes place mostly in seedling and sapling stages in mature forest, is delayed until a later stage in secondary forest.

Associated with this were differences in pattern. Less randomness was apparent in the secondary than in the undisturbed forest and the scale of heterogeneity was less in the earlier stages of the succession than later on. Marked clumping of woody species was often seen in secondary situations. Analysis of joint occurrences of pairs of species indicated greater interdependence in the earlier successional stages. As communities approached stability, they tended to show patterns determined more by interaction than directly by environmental causes.

Other observations showed that secondary sites had lower proportions of localized or endemic species and a higher proportion of widespread species, the newcomers often being more characteristic of drier habitats. Shifts in the proportional representation of certain families were also noted.

In the wetter islands of the Lesser Antilles, for example St Lucia, ferns and members of the *Melastomataceae* and *Piperaceae* are prominent in early stages of sucession following the abandonment of banana plantations (Wardlaw, 1931).

Analytic Characteristics of the Community

We now wish to consider the analytic features of communities as a basis for the next few chapters. The community exhibits a number of features which may be classified into two main groups, namely analytic and synthetic. The analytic group of characters consists of the qualitative and the quantitative types. The qualitative types are usually of a descriptive nature because of the difficulty in measuring them, even though most qualitative data can be quantified secondarily. But the quantitative ones are those features which can be measured readily. Some important qualitative characteristics will be described first.

Qualitative Characteristics

Among the principal qualitative characteristics of the community are: the species of plants and animals comprising the community (floristic and faunal composition); sociability as a broad descriptive statement of the spatial patterns of the components; stratification or layering of the various elements in the community; vitality; life form (vegetative habit or growth form); and periodicity (including phenology).

Floristic and Faunal Composition

No study of a community is possible without a record of all the species of plants and animals present. It is almost always the first step in the study of the community and it should not be limited to one season but should be recorded throughout the year in order to have a complete list of species of the area. The lists should ideally reflect accurate identification and the correct application of scientific names. The importance of the species list lies in the fact that it makes the ecologist aware of the possible interactions among the different species present. The species structure of a community provides an expression of diversity, successional changes and stability of the community.

Sociability

Sociability describes the proximity or space relations of individual organisms. The scale normally used for analysing vegetation in this respect is the one put forward by Braun-Blanquet (1932).

Class 1: Shoots growing singly, one in a place.
Class 2: Small groups of plants or scattered tufts.
Class 3: Small, scattered patches or cushions.
Class 4: Large patches or broken mats or colonies.
Class 5: Very large mats or strands of nearly pure populations which cover almost completely a large area.

The mode of reproduction, the reproductive capacity and the structure and growth of underground parts of the different plant species contribute to the nature of the aggregation and add more detailed information of a qualitative kind if required.

Stratification

Layering or stratification expresses the vertical position of the various elements in the community. Four principal layers are often recognized in a plant community. These are a tree layer, a shrub layer, a herb layer and a ground layer. As has been mentioned, tropical rain forests are remarkable for exhibiting more than one tree layer; they usually have two, and sometimes three. In addition to these are the various plant synusiae, consisting of epiphytes, climbers, saprophytes and parasites, which are somehow associated with the tree layers. The animals are also associated with these tree layers. The birds often feed in the canopies of the emergents, the herbivorous mammals such as lemurs and squirrels feed in the intermediate layers of trees and shrubs, and the large herbivores such as okapis or deer feed only on or near the floor of the forest.

Plant and animal communities in the soil and in water exhibit similar forms of stratification. In general stratification is more pronounced in the tropics than in temperate climates.

Vitality

This is an indication of the degree of prosperity or vigour of the species of the community and in particular it indicates whether or not the species regularly complete their life cycle and produce seed. A number of criteria may be employed in determining the vitality of a species, such as the rate of growth, rate of renewal after grazing or mowing, the quantity or area of foliage, colour and turgidity of the leaves and stems, degree of infection or insect attack, the number and height of flower stalks, rate of growth and size of the root system, the rate of development of new stems and leaves and the extent of dead branches or other portions.

These criteria can lead to a number of levels of vitality. The levels of vitality which have been much used are those provided by Braun-Blanquet (1932).

Class 1: Well-developed plants which regularly complete their life cycles.
Class 2: Vigorous plants which usually do not complete their life cycles or which are poorly developed, and sparsely distributed plants that spread vegetatively.
Class 3: Feeble plants which never complete their life cycles and do not spread vegetatively.
Class 4: Plants occasionally appearing from seed but which do not increase in number, such as ephemeral plants (e.g. herbaceous weeds).

Life Form

By life form is meant the characteristic form of a plant in the vegetative state. Any species may be assigned to a life form class on the basis of its size, shape, mode of branching, method of perennation and longevity, as well as the average area of its leaf or leaflet. The position of the perennating bud is often crucial in this determination. Thus, trees have buds which are held high above the ground and exposed to the wind which may tend to desiccate them, but in herbaceous plants the buds may lie just above the soil where the humidity is high or they may be buried beneath the soil where they do not experience the dry period of the year. It appears, then, that the severity of the dry period in tropical climates or the winter in temperate climates is correlated with the predominant life form of the vegetation. This was made a special study by the Danish botanist Raunkiaer who devised a system of classifying the whole range of life forms found throughout the vegetation of the world. This classification has world-wide application in ecological studies.

Raunkiaer's life form classification is based simply on the position of the highest perennating (resting) bud in relation to ground level as an indication of the manner in which the plant survives the adverse season. He regarded trees as the most primitive since they carry their resting buds high above the ground, a situation that was thought to be suitable for the weather in which flowering plants arose. Other life forms with the resting bud borne closer to the ground or buried in the soil were considered more advanced and adapted to the dry (or winter) periods. The extreme end of this trend is represented by plants which go through the perennating period as dormant seeds. These were considered the most effective means of resisting the adverse season. Temperate species of trees often have bud scales covering the dormant growing parts in winter. Such scales (modified stipules or vestigial leaves) are much less prevalent in tropical species. The life forms of Raunkiaer, their descriptions and symbols to denote them are shown in Table 3.

Biological spectrum is the expression given to the numbers of species in each of the classes of life forms, expressed as percentages of the total number of species in a given vegetation.

Normal spectrum (a random world spectrum) is the biological spectrum applied to the world vegetation as a whole. It was estimated by Raunkiaer by analysing a random sample of 400 species of the world. It is thus properly a random world spectrum. Departures from the normal spectrum compared with the biological spectrum of a temperate forest, a tropical forest and a hot desert are shown in Table 4.

It will be noticed from Table 4 that a high percentage of phanerophytes is typical of moist tropical climates, which also show a high percentage of epiphytes. In the temperate forest hemicryptophytes predominate and no epiphytes are present. The desert is typified by therophytes.

Apart from the use of life form based on the degree of protection of the renewal buds and their height position from the soil surface, as a reflection of climate, it has been shown that leaf size variation can also be used since it is independently related to climate. For it is a common observation that large leaf blades are generally found in the warm and wet tropical climates, while small blades are typical of dry and cold climates. Experimentally too, it is observed that the size which an expanding leaf blade ultimately reaches depends very much on the temperature and moisture of the environment.

Raunkiaer proposed leaf size classes and characterized a given vegetation by the series of percentages of the flora that fell into each of the leaf size classes. Thus he showed that vegetation from the same climate exhibited a similar spread of percentages of categories of leaf size classes. The leaf size classes proposed by Raunkiaer for this purpose were:

Megaphylls – leaves larger than 164 025 mm².
Macrophylls – leaves from 18 226 to 164 025 mm².
Mesophylls – leaves from 2025 to 18 225 mm².

Table 3. Life forms, their symbols and descriptions, according to Raunkiaer

LIFE FORM	SYMBOL	DESCRIPTION
1. Phanerophytes	P	Aerial plants with perennating buds borne above ground level and exposed to the weather. These include trees and shrubs, tropical herbs of tree size and lianes. This class is sometimes subdivided on the basis of height to which the buds are borne, as follows:
(a) Megaphanerophytes	MM	Over 30 m high
(b) Mesophanerophytes	MM	Between 30 m and 8 m
(c) Microphanerophytes	M	Between 8 m and 2 m
(d) Nanophanerophytes	N	Between 2 m and 0.3 m
2. Chamaephytes	Ch	Surface plants with perennating buds borne just above ground level. They may be herbaceous or low woody plants.
3. Hemicryptophytes	H	Plants with perennating buds half-hidden in the surface of the soil.
4. Cryptophytes		Plants with perennating buds buried in the soil or beneath standing water. This group is subdivided as follows:
(a) Geophytes	G	Plants with perennating buds buried in the soil, e.g. plants with bulbs, rhizomes and corms.
(b) Helophytes and hydrophytes	HH	Helophytes are marsh plants with perennating buds in water-logged mud. Hydrophytes are water plants with the perennating buds beneath the water.
5. Therophytes	Th	Therophytes germinate, fruit and produce seeds in brief complete cycles. They survive as seeds during unfavourable seasons. They are cosmopolitan in distribution including the hot, dry deserts but are very rare in forests.
6. Stem succulents	S	These are plants which survive unfavourable weather or drought by living on water stored in their tissues from the previous rains.
7. Epiphytes	E	These are plants growing on other plants (support plants) without drawing food from them.
8. Parasites and saprophytes		See Chapters 4 and 8.

Microphylls – leaves from 225 to 2025 mm².
Nanophylls – leaves from 25 to 225 mm².
Leptophylls – leaves smaller than 25 mm².

To ease the process of determining the leaf size category of different leaves a graphic representation of the maximum sizes of leaf size class as shown in Figure 11 may be used. Lobed leaves, however, present problems in this regard and as such their areas have to be determined by means of a planimeter or by some matching method.

Apart from the use of leaf area, complementary to the bud height system, another aspect of the

Table 4. A comparison of the biological spectra of a temperate forest, a tropical forest and a hot desert

VEGETATION TYPE	NO. OF SPECIES	% DISTRIBUTION OF THE SPECIES AMONG THE LIFE FORMS									
		S	E	MM	M	N	Ch	H	G	HH	Th
Normal spectrum	400	1	3	6	17	20	9	27	3	1	13
Temperate forest (West Germany)	250	0	0	0	27	0	6	39	23	0	5
Tropical rain forest (British Guiana)	220	0	22	0	66	0	12	0	0	0	0
Hot desert (Algeria)	169	0	0	0	0	9	13	15	5	2	56

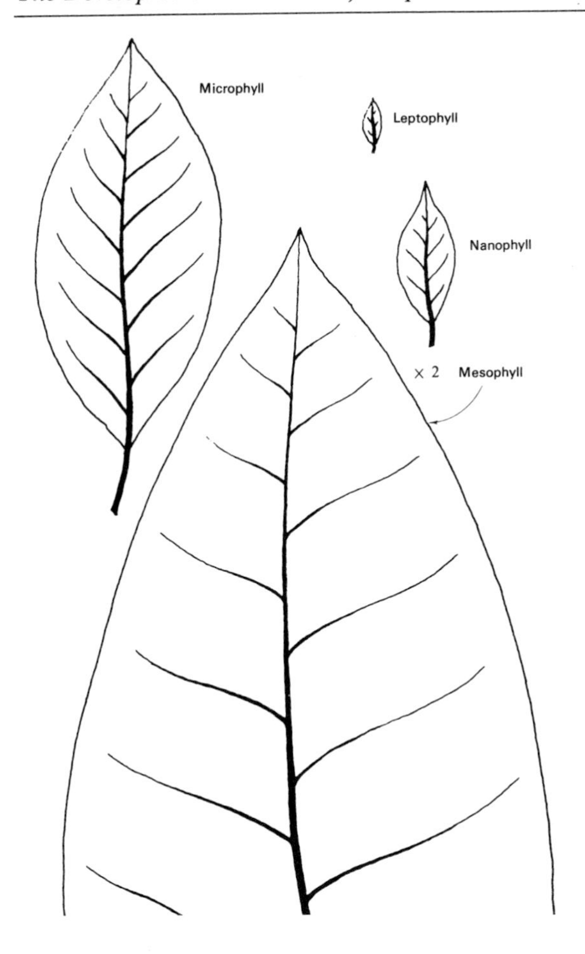

11 Leaf size classes. Graphic representation of the maximum sizes of leaf size classes. Only half of the area of a mesophyll is indicated. Still larger leaves are megaphylls
SOURCE: C J Taylor (1960)

leaf area has become very important in biomass and productivity studies (see Chapter 6). This is the leaf area index. The leaf area index (LAI) is defined as the ratio of total leaf area of the stand or a particular ecosystem as compared with the total ground area covered by it. The ratio is limited to a maximum value in a given situation. Its value may not rise above that maximum value in view of the fact that if that were to happen, it would imply that the lower and overshadowed leaves would not be able to maintain a positive balance of assimilation, a situation that is almost impossible. The maximum value may, however, change only with changes in light intensity or in the supplies of water and nutrients to the ecosystem.

Periodicity

Periodicity refers to the regular seasonal occurrence of various processes such as photosynthesis, growth, pollination, flowering and ripening of fruits and seeds. Periodicity appears to result from inherent genetic characteristics of each species under the influence of a particular combination of environmental conditions. Periodicity means particularly the recurrence at certain times of these processes and their manifestations, while phenology refers more to the appearance of the manifestations at certain seasons of the year.

Since this is a topic for Chapter 5 it will not be discussed further here.

Quantitative Parameters

The quantitative parameters often studied are frequency, density and cover.

Frequency

Frequency deals with the degree of uniformity of the occurrence of individuals of a species within an area. It is measured by noting the presence or absence of a species in sample areas (or quadrats) which are ideally distributed at random throughout the area under study. Frequency is therefore expressed as the percentage of the total sample areas (or quadrats) used in which a particular species occurs. If, for example, a species is found in 15 out of 30 samples, its frequency is 50 per cent.

As already mentioned in Chapter 2 most species tend to have non-random distribution. In order to verify the nature of the combined distribution of all the species in a community Raunkiaer grouped the species into five frequency classes: A, 1–20 per cent; B, 21–40 per cent; C, 41–60 per cent; D, 61–80 per cent; and E, 81–100 per cent. The application of such classification to various communities showed that the normal distribution of the frequency percentages, derived from such classifications, is expressed in the proportions of Classes A, $>$ B, $>$ C, $\geqq$ D, and $<$ E. This has been named Raunkiaer's ratio. The occurrence of this ratio is largely due to the facts that a large number of less common species are widely scattered (Class A) when compared with other frequency classes, and that a significant number of common species are more or less evenly distributed (Class E). The latter are the dominants. The ratio is useful in determining in the field the uniformity or otherwise of the community, an important feature being that Class E should be larger than Class D if species in the community are randomly distributed.

Frequency determinations have been used to show the effects of various treatments such as grazing or management practices on different species in grasslands, to compare different vegetation types, to study the role of micro-topography in causing variations in plant cover and to assess the significance of different species in various communities at one time or in the same community at different times.

Density

Population density has already been discussed to some extent in Chapter 2. Density values are used to show the relative importance of each species in a community when they are similar in life form and size. Where the plants are of different life forms, such as shrubs, grasses or climbers, density alone is not sufficient for comparison, and data on cover (discussed below) will have to be involved. This is because some plants may have low density by virtue of their tufted or matted nature but this will give an incorrect impression of their importance or role in the community because they may have much greater cover which in fact reflects better their importance in the community than mere density.

One of the greatest difficulties in estimating the densities of species in a community arises with those species in which it is difficult to distinguish separate individual plants, particularly species that propagate by rhizomes or runners. One way of overcoming these problems has been the use of estimation scales such as: 1, denoting very sparse individuals; 2, sparse; 3, infrequent; 4, frequent to numerous; and 5, very numerous. Such a scale should properly be based on stated approximate numbers or density values.

Density values have been used to measure the effects of burning, spraying of chemicals on plants and successional changes in communities.

Cover

Cover is concerned with the area of ground occupied by the above ground parts of plants as viewed from above. It is estimated from a number of samples and is defined as the proportion (usually expressed as a percentage) of the ground occupied by any species. In view of the overlapping nature of plant parts, the total percentage cover is often more than 100 per cent. In order to correct this, relative cover is sometimes used, that is, the cover of a species as a percentage of the total area of all the species and bare ground in a given habitat. (In all cover and relative cover statements the proportion of bare ground should be included or taken into account.) In this way the total does not exceed 100 per cent.

As mentioned in the section on density, the importance of cover lies in the fact that it gives an indication of dominant species in the community, unlike frequency and density which merely indicate distribution or numbers and not the overpowering influence of taller species through their cover. The more intrinsic meaning of cover is often competition for light in a closed community. Cover is also a most suitable medium for estimating changes in quantitative relations between species but, no doubt, for single-stalked plants population density may be preferred. Finally, cover can be a useful concept below ground in regard to root competition. Thus scattered plants in a desert may represent a closed community underground.

From what has been said about the value of the quantitative parameters of frequency, density and cover, it is evident that when time is limited for field work, frequency and cover are particularly useful in furnishing much of the information on the nature of the community.

In principle the **species structure** of a community depends on the nutrient level of the soil or substratum. For example, a habitat like a rock face or a mobile sand dune has a low nutrient level and is exposed to such harsh conditions that only a few plants can survive there, so that it has a poor species structure. These communities are more susceptible to the ravages of nature, such as an excessively long dry period, and of man, such as the introduction of a parasite. If the population of one species is seriously affected, there will be fewer species to buffer the damage and this may throw out the stability of community. Communities that are richer in species, on the other hand, are more stable. Any sudden change in the population of one species in a community is compensated for by other species. This is because more ecological niches are generally present. The buffering effect makes communities resilient to sudden changes in the environment.

It is often noticed that during the early stages of succession, the bulk of the community is made up of relatively few common species. The diversity which later stages of succession often bring depends on those species that remain uncommon, and it is they which contribute a reserve of adaptability to the seral community. If a community is damaged by, for example, pollution or by the indiscriminate use of insecticides, a decrease in the diversity sets in with the least tolerant species dying out first, followed by others as the pollution continues. This particular type of problem is at present more common in industrialized countries, but it is gradually beginning to appear in some developing and tropical countries too.

There are a number of methods of measuring species structure. One of the simplest estimates which can suit our purpose at this stage is the **diversity index**, measured by the ratio:

$$\frac{\text{Total number of species recorded}}{\text{Log. of total number of individuals counted}}$$

Graphically, it is the slope of the straight line obtained by plotting the running totals of different species in different sample areas counted against the logarithm of the total number of individuals. It should be noted that the slope (or the diversity index) obtained for a community as a measure of its species structure is independent of sample size.

Synthetic Characteristics of the Community

A number of the characteristics which have been discussed above are used ultimately in distinguishing between different communities. But some of these parameters when combined give expression to features which are more useful for classification. It is now intended to discuss some of these in relation to the identification of communities.

Dominance

Dominance is the characteristic of a community which expresses the predominating influence of one or more species in it, so that populations of other species are more or less subordinated or reduced in number or vitality. Dominants are those species which are so highly successful ecologically that they determine to a considerable extent the conditions under which associated species must live. This obviously means that all populations in a community are not of equal importance in determining the characteristics of the community or the interaction of the community with its habitat.

As shown earlier, cover and population density are the principal parameters which help us to determine the dominant species. However frequency, height, life form and vitality are also important in this regard. It should be pointed out that height alone is not enough. In the tropical dry savannas the scattered trees or tall shrubs are not sufficiently numerous nor sufficiently close together to enable them to dominate the densely growing grasses. The dominants here are the grasses because they exert more influence on the habitat and on other plants and animals. In the relatively wet savannas, however, where shrubs and trees grow close enough together to form a canopy, they then become the dominants in the upper layer while the grasses form the dominants in the field layer. The removal of any dominant species would result in a drastic change in the character of the community, whereas a removal of a non-dominant species would not have such a far-reaching effect.

Communities in temperate climates, pioneer communities generally and agricultural communities have few dominant species and in some of them there are indeed only single dominants. But, as already referred to, natural tropical communities tend to have a large number of species as dominants.

In practice, synthetic tables are constructed in which the names of the species, their density or cover, and frequency, provide the basic data for the qualitative determination of dominants. Usually species occurring in more than 80 per cent of the sample areas of a community and with the highest cover or numerical abundance may be recognized as the dominant species.

Zonation (Gradient Studies)

In certain habitats it is noticed that within a limited range the vegetation occurs in the form of a series of more or less parallel or concentric bands. This is referred to as zonation. The extent of the area covered by any band varies considerably from several kilometres in vegetation belts to a few millimeters on agar in a petri-dish. Zonation will be illustrated by the broad vegetation belts of West Africa, the montane vegetation, zonation around a pond, strand (coastal) vegetation and marine algal zonation.

The cause of zonation is not always clear and it is evident that the cause cannot always be the same in each type of zonation. However, by its very nature, zonation must be due to a gradient of some particular over-riding factor in the habitat.

Zonation of Major Vegetation Belts

The West African vegetation exists in broad zones which run approximately parallel with the equator. These zones are easily appreciated even though the boundaries between them are not always clear. Firstly there is the forest zone which lies near to, but does not necessarily reach, the sea except at a few points. Beyond this (landwards) is the southern band of relatively moist savanna with broad-leaved trees (the Guinea savanna). Next comes the zone of relatively dry savanna with reduced tree cover and fewer broad-leaved trees (Sudan savanna). This is followed by the zone of much drier savanna which has small trees with thorns and small leaves (Sahel savanna). Finally comes the Sahara desert (Figure 12). As stated already, it appears that this type of broad zonation of vegetation in West Africa is due largely to rainfall differences and not to temperature differences, which are rather slight.

Montane Zonation

This type of zonation is related to altitude. Many mountains exist in the tropics which show this type of zonation as one moves from the base to the top of the mountain. On the mountains in the tropical forests, beginning from the base of the mountain, there is the typical rain forest of three tree storeys, followed by a type of forest with only two tree storeys, and then by a forest with heavy mist and a profuse growth of mosses with only one tree layer. This type of zonation evidently results from different temperatures at different heights on the mountain, since a rise of 100 m results in a drop of temperature of about 0.4–0.7°C. As a result of the low temperature at the top of high mountains some of the species there are temperate plants. Actual examples are described in detail in Chapter 10.

Zonation around Lagoons

Tropical coastal lagoons often show zonation of species from wetter to drier land. The first zone often consists of mangrove plants such as species of *Rhizophora* or *Avicennia*. The lowest roots of these plants are permanently inundated. Approaching drier land the water becomes shallower, more prone to drying up and less brackish. In low rainfall areas the mangrove zone gives way to herbaceous swamp with grasses and sedges which are salt tolerant. In high rainfall areas there may be herbaceous swamp comprising less salt-tolerant species or swamp forest with palms, adjacent to the mangrove. In Chapter 10 actual examples from different tropical areas will be described.

Strand Zonation

On the coasts there is a progression from the extreme conditions of the beach, which only a few species can withstand, through much less extreme conditions which more species can tolerate, to more favourable conditions suitable for the climax vegetation of the region. The extreme conditions of the shore line include wave action, strong wind, salt water as well as a sandy or rocky substratum. Only a few species with physiological or structural adaptations for successful living in this belt are found. As one moves away from this area one comes into a zone where there is relatively more soil and here more species are able to live. This zone may have different dominant species which show their own adaptations to it. This may be followed by a transitional zone where some plants from the climax vegetation of the region are found. Finally, normal land conditions are reached and species of the climatic climax of the region predominate. Again, this is more fully discussed in Chapter 10.

Marine Algal Zonation

This is exhibited by the inter-tidal rocks of both tropical and temperate shores. It is noticeable that the living organisms on the rocks are arranged into horizontal zones or belts. The top part of the rock is often exposed at low tide and so this zone tends to become periodically very dry. The organisms that are able to live here often consist of animals which arrange themselves into two or more zones according to the increasing humidity down the rock. First there is the zone dominated by snails such as *Littorina punctata* and *Tectarius granosus*. Next is the zone of small barnacles, often being species of *Chthamalus*. Also present here are *Cithothamnia* and small tufts of red algae such as *Pterocladia pinnata*. From the mid-tide level downwards the rock surface is almost always wet, and in this zone is found a concentration of other algae such as *Cracillăra* sp., *Sargassum vulgare* and *Polysiphonia* sp. (Figure 13).

The different organisms arrange themselves into zones as a reflection of their tolerance to exposure,

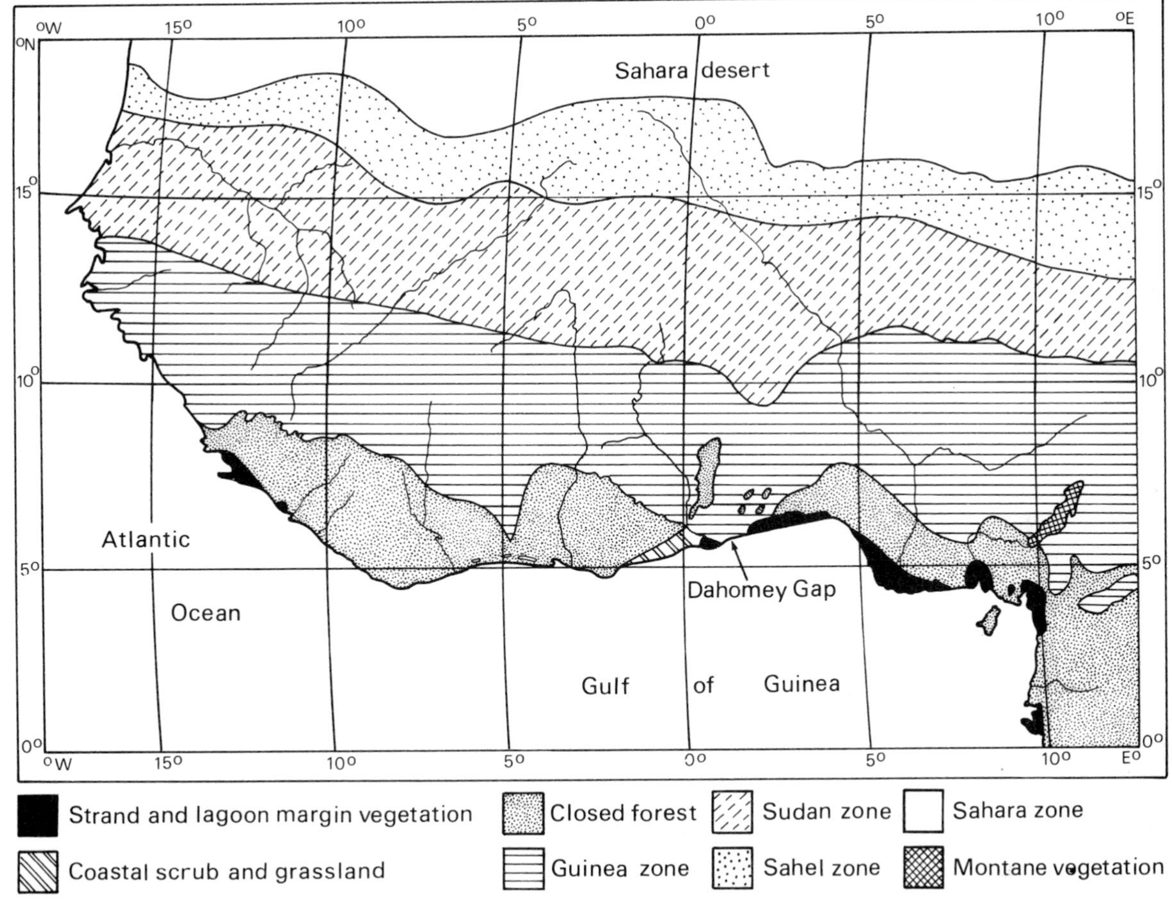

12 Vegetation zonation in West Africa
SOURCE: J Brian Wills (1962)

desiccation, high light intensity and also temperature. Where there is a broad inter-tidal area of irregular configuration the zonation may be spread so widely as to be distinct. A good example is afforded by rocky shore zonation described by Lawson (1955) for the Cameroons (Figure 14).

Zonation in Relation to Succession

Zonation on the ground often appears to mirror plant succession. On sand dunes it may be possible to notice stages of a primary xerarch succession such as is found on the strand. A hydrarch succession may also be found in mangroves (Chapter 10), estuaries of salt marshes or lagoons or in marshes at the edge of a fresh water lake. It should be borne in mind that the similarities cannot be taken too literally. Even though zonation is said to recapitulate succession, it should still be remembered that this is only an approximation and that the real test lies in experimentation. Thus, for example, while the zonation on the sea shore is maintained primarily by tidal exposure, the nature

13 An example of zonation between tide marks on the coast of Ghana. The upper part of the rock is covered by *Lyngbya conferivides* which soon gives way to a white belt of barnacks. Below this is a belt of *Enteromorpha* and, lower still, tufts of *Ulva*, partly covered by sand
SOURCE: G W Lawson (1966)

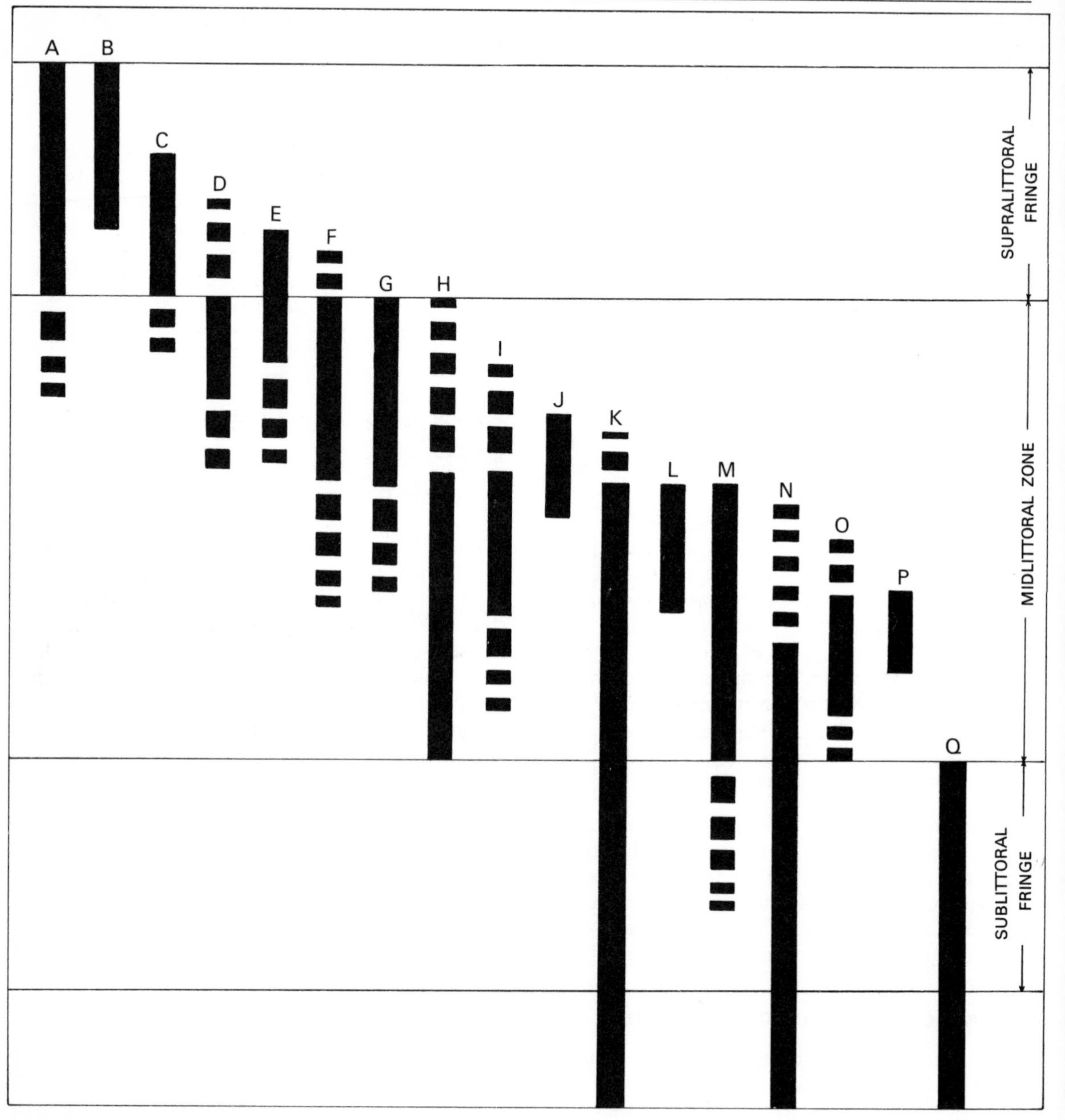

14 Diagram to show the zones occupied by some of the commoner or more ecologically important intertidal organisms of the Cameroons
SOURCE: G W Lawson (1955)

A *Tectarius granosus*
B *Rhizoclonium riparium*
C *Bostrichia tenella*
D *Cladophora camcrunica*
E *Centroceras clavulatum*
F *Nerita senegalensis*
G *Chthamalus dentatus*
H *Gelidium pusillum*
I *Siphonaria pectinata*
J *Ostrea* sp.
K *Lithothamnia*
L *Gymnocongrus nigricans*
M *Pterocladia pinnata*
N *Balanus tintinnabulum*
O *Caloglossa lepricurii*
P *Chactomorpha antennina*
Q *Sargassum vulgare*

of succession at the same site can be obtained only if the tidal exposure is cut off in a portion of the post-tidal dunes. The detailed comparison of zonation and succession remains a fruitful area for research.

The Catena

In savanna, especially in Africa, where the land is undulating, a sequence of somewhat different but related subsoils frequently occurs from the higher ground to the valley bottom and this is repeated several times on the undulating plains. The subsoils of the higher ground are often red, brownish-red or orange-brown in colour, typical of well-drained soils in the tropics. However, further down the subsoils are lighter in colour, such as brown, brownish-yellow or yellow. These are soils which are less well-drained than the red ones. At the flat bottom below the slope, subsoil colours range from greyish-yellow to nearly white. These are poorly-drained soils often flooded during the rainy season (Ahn, 1970).

These soil differences are reflected in differences in the vegetation along the slope, with each soil type supporting plants that are adapted to it. When such a sequence of soil and vegetation is repeated over a wide area of vegetation it forms a pattern of zonation often referred to as a 'catena' (see also Chapter 8).

Growth Rates

The luxuriance of plants in tropical communities would seem to be due to increased growth rates. Recent experiments in some tropical countries have shown that the rate of growth of plants such as maize, sunflower and tomato is much greater in the tropics than in the temperate climates. It has been suggested that the increased rates of growth may be due to the high tropical temperature. This is not yet proved.

Gregarious Flowering and the Flushing of Leaves

As already referred to, there are always some species of plants flowering at any time of the year in the tropics. Some plants respond in subtle ways to changes in temperature, moisture regime and day length. These environmental factors, acting singly or in combination, may be the necessary stimuli to flowering or leafing. Most tropical trees tend to 'flush' rather brightly-coloured leaves which remain soft for some time before they stiffen. These peculiarities in tropical plant communities are discussed in greater detail in Chapter 5.

Chapter 4 The Role of Environmental Factors in Tropical Communities

The environmental factors to be considered are those common physical and biological factors which normally operate in terrestrial communities. These may be classified as shown in Table 5.

Climatic Factors

Rainfall and other Precipitation

Water generally constitutes one of the important factors whose variation is reflected in the nature of the vegetation, and so in the habitat for particular animal communities. Variation in rainfall in the tropics is shown not only in total amount but also in the way it is distributed throughout the year. In the rain forest the total rainfall per annum usually ranges from about 1600 to 4000 mm. Very high rainfall occurs in a few areas, such as Debundja at the foot of the Cameroon Mountains in West Africa with about 10 000 mm, and at Cherrapunji in the Khasi Hills in India with about 12 000 mm per annum. Where rainfall is less than about 1000 mm per annum, woodland savanna type of vegetation is found while in the deserts rainfall may be as low as 300 mm.

The seasonal distribution of rainfall is such that it is only in parts of the lowland tropics that it is virtually continuous throughout the year. Moving away from the equator the forests show certain periods of reduced rainfall during the year usually lasting for about three or four consecutive months. Thus even at Cherrapunji with its very high rainfall total there are some four consecutive months with less than 100 mm each. In general, the pattern of seasonal distribution of rainfall in the tropics is related to the distance from the equator. Between the equator and the latitudes of the tropics (Cancer at $23\frac{1}{2}°$N and Capricorn at $23\frac{1}{2}°$S) there can be two dry periods and two wet periods in each year. Rain follows the sun in its apparent movements northwards and southwards because, where the sun is overhead, greater convectional air movement results in the formation of more cloud. Between the tropics and the poles the sun is never overhead but reaches its highest elevation once a year at midsummer. In temperate regions there is only one warm and one cool season in each year. It follows that from the equator to either tropic, one of the intervals between the two occasions when the sun is overhead gets longer while the other gets shorter closer to the tropic latitude. At and near the equator wet and dry seasons are of about equal length and may be less intense. This general picture is illustrated by Figure 15. As may be expected, various other factors modify the expectation in certain areas. Among such factors are irregular distribution of land and sea, ocean currents, the deflexion of air currents by mountains, and the incidence of monsoon winds. Local conditions may also affect not only the pattern of rainfall but also the total for the year.

The West Indies, the Malay archipelago and the Philippines have tropical island biota which differ in some respects from those of the continental masses of Africa, India or South America. The climates of islands are determined by their size, height and the prevailing wind. The prevailing wind in most inter-tropical latitudes is the easterly trade wind and where this wind has passed over an ocean it is saturated with water vapour. Low islands may have very low rainfall but mountainous islands, like continental mountains, have higher rainfall and shorter dry seasons. Some island systems are like continental systems in miniature so that, for example, in Jamaica and Cuba the mountains receive high rainfall and support rain forest, with moisture-loving species such as tree ferns, especially on the eastward or windward slopes of their mountains; but to the leeward a rain shadow is created and within a few kilometres of the humid forest there can be an area of very low rainfall capable of

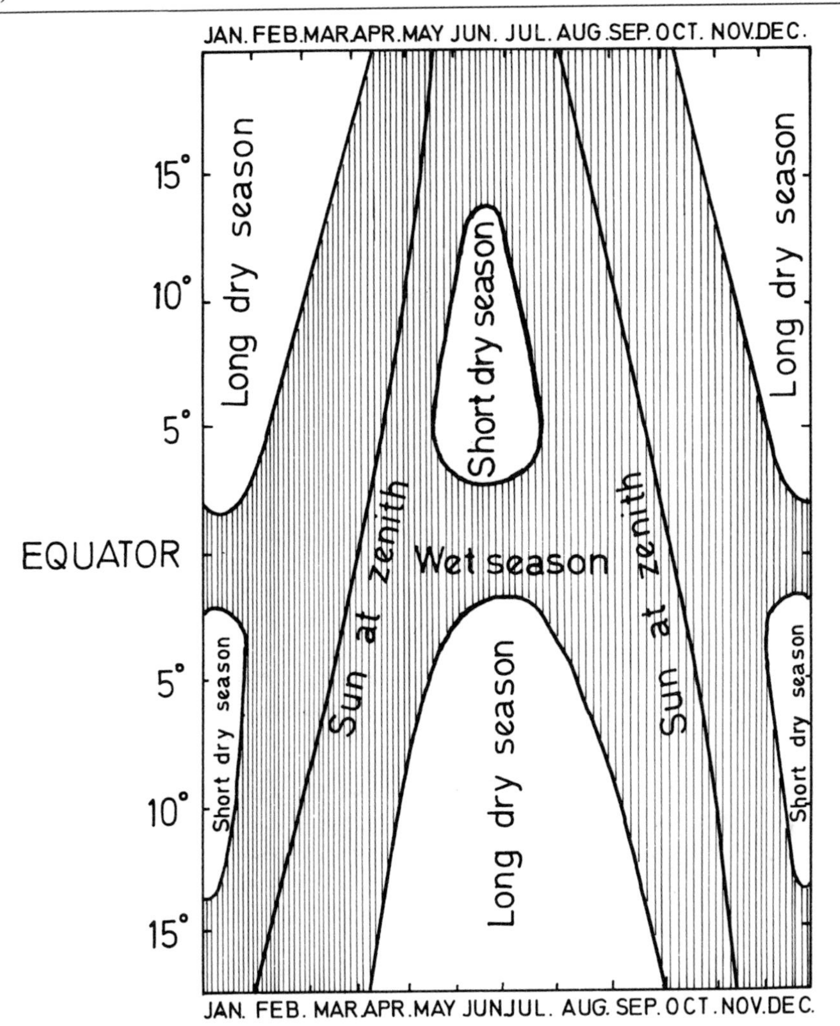

15 Wet and dry seasons in the tropics in relation to latitude near the equator.
SOURCE: P W Richards (1952)

Table 5. Classification of common ecological factors

PHYSICAL (ABIOTIC) FACTORS	BIOLOGICAL (BIOTIC) FACTORS
Climatic Factors 1. Rainfall and other precipitation 2. Temperature 3. Atmospheric humidity 4. Wind 5. Light 6. Energy balance Physiographic and Edaphic Factors 1. Topography 2. Edaphic (soil) factors 3. Geological substratum	1. Green plants or other plants or animals 2. Interactions between organisms Predation Scavenging Symbiosis Parasitism 3. Animals or plants or other animals 4. Man

supporting only a near desert with cacti and thorny shrubs.

Tropical rain forest can be found also in areas with two distinct rainfall seasons as well as in areas where local conditions help to increase the rainfall during any relatively dry season. In some areas there are compensating factors like high humidity in the dry periods, as at Akilla in southern Nigeria or on the Accra plains in Ghana, or edaphic moisture as is generally found with the post-climax riverine (gallery) forest and forest outliers. Some areas have total rainfall sufficient to support tropical rain forest but long and severe periods of drought prevent its development. In the regions of two rainfall maxima there are deciduous forests marked by a certain proportion of deciduous trees and tuberous geophytes. The length and severity of the drought period in this way modify the climatic climax of tropical forest. When the rainfall decreases further the vegetation alters again. Where the rainfall is less than 1600 mm there is woodland savanna vegetation and where it falls below 300 mm there are desert conditions. This is depicted by the zonation of vegetation in West Africa shown in Figure 12.

Rainfall may also modify soil salinity and thus affect the vegetation of an area. In a region with tropical rain forest as the climatic climax the extent to which the forest encroaches upon the strand vegetation is a function of the total rainfall. In Ghana, for instance, the strand vegetation is widest at the eastern end of the coast-line which has the lowest rainfall. Towards the western end, where rainfall is heaviest, forest extends in some places almost to the high tide mark.

The effects produced by seasonal changes in rainfall on some plant species are discussed in Chapter 5.

Temperature

Temperatures in the tropical lowlands never drop to freezing point, and range mostly between about 20 and 28°C. The high tropical temperatures are due to the high angle of incidence of solar radiation, the small annual change in day-length and the heat capacity of the oceans and the soil. The mean varies only very gradually from one season to another. The difference between the hottest month, in the dry season, and the coldest month, which occurs in the wet season, can be only about 1°C or less on tropical islands and about 5°C on the equator in continental areas. Seasonal variation, however, increases progressively as one moves away from the equator, but even here a difference of 13°C is about the greatest often encountered. The high temperatures of the lowland tropics are due largely to the higher minima and not so much to the maxima which, near the equator, reach about 33°C. High temperatures are in fact lower than the maxima found in southern parts of Europe. It is only in the hot deserts nearer the limits of the tropical latitudes that temperatures exceed 45°C in the middle of the day in the dry season.

The daily range of temperature varies as much or more than the range experienced throughout the year. The greatest differences between day and night temperatures are found during the dry periods in inland savanna areas. These differences naturally affect the physiological processes of plants as well as their anatomy.

As already mentioned in Chapter 3, temperature decreases with altitude on high mountains. In the tropics the mean temperature decreases by about 0.4–0°C for every 100 m rise. In this way, low temperatures are found on the tops of tropical mountains. As the temperature drops with elevation, the vegetation is zoned and the species change. At these higher altitudes a number of temperate species may be found (see also Chapter 3). For example, on the Cameroon Mountains of West Africa (about 4000 m high) are found common northern temperate genera like *Galium, Veronica* and *Ranunculus*. On Mount Kilimanjaro in East Africa three species of *Collembola*, three species of mites and a fly are found which are otherwise found in Europe and North America.

Temperature decreases with depth in water. For example, it is estimated that at a depth of 1500 m in the Gulf of Guinea the temperature has fallen from the normal surface figure of 30°C to about 5°C. However, the fall in temperature is not proportional to the depth all the way down to the seafloor. In the oceans and in shallow lakes it has been found that at a depth of 120 m the temperature of the water begins to drop more rapidly until at about 175 m the decrease becomes slow again. This zone of rapid temperature change forms a barrier, known as the **thermocline**, between the upper body of water of high temperature (the **epilimnion**) and the lower body of low temperature water (the **hypolimnion**). (See also Chapter 7.) This situation is, however, not permanent for, when the atmospheric temperature falls at night, so does the temperature of the water of the epilimnion, which tends to sink to the bottom as its density increases. The convection currents thus set up help to mix up the water. The mixing can also be effected by strong winds.

Tropical and subtropical temperatures are modified more along the western coasts of tropical continental land masses than they are on the eastern coasts. This is because of the upwelling of deeper cold ocean water or because of polar ocean currents. Thus tropical littoral animals are found from the northern coast of Peru at only 5° or less south of the equator and northwards to northern Mexico or southern California.

Recent examination of the flora of the ocean floor of the Gulf of Guinea has shown that many of the species of algae at the bottom of the seas are the same as those found nearer the surface of the sea in temperate climates. This finding thus parallels the presence of typical temperate plant species on high tropical mountains and demonstrates in a most convincing manner the role of temperature in the distribution of plant species.

It has been shown also that major differences in the range of tolerance of plants to temperature accounts for their different distributions. It is reported (Evans *et al.*, 1964) that temperate festucoid grasses and the more tropical non-festucoid species differ markedly in their response to air and soil temperatures in temperate and tropical climates respectively. But even more significant is the difference in their response to lower temperature. The more temperate (festucoid) genera such as *Poa, Dactylis*, etc. grow actively in temperatures of 15°C and below, whereas many of the non-festucoid genera such as *Paspalum,* and *Sorghum* stop growing or are killed by temperatures between 10°C and 15°C. The inability of many tropical species to survive moderately low temperatures accounts for their distribution, but whereas the panicoid grasses are not well represented in temperate regions, some festucoid grasses, including *Eragrostis, Chloris* and *Eleusine*, are well represented in the tropics.

These temperature effects are naturally reflected also in seed germination. The optimum temperatures for the germination of most seeds are between 15° and 30°C, but higher optima of 35–40°C have been reported by Williams and Webb (1958) for tropical species of *Paspalum* and *Saccharum* (sugar cane). In general the temperature range for germination of tropical species is shifted to higher temperatures, with a minimum between 10° and 20°C, whereas many temperate species such as wheat and pasture grasses germinate at temperatures as low as 0°C.

Atmospheric Humidity

Atmospheric humidity, which is a function of the amount and duration of rainfall, the presence of standing water, and the temperature, is an important environmental factor which may determine the presence or absence of some plants and animals in particular habitats, Evaporation from rivers or standing water, the sea and the soil, and transpiration from plants, constitute the major sources of water in the atmosphere.

Atmospheric humidity may be measured by a number of methods. The most useful measure for ecological work is the evaporating power of the air. This is the saturation deficit and it is calculated from simultaneous readings of wet and dry bulb thermometers. The other measures of atmospheric humidity are percentage relative humidity (also calculated from wet and dry bulb thermometer readings or by means of the hair hygrometer) and the rate of evaporation as measured by the evaporimeter. The latter should provide the more direct method, but in practice measurements of rate of evaporation have proved difficult to compare from one place to another. Relative humidity is variable with temperature, so it is advisable not to use relative humidity figures in environments with wide fluctuations in temperature. Saturation deficit is independent of temperature.

Seasonal variation in relative humidity generally parallels variations in rainfall. In tropical rain forests the mean relative humidity in the morning may vary from 95 to 75 per cent, and in deciduous forests from 85 to 75 per cent, but it can fall to about 55 per cent and often much lower in the dry season. In the tropical savannas it may be about 60 per cent but can fall to as low as 10 per cent in the dry season. In the hot deserts relative humidity is less than 50 per cent and can drop to 5 per cent in the dry season.

Atmospheric humidity tends to decrease from the coast inland. This is easily observed as one of the important factors resulting in the climatic gradients. It has been shown, for example, that the distribution of evergreen and deciduous forests in the southern part of Nigeria seems to be determined by the distance inland reached by moist air from the sea. On tropical mountains, atmospheric humidity appears to rise with increasing elevation up to a certain height (which varies in different situations) where it reaches saturation. Here forests are engulfed in perpetual cloud and drizzle. Often above this saturated level on high mountains, atmospheric humidity falls and savanna-like montane grassland exists.

Figures of annual average and annual minimum relative humidities (percentages) for a few localities in the tropics are shown in Table 6.

Table 6. Annual average and annual minimum relative humidities for some localities in the tropics (%)

LOCATION	ANNUAL MEAN (MORNING)	ANNUAL MINIMUM (MIDDAY)
Douala, Cameroon	95	94
Singapore	83	82
Mazaruni Station, British Guiana	94	93
Lagos, Nigeria	85	81
Eala, Zaïre	73	70
Port of Spain, Trinidad	93	91
Djakarta, Java	87	81
Kumasi, Ghana	96	95
Rangoon, Burma	85	77
Sokoto, Nigeria	52	24

The high humidities are reflected in the wet soil surface and the rate at which various materials became mouldy in the forest.

The high relative humidities of the humid tropics contrast with those of temperate climates where, at an average temperature of 15°C, the relative humidity is often as low as 6 per cent.

In spite of these apparently high relative humidity figures it is remarkable that during the day the humidity can fall to very low figures for short periods. Thus, for example a daily minimum of 51 per cent at Singapore and 56 per cent at Eala in Zaïre have been recorded. Relative humidity is almost always at saturation point at night resulting in deposition of dew on plants, and it reaches its lowest point near midday. Evaporating power in terms of saturation deficit is lowest at night and at its highest near midday. The best results of this are given by Evans (1939) from a tropical rain forest in the Shasha Forest Reserve in Nigeria, as shown in Figure 16.

The importance of particular levels of humidity at the micro-habitat level for the lives of certain organisms has already been illustrated in Chapter 2 by the upward and downward movement of two species of mosquitoes in response to changes in humidity during the day.

Wind

In tropical regions wind velocities are generally lower than in temperate regions which have frequent violent winds. It is in a belt in the West Indies and Western Pacific that violent winds (hurricanes and tornadoes) are known to be frequent in the tropics. They occur to some extent also in the Indian Ocean. Thunderstorms are, however, common in many tropical areas. Strong winds known as line squalls which last for only a few minutes often precede these thunderstorms. One effect of the squalls is the destruction of trees covering sometimes many hectares, with the worst cases known in Nigeria and some parts of the West Indies. When hurricanes and cyclones occasionally occur in tropical regions they are equally destructive to trees and plantations. It is reported (Vaughan and Wiehe, 1937) that in Mauritius the cyclones appear to do relatively less damage to local natural vegetation than to plantations, and this is presumed to be due to better adaptation of species as a result of selection in course of time. Certain areas in the West Indies with frequent hurricanes and steep slopes have not been able to develop a forest climax vegetation but support only a type of sub-climax vegetation that is poor in large trees. The wind has thus been able to influence the structure and species composition of such vegetation.

The direction of strong winds on exposed sea coasts is also known to exert a marked effect on the physiognomy of the trees and shrubs. The common effect of this is the exhibition of asymmetrical growth by some of the trees and shrubs. This is largely due to the drying effect of the salt-laden wind on the development of buds on the windward side of the trees while buds on the leeward side are shielded by the shoots and are thus able to grow. Characteristic woodland or scrub vegetation with a trimmed appearance, rising towards the land, is thus developed (Figure 17).

The vegetation on the windward slopes of isolated peaks and ridges on tropical mountains also shows stunted growth and forms a tangle of compact vegetation. The mechanical effects of strong harmattan winds in West Africa on the

16 Daily march of saturation deficit in the undergrowth (broken line) and tree tops (continuous line), Shasha Forest Reserve, Nigeria. The top diagram shows the relationship during the dry season and the bottom, the wet season
SOURCE: P W Richards (1952)

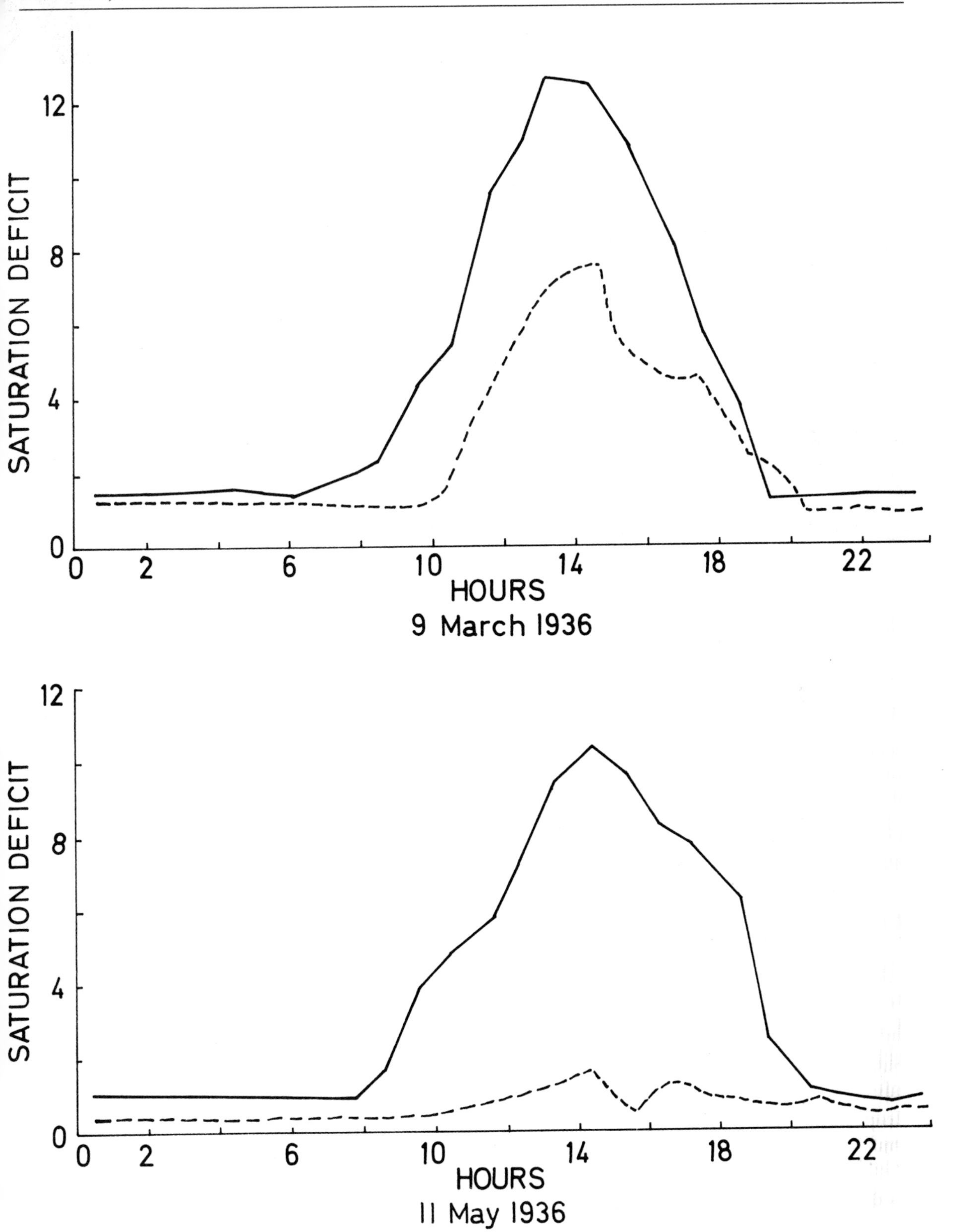
12
8
4
0
SATURATION DEFICIT
0 2 6 10 14 18 22
HOURS
9 March 1936
12
8
4
0
SATURATION DEFICIT
0 2 6 10 14 18 22
HOURS
11 May 1936

17 Coastal scrub vegetation with windcut effect

vegetation of mountain tops are responsible for the development of distinct windward and leeward forms of tree steppe and savanna thicket respectively (Jenik and Hall, 1966).

Desiccating winds in some parts of the tropics, which prevail for short periods each year, reduce the humidity of the vegetation. An example is the dry dust-laden wind of West Africa referred to above, the harmattan, which blows from the northeast from the Sahara Desert towards the Guinea coast. This further increases the dryness of the vegetation during part of the normal dry season and thus facilitates burning of much of the vegetation during that period. On high hills the harmattan favours communities of xeromorphic species resembling semi-desert forms (Jenik and Hall, 1966).

Light

Although the variation in the length of day in the tropics is up to a maximum of only about two hours, so that the daily possible amount of sunshine is never less than eleven hours, yet there is often a high degree of persistent cloud in the tropical zone and the period of bright sunshine is much reduced. Typical figures from a station in Guiana show a general average of about 5.5 hours a day with 6.3 hours a day as the average for the sunniest month and 4.4 hours for the least sunny month. The amount of sunshine in the tropical regions, particularly near the equator, can be surprisingly small.

Light also plays an important part in the distribution, orientation and flowering of plants. In the tropical forest light is a limiting factor and the amount of light penetrating through a forest canopy would appear to determine the layers or storeys formed by the trees. This is a reflection of the plants' demand for or tolerance of different amounts of light in the forest. It has been shown that the seedlings of emergents (the tallest trees) of the tropical forest require a lot of light and so have good early growth when they are not shaded. If they happen to be shaded their growth is considerably slowed down, but growth is very much accelerated when a gap occurs in the canopy and the seedling whose growth has earlier been arrested for lack of light receives more light.

It is evident that some plants are light demanding (or 'sun plants') while others are shade tolerant (or 'shade plants'). In physiological terms, 'shade plants' are plants with a low **compensation point**, and 'sun plants' are plants with high compensation point. (Compensation point is the light intensity at which the plant's energy income through photosynthesis is balanced by its expenditure in respiration.) In the tropical climate most plants may be deemed to be sun plants, but shade plants are found under canopies of the forests or in similar dark places. Examples of shade plants are species of *Costus* and *Geophila* found on tropical forest floors. In fresh water too, there are certain species of water plants which are often found in dark places and may thus be considered 'shade plants'. Examples are *Ceratophyllum* and *Chara*.

The concept of compensation point as a definite light intensity is difficult to apply in the field because the duration of daylight and darkness is always changing from day to day, and in any case the plant must receive a little more light above its compensation point during the day in order to grow. A better field parameter therefore is **extinction point**, which is the lowest percentage of full light in which a species is found under natural conditions. Here the measurement is in terms of the percentage of full daylight which the species requires in the field to keep it alive. Extinction point indicates the degree of shading that the species will tolerate, so that it will not be found in places with lower light intensity than the extinction point. In the undergrowth of climax forest there are many small plants which, by their short internodes and numerous leaf scars can be seen to be of great age. These are often saplings of trees whose growth has been arrested by inadequate illumination. When a gap in the canopy occurs, their growth accelerates.

In the aquatic medium, because carbon dioxide is more plentiful than it is in the air, the rate of photosynthesis depends more on light intensity than on carbon dioxide availability. Here the light reaching a submerged plant is reduced by reflection, through its absorption by the water itself, by floating water plants and by overhanging leaves of trees. A thick cover of the water surface by plant plankton, for example, can act as a colour filter through which predominantly green light, which is normally not useful for photosynthesis, would pass. Beyond a certain depth, depending on the light requirements of the species, submerged leaves may receive light intensities below their compensation point. At such depths all food reserves are used up, and the species may not survive, and certainly cannot colonize other situations. Where seeds can be dispersed and contain enough food reserves to enable them to germinate and grow above the critical depth, fresh colonization of the species becomes possible. In the deep-water plant *Crinum natans* it is observed in Ghana that flowering takes place in March–April when the level of the water is low enough to permit enough light to reach the plant.

In the sea there seems to be a general correlation between pigmentation and the depths at which different marine algae are found. For example, red algae are found in deeper parts than other groups of algae; here the red pigment which such algae have in their chlorophyll enables them to absorb the prevailing light much more efficiently because the red colour is complementary to the green light which succeeds in penetrating to those depths.

The demand for light by parts of a single plant results in what is often referred to as a **leaf mosaic**. In order to get enough sunlight for photosynthesis, the leaves of many plants tend to be adjusted in such a way that each of them shades the neighbouring leaves as little as possible. In this way they appear to be fitted together into a kind of mosaic. The plants achieve a minimum of overlapping of the leaves, which in effect means a maximum use of the sunlight falling on the plant. Some plants, of course, exhibit this feature more than others. Some of the tropical plants exhibiting this phenomenon more clearly are *Acalypha*, *Boerhaavia*, *Oxalis*, and *Artabotrys*. Densely-leafed climbers on walls and in the tropical forest also exhibit the same phenomenon.

Different tree species cast different degrees of shade. In deciduous trees the duration of this shade also varies according to the time of the year, that is, whether the leaves are shed in the rainy or dry period of the year and whether this corresponds with the flowering and fruiting period of the species (see Chapter 5).

As stated above, the length of time in a day of twenty-four hours for which certain plants are exposed to light may determine their flowering or continued vegetative growth. Contrary to an earlier popular belief, it has been shown that many tropical plants are sensitive to very slight changes in day-length, with respect to their flowering (see section on photoperiodism in Chapter 5). Light also plays a part in the germination of the seeds of some species; while the seeds of some species require light for germination (photoblastic seeds), such as tobacco, capsicum peppers, *Musanga* sp., *Terminalia superba* and *T. ivorensis*, the germination in others, for example garden egg or egg plant, is rather inhibited by light.

Since light or visible radiation forms the predominant part of the solar radiation received at the earth's surface, and since plants receive their energy for photosynthesis in this way, the energy budget of a habitat can be determined only in terms of energy units of radiation.

Energy Balance

Both plants and animals are irradiated from all directions with solar radiation, that is direct sunlight as well as light reflected from the ground, from other objects around them and from clouds. Solar radiation is important to plants because it is the only source of energy for photosynthesis. It also indirectly provides energy for all the life processes in the biosphere.

The surface of the earth re-radiates long-wave infra-red radiation to the atmosphere. This radiation goes beyond the atmosphere into outer space. The atmosphere itself also emits infra-red radiation into outer space as well as to the surface of the earth. Thus plants and animals are subjected also to thermal (infra-red) re-radiation from the surface of the earth, from surrounding objects, the clouds and the atmosphere. Plants and animals also loose heat in all directions into space, into the sky and to the ground. They lose heat continously by the process of convection, which can take place even in the absence of any wind. Plants also lose heat through transpiration (evaporation) while some animals may do the same through the evaporation of sweat. Solar radiation is received on the earth only during the day. Long-wave infra-red radiation operates both during the day and the night. This has important ecological significance to plant and animal life.

Of the gases in the atmosphere, oxygen, nitrogen and the inert gases, which make up over 99 per cent of it, readily transmit sunlight as well as the long-wave radiation from the surface of the earth to outer space. This is because these elements do not have any significant absorption bands in the infra-red wavelengths. Thus, if only these gases made up the atmosphere, the temperature limits on the earth would be much wider; that is, it would be much hotter at day and much colder at night. However, the minor constituents of the atmosphere, chiefly water vapour and carbon dioxide, have absorption bands which are located in the wavelength region where radiators at the temperature of the earth or the atmosphere emit most of their energy. Thus these minor constituents of the atmosphere exercise a measure of control on the radiation exchange between the earth, the sun and outer space which results in a reduction in the temperature limits of the biosphere. The amount of carbon dioxide by volume does not vary all over the biosphere, as a result of breathing, burning and decay, so that the flux of energy due to carbon dioxide will not reflect any minor variations. However, the water vapour content of the atmosphere varies considerably over the biosphere from about 0.1 per cent in the case of dry air to 1 per cent volume in the case of warm saturated air.

This means that in habitats such as deserts and in other situations when the atmosphere is dry, that is, when negligible water vapour exists and the sky is clear of clouds, any living organism is subjected to extremes of temperature. During the day the organism radiates heat and at the same time receives almost direct, and with no significant absorption by the atmosphere, the sunlight and radiation from the atmosphere, resulting in rather high temperatures for the organism. At night on the other hand, while the organism radiates heat, it does not receive any sunlight or thermal radiation, resulting in a much lower temperature for the organism. If during the day a layer of clouds moved in overhead, resulting in an increase in the moisture content of the atmosphere of the locality, the increased water vapour would absorb some of the sunlight and thus reduce the radiation reaching the organism. If this happened at night the increased water vapour of the clouds would emit more thermal radiation to the ground than was being emitted by the organism, resulting in a feeling of more warmth by the organism than the air temperature would indicate. This explains the hot days and cold nights experienced by organisms in tropical deserts (see Chapter 11).

Physiographic and Edaphic Factors

Topography

Physiographic factors are inanimate factors peculiar to the precise location of a habitat. One of these factors is topography which deals with features of land surface and includes altitude, slope of land, geological substratum affecting drainage, erosion and silting. These features of land surface have their influence on the nature and distribution of plant communities.

We have already seen the affect of altitude in relation to zonation on tropical mountains and in the occurrence of some temperate plant genera there. High elevations are also more exposed to extreme expressions of climate like wind, low temperature and humidity. Topographical effects are also shown with respect to rainfall as has already been discussed. Forest outliers, too, appear in savanna areas where the presence of more water in depressions results in post-climax forests. The shape of the landscape may partly determine the amount of radiant energy that reaches the ground from the sun. This explains the special types of community which exist in cliffs, caves, gullies and on steep hillsides. In inter-tropical latitudes sunlight can fall on both northern and southern aspect slopes at different times of the year. In northern latitudes, and conversely in southern latitudes, a steep slope facing the pole may never receive direct illumination.

The degree of slope also affects the movement of water and soil, so that erosion is greatest on hilltops and least in the valleys. Serious erosion on hills in the tropics produces gullies on the hillsides. In some tropical countries one of the effects of this on the undulating plains of savannas is the catena. Further details of soil catenas will be given in Chapter 8.

Edaphic (Soil) Factors

These are considered in Chapter 8.

Geological Substratum

Rocks that have an excess of or lack of certain minerals tend to cause restriction to growth of certain plant species. Within a given climate the geological substratum is responsible for much of the variation in soils and vegetation. The geological substratum consists of the different sorts of rocks which make up the earth's crust. Each of these rocks has its own particular mineral composition and characteristics. These rocks, serving as the parent materials of the soil, often lie immediately below the soil. Particularly where these rocks are overlain by alluvium, sand dunes or other thinner layers of material, the chemical nature of the underlying rock greatly influences the nature of plants that grow in those areas, to the exclusion of others that are not tolerant of that particular chemical composition. This is more marked in dry climates, and in West Africa the 'hard pan' of laterite in savanna areas exerts much influence over the vegetation in those areas (see examples in Chapter 8).

Biological (Biotic) Factors

Lowe-McConnell (1969*a*) has demonstrated from her studies of some tropical fishes that in the tropics biotic factors are relatively more decisive than they are in temperate regions. Dobzhansky (1950) suggested that the nature of selection differs in tropical regions from that in high latitudes. His contention, as put by Lowe-McConnell, was that

> 'Where the physical environment is harsh and indiscriminate (density-independent) and catastrophic mortalities are likely to occur, selection will take the form of physiological adaptations to the controlling factor (such as cold or drought) and to increased fecundity. But under most tropical conditions the individuals that survive and reproduce will be those that are most attuned to the complex inter-relationships of the organic community. Thus selection is here a more creative, moulding, process capable of producing subtleties such as concealment behaviour and mimicry. It would therefore appear that at temperate latitudes the environment creates or selects the creatures, whereas in the tropics the creatures play a much more active part in creating the environment.'

Green Plants (Photosynthesis) and Competition

Green plants play an indispensable role in the ecosystem through photosynthesis which they carry out with the aid of water, carbon dioxide and light. The food they manufacture serves not only themselves but also all non-green plants as well as animals in the habitat. They also exert influence in their habitat through **competition**, by modifying many of the environmental factors, and thus help to determine the local climate (micro-climate) through such means as shading, use of water or minerals and the addition of leaf litter to the habitat. In the terrestrial habitat, these influences are even greater since the trees, shrubs and grasses are often large plants. Shading by large trees in the tropical forests can reduce the temperature in the canopy by as much as 5°C.

The effect of shading depends on its intensity, the height of the leaves casting the shade and the variation in the quantity of leaves borne during the year, that is whether the leaves are shed at one period or not. The nature of the sunflecks which penetrate the canopy to the ground may also be important to the ground vegetation. For a species to be successful in competition it should be able not only to shade out competitors but also to maintain its numbers and spread to new ground at the expense of its rival species. The latter is achieved through a complete cycle of the life of the plant; that is, pollination must be effective (both self and cross-pollination), seed dispersal efficient, germination of seeds good, and the growth of seedlings rapid. For example, the grass *Andropogon* spp. often replaces *Imperata cylindrica* (Lalang grass) because of the former's greater height and the shade it casts on the *Imperata* as well as its high degree of aggressiveness in spreading. *Paspalum vaginatum* grass can also replace *Sesuvium portulacastrum* through competition in a lagoon or a mangrove basin.

Unlike an aquatic habitat, water is often in short supply in the terrestrial environment and the influence exerted by the large plants includes the use of water. In the aquatic environment, however, there are often no large plants like trees, and in any case water is not in short supply, so that a modification of the environment through the use of water does not arise. However, in some tropical swamps where oxygen-consuming micro-organisms are abundant, these compete for oxygen which is easily reduced in amount owing to high temperature, the limited movement of water and low photosynthesis resulting from the limited light penetration.

All plants whether large or small compete for light, minerals or space and, in the terrestrial habitat, also for water which may be limited. The most acute competition is between individuals of the same species, since their requirements are essentially the same. Species with different requirements may often not compete for certain factors. For example, shade-tolerant species do not compete with light-demanding species for light. Competition for the same general environmental requirements results in diversity in the structure of the plants of a population. Such competition is demonstrated when individuals of the same species are grown in different densities in pure culture. Competition among the root systems of plants can be quite intense. Thus the possession by a species of suitable adaptations to the main environmental factors of a given habitat is no guarantee that a particular species will succeed in the community.

In some types of relationships complex physiological reactions take place involving the production and excretion of organic compounds by one plant which may inhibit or retard its own growth or the growth of other plants in the habitat. Recent studies have shown that the organic compounds concerned are often phytotoxins, mainly phenols and terpenes which are released into the environment as leaf leachates or root exudates by certain plants. The presence of plants which exude such chemicals is inhibitory to the growth of other species in the vicinity. The effects of the chemicals may result in complete inhibition of growth or in stunted or retarded growth. When complete inhibition occurs this may be noticed in the form of bare areas around the trees which exude the chemicals. This phenomenon is now known as **allelopathy**. It should be obvious that allelopathy can be a factor in plant succession, in single-species dominance (see Chapter 9) and in the patterning of vegetation generally. Most of the well-known examples of this phenomenon come from temperate vegetation. Among the known subtropical examples is the non-gregarious subtropical forest tree *Grevillea robusta*, whose phytotoxins, in the form of rainwash from leaves, have been shown to be toxic to the seedlings of the species itself but not to those of other species. This example illustrates how allelopathy may contribute to the mixed nature of tropical forests.

It has been shown that many animals exhibit growth inhibition when kept in overcrowded conditions in laboratories. Among these are aquatic snails and tadpoles, larger individuals of which appear to release more growth inhibiting substances which suppress the growth of the smaller members of the population. An example is the aquatic snail *Biomphataria sudanica* found in tropical Africa (Berrie and Vissler, 1963).

Finally, it should be mentioned that some green plants grow while standing on other green plants. These are called **epiphytes**. They grow in the soil and other debris which may be collected at the junctions of branches and on the main trunk of forest trees. They naturally intercept some of the light reaching the tree on which they grow and thus modify the environment of such trees. Examples are *Platycerium* (the staghorn fern), *Polypodium, Microsorium* and a number of epiphytic bromeliads and orchids. Epiphytes and the arboreal ecosystem are discussed in detail in Chapter 10.

Interactions between Organisms (Some Inter-specific Relationships)

As has been pointed out, animal food is one of the central agents which triggers off most of the relationships between organisms in any community. Some organisms may struggle and compete for food, others may feed on other organisms, while still others have evolved ways of avoiding their prey or of establishing other special relationships with other organisms. Relationships among organisms in a community may be considered in four categories, namely, predation, scavenging, symbiosis and parasitism.

Predation

This is the phenomenon by which one animal feeds on another animal after catching and killing it. The animal that feeds on another is the predator (carnivore) and the animal fed on is the prey. A relationship is thus established in a community between these groups of animals. It is estimated that lions, for example, kill about 8 per cent of game animals a year in the East African national parks.

Predators are aided in their mode of life by the high development of their senses and their organs for catching and holding. Predator birds, like the owl or the eagle, have hooked beaks and acute eyesight. The owl is able to locate the source of an echo and a number of predator mammals often have keen senses of smell. Also special structures assist predation. Some predators have strong jaws and cutting teeth as well as sharp prehensile claws, toads have long tongues, rattle snakes have fangs and insects have mandibles. The prey have also evolved methods of avoiding being preyed upon. Like the predators some of them have sharp eyes and an acute sense of smell, but often they are adapted to high speeds in running, flying or swimming. They also exhibit various forms of protectice colouration (which is discussed in detail below).

The only predaceous plants are the insectivorous ones among which the tropical examples include *Nepenthes*, species of *Utricularia* and *Drosera* (see Figure 18). These plants have morphological and physiological adaptations for trapping and digesting insects. Small crustaceans may also be trapped and digested in water by *Utricularia*. The particular adaptations of each of these are commonly described in most biological textbooks and will therefore not be described here.

Some animal prey have various devices whereby they create an optical illusion in order to disguise their true identity in their communities. These devices are referred to as **camouflage** and can be rather elaborate and effective since it makes the animals concerned resemble other organisms. One of the commonest ways of camouflaging is by imitation of the background colour. Here the animal blends with its surroundings to such an extent that it is difficult to see it. Examples include the grey-brown insects on the bark of trees or the green insects which inhabit grasses and leaves. The African lion, whose hair colouring resembles the surrounding savanna vegetation, is able to creep up to its prey or lie in wait for it without being noticed. Other animals are able to change the colour of their bodies in conformity with a change in the background. Among these are the well-known chameleon and some species of lizards and fishes.

Another method of camouflaging is by what is generally referred to as confusion in colouring, which arises from distortion or disruption in colouring. This breaks the solidity of the body of the animal. Examples are found in the zebra, the giraffe, toads and butterflies. Other animals camouflage by what is described as counter-shading. As light falls on any surface from above, the natural environmental shading is that of lighting up the upper part and leaving the lower part dark. It is found that most animals bear their shading in an opposite fashion to this environmental shading. To name a few examples one could refer to the toad, snake and antelope and fishes in water, which are all lighter in colour below and darker above.

Another method by which certain harmless animals avoid falling prey to other animals is by having form and colour similar to those of dangerous animals of their kind. Thus some moths and flies look like bees in form and colour, some spiders like ants, some grasshoppers like wasps and some harmless kinds of snake may look like poisonous ones.

Finally, misdirecting colours are used by some prey to divert the attention of the predator to the less vulnerable part of their body. Examples are the conspicuously coloured tail of some lizards and the 'eye-spots' on the wings of some moths and butterflies. The tail of the lizard and the wings of the moths and butterflies are less vulnerable than the main body or abdomen.

Scavenging (Decomposers)

There are some organisms on land and particularly in soil communities which live on dead bodies of plants and animals. On land the common ones include vultures and worms. In the soil are the decomposers consisting of a wide range of soil bacteria and fungi. These organisms are important

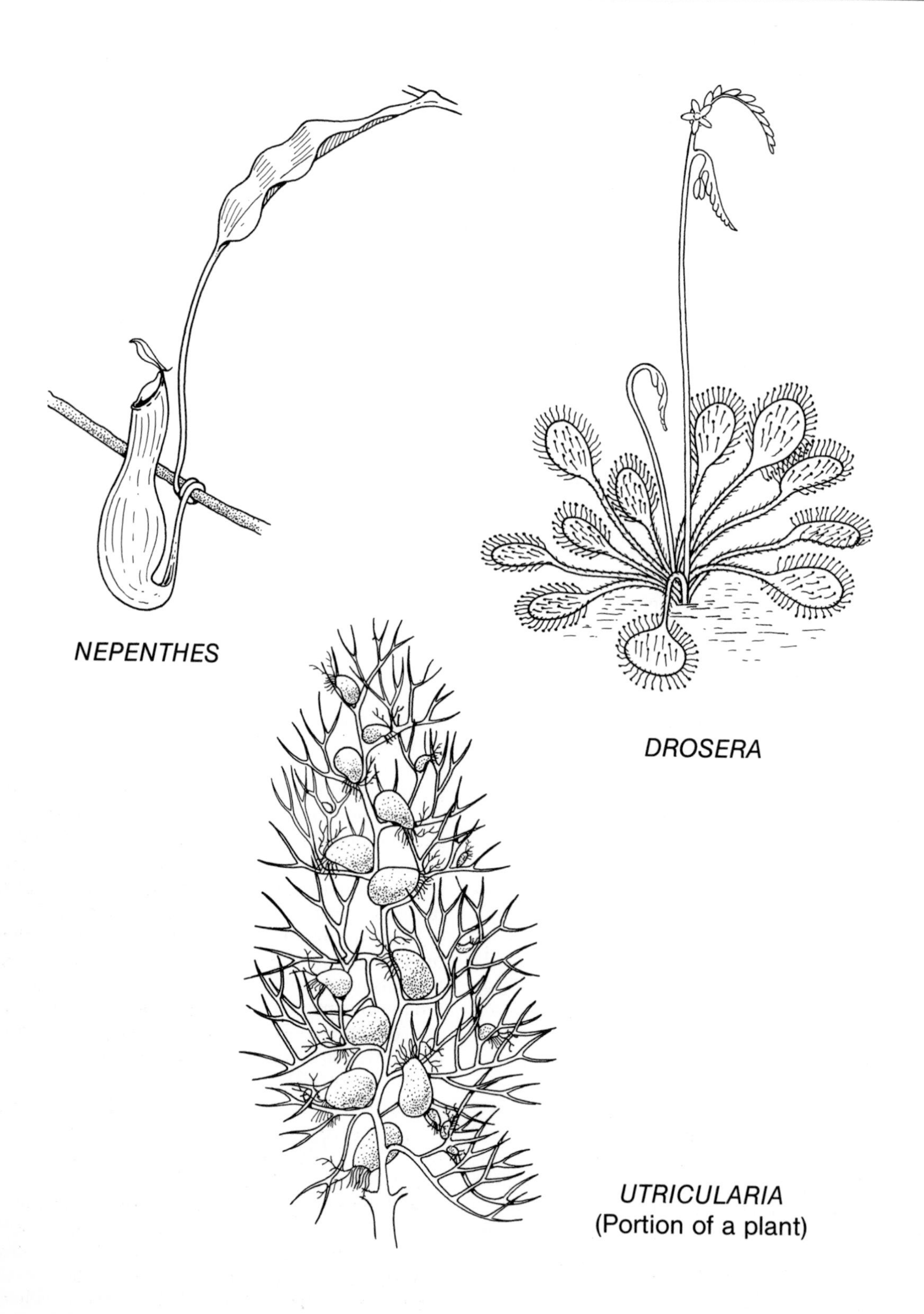
NEPENTHES
DROSERA
UTRICULARIA
(Portion of a plant)

in the cycling of inorganic nutrients in nature. The bacteria and fungi are assisted by enzymes in decomposing dead material. Without them organic matter would continue to accumulate over the years, with the minerals locked up, leading to a complete depletion of essential minerals from the root zone of the soil. If that happened, green plants could not obtain any more of these minerals and new protoplasm would therefore not be produced by green plants. By their action the scavengers provide themselves with food and also return to the soil or water essential elements which are constantly being absorbed by the roots of plants. Their action thus terminates one food chain and prepares the conditions for another one to begin.

In tropical climates the high temperature in conjunction with the high humidity accelerates the activities of the decomposers in breaking down organic matter, as a result of which not very much of it accumulates. This is discussed in detail in Chapter 8.

Symbiosis

By definition symbiosis covers the situation in a community where two organisms live in close association, each of them deriving some benefit from the other partner of the association. One of the well-known examples of symbiosis is afforded by the association of the nitrogen-fixing bacteria (*Rhizobium*) with the tissues of the roots of leguminous plants in the form of root nodules (Figure 19). These nodules are swellings of the root tissues of the plant which contain the bacteria. The bacteria feed on the carbohydrates in the root tissues. The plant also makes use of some of the nitrogenous matter which the bacteria are able to build up from the nitrogen in the air between the soil particles. Root nodules may also be found on some non-leguminous plants inhabiting sandy places. Such symbionts play an important part in the nitrogen cycle in nature. They make the plants with which they are associated valuable sources of nitrogen for the soil and for that reason legumes are much used in crop rotations by farmers.

Lichens constitute another example of symbiosis. Lichens are associations of a fungus and an alga. The fungus obtains carbohydrates from the alga, while the alga is protected from dehydration by the tough gelatinous wall of the fungus. Some of the common lichens are observed as thin, round and greenish patches on trunks of trees. Among the flowering plants *Mycorrhiza* provides the best example of symbiosis. It is found in many forest trees in saprophytic phanerogams and in orchid roots. The seedlings of the latter will not grow unless their roots become infected with a particular fungus or are artificially supplied with soluble organic nutrients. In *Cycas* also one often finds algae and bacteria inside the cortex. Saprophytes like *Monotropa galeola*, *Leiphaimos*, *Afrothismia* and species of *Burmannia* have their roots associated with a filamentous mass of a fungus which causes decay of dead organic matter and acts like root hairs in absorbing food material from decomposed organic substances and passing them on to the saprophytic plant, which itself may lack green colour entirely.

An interesting symbiotic relationship is found in tropical ant-plants. In tropical America and Africa, many species of *Acacia* have hollow stipular thorns, developed near the bases of the leaves. Ants live in the hollows of these thorns which

18 Some tropical insectivorous plants: *Nepenthes*, *Utricularia* and *Drosera*
SOURCE: W H Brown (1935)

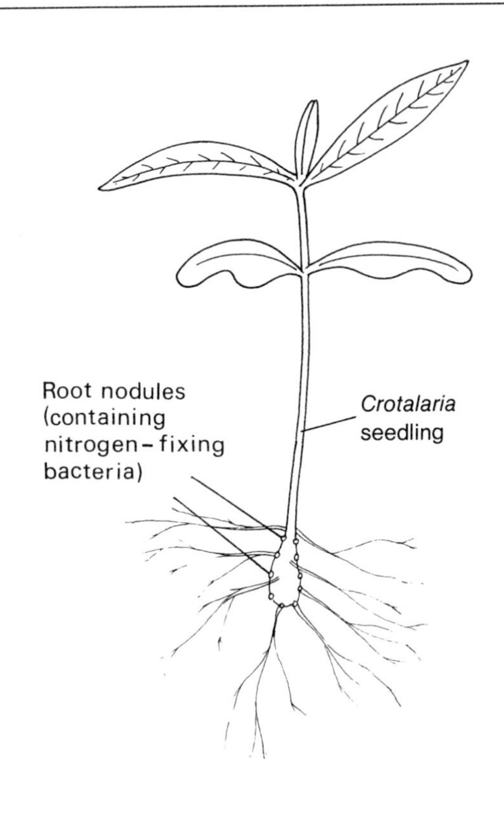

19 Root nodules on a leguminous plant
SOURCE: J Y Ewusie (1974)

also produce glandular secretions on which the ants feed. In return the ants drive away leaf-cutting and plant-eating insects as well as browsing animals. Finally there are also cases where algae inhabit coelenterates, as evidenced by green *Hydra*.

Before concluding on symbiosis, there is a similar type of association in which only one partner benefits, while the other apparently obtains no known benefit but does not, however, seem to suffer from the association. This relationship is often referred to as **commensalism**. Generally one of the partners consumes the unused food of the other. An example involving two animals is afforded by the *Remora* fishes and sharks.

In plants it is found that innumerable micro-habitats are provided for a number of animals. Thus many small animals use plants as shelter for protection from their enemies, from wind and evaporation, from extremes of temperature and from other adverse weather conditions. For example, insects and spiders often rest within flowers and fruits. In tropical forests it is found that the bases of palm fronds are inhabited by a host of small animals many of which have evolved a flattened shape to suit their peculiar micro-habitat. Also the spaces between the leaves of epiphytic orchids and bromeliads harbour a large number of insect larvae, earthworms, snails, centipedes, millipedes, grasshoppers, ants, scorpions, tree-frogs, lizards, snakes and other animals, some of which may have flattened shapes. One of the extreme forms of micro-habitat in plants is occupied by leaf-miners which are tiny insects whose larvae live and feed between the upper and lower epidermal layers of leaves. They have very flattened bodies and reduced legs. Some ant-plants of the tropics also exhibit the same phenomenon. One ant-plant in Guiana has been found to have as many as 20 species of ants associated with it, together with 30 different kinds of other animals.

Parasitism

By definition parasites benefit from the other partner of the association while the host suffers some loss or damage. There are various grades of parasitism in plant and animal communities. Parasites can be either partial or total parasites depending on how much of their food requirements is obtained from the host. They can also be internal (endoparasites) or external (ectoparasites). The discussion is confined here to partial and total parasites.

20 Mistletoe. A typical specimen of *Tapinanthus bangwensis*
SOURCE: *Cocoa Growers' Bulletin* (1972)

Partial (or Semi-) Parasites. A well-known example of a partial parasite in the tropics is afforded by the mistletoes found on wild trees and on crop plants such as the cocoa tree. Most partial parasites have no normal roots but some species of *Phthirusa* and *Dendropemon* (*Loranthaceae*) in tropical America have roots which not only curl around the branches of the host but develop penetrating haustoria. They may represent an evolutionary stage in the development of more complete parasitism where roots have been disposed of altogether, but they have the advantage that small pieces of stem, carried from tree to tree by birds, can establish themselves on a new host. As such they are less completely dependent on becoming established by seed. Partial parasites obtain only their water and mineral salts from the host but they contain chlorophyll themselves and they are thus able to manufacture their own carbohydrates by means of photosynthesis. Among these mistletoes (*Loranthaceae*) *Tapinanthus bangwensis* is a common African mistletoe found in the top of the canopy of unshaded branches of the host or on twigs lower down which are exposed at the side of a clearing or a road. It has red candlestick flowers, red berries and a globular haustorium. It differs from the European mistletoe by having normal non-fleshy ovate leaves. As soon as a seedling of mistletoe penetrates a branch of the host it develops a disc between and in contact with both the phloem and the xylem of its host. The disc gradually expands across the full cross-section of the branch as it grows into a roughly spherical mature haustorium. As the xylem vessels of the host branch become blocked by the haustorium, the part of the branch distal to the mistletoe sometimes gradually dies and rots away (Figure 20).

It is remarkable that, in spite of these effects, certain observations on the mistletoe on cocoa suggest that the association may be beneficial to the host under certain circumstances. Thus, when equal numbers of cocoa seedlings with and without mistletoe infestation were subjected to severe water stress (Room, 1972) those with mistletoes retained more leaves and generally survived better than those without. Similar observations have been reported for other mistletoe – host relationships (Gill and Hawkesworth, 1961) which suggest that mistletoes are relatively refined parasites which may help indirectly to keep their hosts and themselves alive during difficult weather conditions. While of some biological interest these observations should not lead to any conclusion that it is desirable to leave vascular parasites on valuable trees. Parasite infestations reduce yields of citrus and other fruit trees and may cause the death of the tree.

Another partial stem parasite is *Cassytha* which has very slender stems which cannot support themselves. They climb their host trees by twining and produce haustoria at short intervals, which support the stem and absorb water from the host plant. It is not certain whether the haustoria also withdraw manufactured food substances since the *Cassytha* stem contains a fair amount of chlorophyll in addition to its minute leaves.

Other tropical root semi-parasites include species of *Striga* with greenish wiry stems and a few green leaves for photosynthesis. It grows on the roots of grasses. There are numerous semi-parasites among the *Scrophularaceae* in all grasslands of both temperate and tropical regions. They cause serious loss to economic grasses such as sorghum as well as to pasture. In recent years Williams (1960) has described *Sopubia ramosa* as a root semi-parasite on *Imperata cylindrica*. This parasite could be used to eliminate that most undesirable grass.

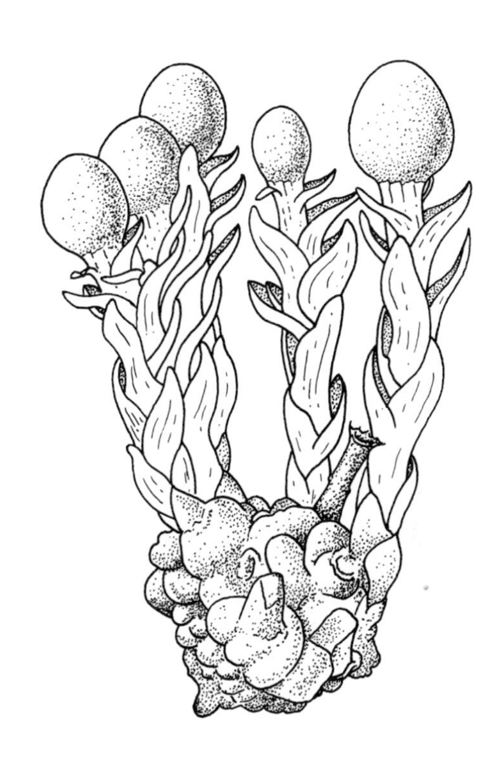

21 *Balanophora*
SOURCE: W H Brown (1935)

Total Parasites. The total parasites are represented by pathogenic fungi and bacteria, some angiosperms, worms and protozoans. Details of adaptation of parasites to their mode of living vary with reference to the host, the habitat and other factors. Parasites damage their hosts either by consuming the tissues or by releasing toxins which are poisonous to the hosts. In general parasites may eventually kill their hosts, or the host may establish a balance with the parasite and thus tolerate the parasite for an indefinite period, or the host may succeed in destroying the parasite. Many botanical and zoological textbooks contain examples of fungal, bacterial and animal parasites which makes it unnecessary to describe them here. However, brief mention could be made of the common flowering plant parasite, the yellow dodder (*Cuscuta*). This wiry yellowish parasite is found growing on garden decorative shrubs and small trees. Except for a short period after germination of the seed, this parasite has no roots, but develops haustoria which penetrate into the tissues of the host to obtain its carbohydrates and water.

There are also total root parasites like the *Orobanche* (broomrape) which is parasitic on the roots of tomato and tobacco among other crops. There are also *Balanophora* (Figure 21), parasitic on tropical forest trees, and *Rafflesia* (Figure 22) which is parasitic on the roots of figs and *Vitis*. *Rafflesia*

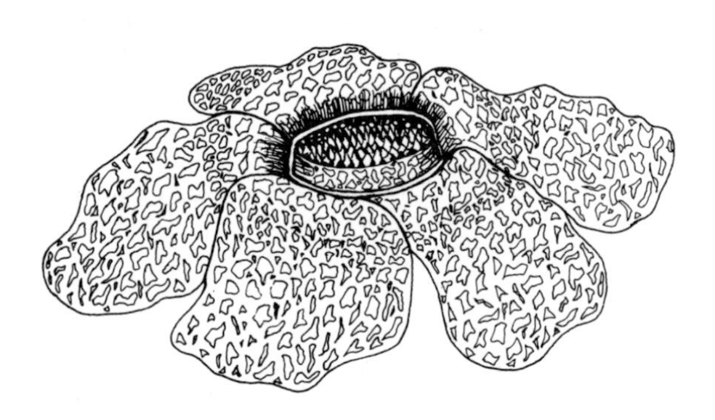

22 *Rafflesia*

is interesting for bearing the largest flower in the world. The flower may measure 1–2 m in diameter and weigh as much as 8 kg, while the stem is reduced to a threadlike structure which passes through the root of the host plant. Vascular parasites exist in all tropical terrestrial ecosystems.

Animals (Other than Man)

The role of animals in vegetation is quite varied but basically animals use plants as a source of food. While many of the effects of the activities of animals in a plant community may be destructive some are beneficial to plants.

Activities of Animals Destructive to Plants

Among the destructive activities of animals on a tropical vegetation the following may be mentioned here.

1. **Physical Destruction of Vegetation**. This includes the physical destruction of woody parts of plants in a vegetation and crops through trampling by bigger animals like the baboons, elephants, antelopes and buffaloes. Elephants dig water holes and push down trees. By such action elephants, in particular, make the forests habitable by other game. In the long run they turn forest into grassland. Elephants in National Parks in East and Central Africa, for example, have in fact reduced some forests to artificial deserts.
2. **Destruction of Vegetation through Feeding**. Here destruction is caused by smaller animals. This includes the infliction of heavy losses on cereal crops like sorghum and millet by a flight of *Quelea* birds; destruction of vegetation by a swarm of insects such as the migratory locust; and the feeding of rodents on food crops like cereals, groundnuts and pulses.
3. **Change of Vegetation through Grazing**. This is effected by game and domesticated animals like cattle, resulting in marked changes in the species composition of the plant communities present. Exotic animal species are known to be more destructive in their grazing than indigenous animals. One of the reasons for this is probably that the different species of wild game complement each other in the selection of browse. A good illustration of this is given by Vesey-Fitzgerald (1960) who showed that in the Rukwa Valley in Tanzania the various species of grazing animals use the habitat successively, each species leaving the vegetation in a state suitable for another.

It is also evident that wild game which are grazers and browsers manage to feed on the animal food production of the plants without cutting into the necessary reserves for plant maintenance and growth. In this way they are in good balance with the habitat. Dasman (1964) has shown that game cropping can be more profitable than cattle ranching. On 130 km^2 of Rhodesian thorn savanna, the annual profit from cattle was £500, whereas that from game was £2300.

The changes produced by intense grazing include the destruction of many of the original plants by trampling and decapitation. A number of other changes occur in the soil on the grazed area, leading to increased erosion. It is, however, known that moderate grazing stimulates growth and that young growth is highly nutritious. Thus, it is reported that the crude protein level of *Cynodon dactylon* in Kenya in moderate rainfall can be trebled by mowing (Dougall and Glover, 1964).

Grazing animals differ in the closeness of their grazing and this can influence the balance of competition between different species in the vegetation. For example, sheep and rabbits graze quite closely by nibbling the plants with their incisor teeth but cattle do not actually nibble the grass. Instead they twist their tongue around it, then grip it against the horny pad of the upper jaw and tear it off with a jerk of the head. Squirrels and the smaller rodents look for a varied diet and so do not nibble closely. They exert their effect on vegetation also by eating seedlings or seeds of trees and thus check the regeneration of trees.

Some shrubs and trees can withstand heavy browsing, either by virtue of the vigour of their growth or by producing a protective thicket around the growing parts, while others can withstand only a little browsing, perhaps up to 20 per cent of the annual growth. It is presumed that some thorny shrubs and trees have evolved spines as a result of which only their soft parts are browsed. When trees and shrubs grow above browsers, heavy browsing produces browse lines. It is not known how far this harms the trees, but it often affects the use of the area if the animal populations change.

Usually grazers first eat the more palatable species and this action removes competition from the less palatable species which may begin to increase in density. An example of this is seen in northern Ghana where the unpalatable *Anogeissus leiocarpus* left by cattle has become more abundant. In East Africa there is often a heavy concentration of grazers on *Sporobolus spicatus*, while there is almost complete neglect of the highly lignified *Pennisetum menzeanum*. By selective feeding, and aided by fire, elephants are known in a short period of time to change forest or wooded savanna to grassland (Glover, 1963).

4. **Transmission of Plant and Animal Diseases.** This if often carried out by insects. An example is the mealy-bug which sucks the juice of the cocoa plant and thus transfers the virus of swollen-shoot disease from a diseased plant to a healthy one.
5. **Reduction of Vigour of Plants.** This is the result of the activity of leaf-eating animals such as beetles and snails or of leaf-cutting ants, the latter being common in South America and Trinidad.

Activities of Animals Beneficial to Plants

The activities of animals which are useful to plants may now be considered. Among these are the following:

1. **Pollination of Plants.** Various types of animals assist in the pollination of plants, and in fact many plants depend absolutely on these animals to provide this service. Most orchids and asclepiads are examples. Humming birds, bats and insects are among the animals which bring about pollination of certain plants when they visit one flower after another for nectar or other food.

   In tropical Africa birds often pollinate flowers which are red in colour and which open during the day, such as those of *Bombax*. Bats also pollinate a number of plants such as *Ceiba pentandra* (the silk cotton), *Kigelia*, *Adansonia digitata* (baobab) and *Parkia*. Here the flowers are light cream or white and the bats are often attracted by the peculiar odour that is given by these flowers (Harris and Baker, 1958 and Baker and Harris, 1959).
2. **Dispersal of Seeds and Fruits.** Effective dispersal of seeds reduces competition between the plant and its offspring and allows the species to spread into new places; monkeys and squirrels are among the animals which bring about dispersal of seeds and fruits. The seeds of certain plants are dispersed by birds along with their droppings since the fertility of the seed is not affected when the latter passes through the digestive tract of the bird. Natural regeneration of *Pimenta dioica* (allspice) often occurs along fence lines where birds perch and defaecate. It has been shown that germination of this species is inhibited by its own fruit coat. The pericarp is removed during passage through the gut of the bird and germination is facilitated. Another example is *Calvaria major* of Mauritius of which tree no seeds had been known to germinate since the extinction of the dodo. Seed did finally germinate when the fruits were first fed to turkeys, a bird of comparable size to the dodo.

   Bats suck the juice of some fruits and spit out the seeds as they fly from tree to tree. In the mistletoe, pollination is affected by sunbirds, bees and flies. The ripe fruits are eaten by a number of species of birds. The seeds pass through the gut of a bird very quickly, often in less than a minute, but they retain an adhesive coating which ensures that the bird has to wipe them from the anal feathers onto a suitable twig. In this way mistletoes spread around a single parent plant. Monkeys, baboons and squirrels either carry fruits with them for long distances or collect them from various places and hoard them. They eat the flesh of the fruits and discard the seeds. The introduction of *Samanea saman* (rain tree), which is a central American species, to the Caribbean was attributed to the use of the palatable fruits of this tree as cattle feed on sea journeys from the mainland to the islands. The seeds pass through the animal undamaged and germinate in its faeces.

   In other cases the adhesive hooks or glands of

fruits or seeds make accidental contact with animals which carry them for long distances until they happen to fall off. Various biological textbooks have good descriptions of pollination and dispersal mechanisms.

3. **Enrichment of the Soil.** It has already been mentioned that scavengers and decomposers play a part in the recycling of nutrients. Termites by their action on woody material and in their building of termite mounds circulate fresh nutrients that may otherwise be unavailable to plants. In wooded tropical savannas the clumps of trees and shrubs are almost always associated with old termite mounds. By transporting large amounts of dead vegetation to the underground parts of the mounds the termites help to concentrate rich soil which provides suitable sites for trees and shrubs. Thus termites play an important role in the distribution of woody vegetation in the savanna. Crabs bore many holes in the mangrove swamp and thus make oxygen available to the roots of the mangrove plants. Finally various animals increase the nutrients of the soil as a result of their droppings which manure the soil.

Relations among Animals

Animals exert influence in many ways amongst themselves. Some of these have already been discussed and it is obvious that predation is the most spectacular of the effects of one animal on another. Studies on prey preference ratings in Kruger National Park in South Africa (Petrides and Pienaar, 1971) illustrate the selective effect of predation on different prey. The lion, the leopard, the cheetah, the hyena and the hunting dog select the waterbuck for food. Thus waterbucks are reported as normally killed from 2.2 to 15.7 times as readily as would be expected if hunting by predators was conducted on a random basis.

The kudu is also sought after by these carnivores as their prey. In view of the fact that the kudu and the waterbuck are sought after, dense populations of the predators seriously limit their numbers. Though wildebeest and impala comprise the major foods of these predator species they are not their preferred foods, and so their numbers are not much limited.

Man

Ecologically man is just one of the animal species which operate in the ecosystem. But man exerts considerable influence on vegetation and his activities result in quite spectacular changes. Among the principal activities of man affecting vegetation are the following.

Felling of Trees

In forest areas man fells and removes the largest trees for use as timber. This activity is common in almost all tropical countries, where it leaves gaps in the forest canopy thus encouraging the rapid growth of secondary forest within those gaps. In most areas trees are felled for use as fuel, either as firewood or converted through controlled burning into charcoal. Important and interesting types of vegetation have been destroyed in tropical countries in this way. Apart from depletion of virgin rain forests in many tropical areas, some habitats, like mangrove, which have high productivity and provide spawning sites for so many useful marine animals, have been destroyed. The destruction of vegetation has the consequence of destroying the habitats for certain animals and the soil. Much of this is discussed in Chapter 12.

Farming

In many tropical countries farming is often preceded by the clearing of the land and this removal of vegetation considerably upsets the natural balance of the ecosystem, including the soil. The heavy rains beat the exposed soil, resulting in increased erosion. This leads to loss of soil nutrients, and a different flora often results through secondary growth or regrowth over a number of years. The land may be left for a length of time after farming. This practice, known as **shifting cultivation**, provides a fallow period. If the fallow period is long enough the regrowth leads to the re-establishment of the woody forest species typical of the forest zones. If the fallow period is too short, and there is over-intensive cultivation, the secondary growth may not regenerate beyond the grass stage, and such vegetation in the forest areas is referred to as derived savanna.

Preservation of certain trees in farmlands also changes the floristic composition of the vegetation.

Fire

Fire, both accidental and intentional, exerts varied effects on species populations and has been responsible for the creation of a number of biotic climaxes in savanna types of vegetation in the tropics.

Results of research in this field appear to have produced contradictory results and recent work has shown that some of the apparent contradictions have resulted from the fact that the experiments have not had the same basis for comparison. In particular, the time of application of the fire seems crucial in the response of a species to it. Since the role of fire is more fully discussed in Chapter 12, the reader is advised to consult that chapter for a detailed account of the effects of fire on vegetation and other organisms.

Plantings

The planting for useful purposes or as forest reserves of special species of plants not common in or typical of an area greatly alters the floristic composition of the vegetation. This is the case with artificial forest reserves in places where enough natural vegetation is not available for preservation. This is because some introduced trees may be easier to grow and provide better yields for certain purposes (such as firewood) than native trees in that area. Examples are the neem and *Cassia siamea*.

Other Activities

Other equally important activities of man influencing vegetation include the keeping of domestic animals, drainage of swamps, irrigation, urbanization and the pollution of the air, rivers and the soil attendant upon industrialization.

With the rapid increase in the populations of tropical countries and modern advances in technology various habitats in these countries are changing. Some of these changes are undesirable because they result in the destruction and loss of valuable natural assets. Only a sound knowledge of ecology can enable us to use our natural resources judiciously. This is discussed in Chapter 13.

Chapter 5 Periodicity in Tropical Populations

Periodicity is a term that is used to describe the occurrence at fairly regular intervals in time of various biological processes in plants and animals and the manifestation, also at regular time intervals, of these processes as significant developmental changes. The processes are often internal, such as active cell division, cell enlargement, increased or decreased photosynthesis or a change in hormone or enzyme action. The manifestations may be expressed as loss of leaves, formation of new leaves, flowering, fruiting and dispersal of fruits and seeds, as well as by the breeding and migration patterns of certain animals. It has been suggested that periodicity results from intrinsic genetic characteristics of species populations under the influence of a particular combination of environmental factors. The periodic or rhythmic phenomena in plants and animals are classified according to the length of the period they show. These are (a) annual rhythms (phenology), (b) lunar rhythms of 28 days period, (c) tidal rhythms of about 12.8 hours period and (d) circadian rhythms of about 24 hours period.

Phenology

Phenology refers only to the manifestations of important periodic phenomena at certain times in a calendar year. Hence it refers to periodicity within an annual cycle. The term was used by Shelford (1929) to embrace the study of correlation between periodic phenomena, such as the flowering of a plant species and the arrival of migratory birds.

In the tropics phenology seems to present a number of features that are peculiar when compared with those of the temperate regions. The nature of these features has not yet been sufficiently studied. The lack of information is probably due to an original mistaken assumption that plant and animal behaviour in this regard is very much like that of temperate organisms. Among the few studies on tropical phenology are those of Ewusie (1968, 1969), Medway (1972), Frankie *et al.* (1974) and Groat (1969).

In temperate climates the seasons are well-marked in terms of temperature and day length, and most plants and animals react in distinct ways in response to these changes during the year. In the tropics, however, the situation is different. Temperature and day length do not vary nearly as much throughout the year. In spite of this, different species of plants and animals show distinct periodicity at specific times of the year. As one moves away from the equator the annual total, as well as the pattern of distribution, of rainfall varies and this becomes one of the significant hubs around which much of the phenology of tropical species populations revolves. The remarkable thing about tropical communities is the apparent absence of synchronization between many different species in most of the phenological phenomena. At almost any time of the year a few species may be flowering, fruiting, losing leaves or producing a flush of leaves; some species of birds or fishes will be migrating or breeding. It is this continuity and overlap which makes tropical phenology different from elsewhere.

It follows that phenology in the tropics is more a population than a community phenomenon, unlike the situation in temperate climates where most of the phenological activities of plants and animals appear to be synchronized. **Aspection**, which is the appearance or the aspect of a whole community at different seasons, is less obvious in the tropics since, with the staggered phenological behaviour of different species, the aspect of a whole community is not markedly altered at any time. Even though Medway (1972) reports that, despite specific variations in phenology, the dipterocarp rain forest he studied in Malaya as a community exhibited regular seasonality, with distinct annual peaks of flowering and fruiting and a double peak of leaf production, the vegetation as a whole did not change in aspect.

Tropical communities are rich in species. If this fact is seen in the light of the non-synchronized nature of phenologic behaviour in species, it becomes evident that genetic constitutions of individual species populations are operating to reduce competition and to facilitate co-existence. Medway (1972) reports that inter-specific differences in periodicity in the Malayan forest trees maintained a minimum incidence of flowering at 44 per cent of the total species present from year to year. The total number of species flowering each year was made up by a changing assortment of species which flowered at longer or less regular intervals, probably in response to different species-specific requirements. He also showed that diversity among the species served to maintain a minimum of 27 per cent of the species populations fruiting in the tree community each year.

An account will now be given of some of the prominent phenological phenomena found in the tropics, such as the flowering and fruiting cycles, defoliation and the subsequent growth of new leaves, breeding seasons and migration.

Flowering and Fruiting Cycles

Flowering and Fruiting of Tropical Trees

A flowering period is taken as the period from the first opening of the flower buds to the time when the last flower ceases to present anthers or stigmas. The fruiting period extends from the initiation of fruits on the tree to the time when only mature fruits are present on it. In cases where the fruits may stay on the plant for as long as a year after their formation and growth to normal size, it sometimes becomes difficult to determine the effective end of the fruiting period. Examples of this situation are found in *Milletia thonningii* and *Avicennia africana*. In the latter case the apparently prolonged fruiting period would be related to the fact that the fruit stays on the plant till the seeds germinate (vivipary) (see p. 161).

Normally the fruiting period would slightly overlap or follow closely the flowering period. The period involved also varies considerably. Flowering and fruiting periods in trees vary widely and can take as short a period as one month or longer than six months. In some varieties of West Indian cherry (*Malpighia punicifolia*) a single tree will complete its flowering in two or three days; fruits begin to develop immediately and are ripe in about 14 days. In *Antiaris africana* flowering and fruiting take about one month, while in *Sesbania grandiflora* they take as much as six months. However, where the flowers stay on the plant for a relatively long period before fruit formation begins, there can be an appreciable gap between the flowering and fruiting periods. An extreme case is found in *Pycnanthus angolensis* where fruiting occurs a year after flowering. The peak period for fruiting is often two months after the peak period for flowering in a community (Ewusie, 1969; Medway, 1972).

Flowering Frequency

As already stated, there has been very little work done in tropical phenology. Of the detailed studies so far made, there are the studies on some 100 West African woody species reported by Ewusie (1968) and the observations on 45 species of Malayan rain forest trees by Medway (1972). In his work Ewusie found that about 48 per cent of the species studied flower once a year, about 44 per cent flower twice a year, about 5 per cent flower three times a year, about 1 per cent flower four times a year and another 1 per cent flower throughout the year.

Table 7 gives a list of the species concerned and their frequency and time(s) of flowering. It should be noted that it is not normal for a tree to flower more than once a year in temperate climates, but it appears to be rather common in tropical vegetation. There are of course species which flower infrequently such as about once in every two or three years, as in *Flacourtia flavescens*, or once in about five to seven years, as in *Triplochiton*.

In line with terminologies in rhythmic studies, it is proposed to use the expression **flowering frequency** ('Fl.f.') to refer to the maximum observed number of times a species flowers in a year. Thus the following flowering frequencies of some West African plant populations emerge from the work under discussion.

Species flowering once a year have a Fl.f of 1
Species flowering twice a year have a Fl.f. of 2
Species flowering thrice a year have a Fl.f of 3
Species flowering four times a year have a Fl.f. of 4
Species flowering throughout the year have a Fl.f of ∞

Also Species flowering once in two years have a Fl.f. of $\frac{1}{2}$
Species flowering once in three years have a Fl.f of $\frac{1}{3}$
Species flowering once in seven years have a Fl.f of $\frac{1}{7}$

Species which do not flower at all or are not able to do so in particular environments will have a flowering frequency of zero (0) in those environments. Trees which flower rather infrequently in different years would not have a defined flowering frequency.

From the reports of Medway (1972) only 10 out of the 45 species he studied in the Malayan rain forest exhibited regular periodicity in flowering while 21 species showed regular foliar cycles. Thus it would appear that a great many rain forest trees flower rather infrequently over the years. Trees in more seasonal (deciduous) forests would appear to show greater regularity in the flowering of particular species than in the true non-seasonal rain forest. Medway (1972) also reports a peculiar behaviour in *Gluta renghas* in which separate branches flowered, and subsequently bore fruit, at different times. Silk cotton tree (*Ceiba pentandra*) and mango (*Mangifera indica*) may also have flowering, fruiting or leaf fall out of phase on different parts of the same tree.

In view of the variation in the frequency of flowering found in tropical species it would be useful and instructive if tropical floras always indicated the flowering frequency of each species described. From the above, it can be said that

Table 7. Woody species of West Africa observed to be flowering and fruiting twice, thrice and four times a year, with period(s) of flowering indicated

SPECIES	FLOWERING PERIOD	SPECIES	FLOWERING PERIOD
TWICE A YEAR		TWICE A YEAR	
Anacardium occidentale	Sep.–Dec.; Feb.	*Mimusops kummel*	Jan.–Mar.; Oct.–Dec.
Dialium guineense	Aug.–Nov.; Apr.–May	*Guainacum officinale*	Oct.–Dec.; Apr.
Erythrophleum guineense	Jan.–Feb.; May	*Anacardium occidentale*	Sep.–Dec.; Feb.
Griffonia simplicifolia	Aug.–Nov.; Feb.	*Magnifera indica*	Jan.–Mar.; Aug.–Nov.
Tamarindus indica	Nov.–Dec.; Feb.–Mar.	*Uvaria ovata*	Oct.–Nov.; June–July
Peltophorum pterocarpum	Nov.–Dec.; Feb.–July	*Carrisa edulis*	Nov.–Dec.; Mar.–May
Capparis erythrocarpos	Oct.–Jan.; Mar.–Apr.	*Strophanthus hispidus*	Feb.–Apr.; July
Dichapetalum guineense	Oct.–Feb.; Apr.–Sep.	*Calotropis procera*	June–July; Dec.–Mar.
Byrsocarpus coccineus	Oct.–Nov.; Jan.–Mar.	*Spathodea campanulata*	Apr.–June; Sep.–Nov.
Dyospyros abyssinica	May–July; Oct.–Nov.	*Crescentia cujete*	Jan.–Feb.; Oct.
Eleophorbia drupifera	July–Oct.; Nov.–June	*Bombax sessilis*	Apr.–May; Oct.–Mar.
Hura crepitans	Mar.–Apr.; Jan.–Oct.	*Ehretia thomningiana*	Apr.–May; Oct.–Mar.
Oncoba spinosa	Nov.–Dec.; May–July	*Cassia nodosa*	Oct.; Apr.–July
Hoslundia opposita	Oct.–Jan.; Mar.–Sep.	*Cassia siamea*	Nov.–Jan; Mar.–June
Strycnos nux-vomica	Oct.–Mar.; May		
Lagerstroemia speciosa	Oct.–Jan.; Apr.–June	THRICE A YEAR	
Acridocarpus alternifolius	Aug.–Oct.; Mar.–Apr.	*Cassia sieberiana*	Nov.–Dec.; Feb.; June–July
Acacia nilotica	Oct.; Apr.	*Parkinsonia aculeata*	Jan.–Feb.; May–June; Oct.–Nov.
Acacia farnesiana	Oct.–Nov.; Apr.	*Drypetes floribunda*	Nov.; Feb.; June;
Albizia lebbeck	Nov.–Dec.; May	*Securinega virosa*	Feb.–Mar.; May; Oct.–Nov.
Parkia clappertoniana	Nov.–Dec.; Apr.–May	*Azadirachta indica*	Feb.–Apr.; June–July; Nov.–Dec.
Samanea saman	Nov.–Dec.; Feb.–July	*Balanities aegyptiaca*	Dec.–Jan.; Apr.; July
Ficus umbellata	Nov.–Dec.; Mar.–July		
Moringa oleifera	Apr.–June; Oct.–Nov.	FOUR TIMES A YEAR	
Psidium guajava	Nov.–Dec.; May–June	*Terminalia catappa*	Oct.–Dec.; Feb.; Apr.; June–July
Ochna kibbiensis	Dec.–Jan.; Apr.–June		
Ximenia americana	Nov.–Dec.; Apr.–June		
Gardenia ternifolia	Nov.–Feb.; May		
Pavetta corymbosa	Oct.–Feb.; Mar.–June		
Canthium horizontale	June–July; Dec.		

temperate species typically have a flowering frequency of one while a number of tropical species have a flowering frequency of two or more.

Medway (1972) found that invariably some of the Malayan forest species which flowered failed to fruit. He estimated this to be between 15 and as much as 43 per cent in different years. Some of the species dropped their fruits before they were mature in certain years. He noticed that while *Cynometra malaccensis* flowered regularly each year from 1963 to 1969, it set fruit only in the exceptionally dry year of 1963. *Elateriospermum tapos* also flowered every year but fruited abundantly only in 1963, while the fruiting was poor in 1968 which was another exceptionally dry year. It was only in 1964 that he found *Palaquium hispidum* carried its fruits till they ripened, these usually falling prematurely. He also found that *Bhesa paniculata* flowered regularly each year, but that it could not carry successive crops of fruit until the season was advanced. These observations seem to show that conditions which stimulate flowering do not necessarily also ensure subsequent successful fruiting.

Ecological Significance of Flowering Periods and Flowering Frequency

It appears that in tropical woody species which flower twice a year the flowering is not at the same level of importance on both occasions. It is argued that the more important (primary) period may be taken as the one which lasts for a longer time while the second and shorter period is regarded as a secondary or subsidiary one (Ewusie, 1969). The longer-flowering period is regarded physiologically as the first flowering of the species in a year. This is because the longer period of flowering must go with the presence of a greater store of energy, which might take place during the rainy period. This would explain why such flowering often takes place at the beginning of the dry period which is between September and December in West Africa. It is remarkable that in nearly all species with a flowering frequency of two, the one which is here described as the first flowering takes a longer time than the second. The reverse situation has so far not been found, and this may well justify the hypothesis put forward here.

This observation seems to imply that here the main rainy season of May to July constitutes a relatively dormant period as far as flowering is concerned and it would follow that it is during this period that more energy is conserved. Only a few species were found to be able to initiate flowering during the rainy season. This is to be expected as the heavy rain can be injurious to the survival of young buds by germinating the pollen grains prematurely, by rotting them through prolonged saturation or by physically washing the young buds off the plant. All these would result in the reduction of their reproductive capacity. The few plants which are able to initiate flowering during the rainy season seem to have rather robust inflorescences and they flower when they are in full leaf. The leaves would help to break the impact of the raindrops. Among such plants are *Mangifera indica*, *Hura crepitans*, *Anacardium occidentale* and *Dialum guineense*.

Flowering frequency is evidently an attribute of species populations, for there are situations where two species of the same genus may have different frequencies of flowering. Thus, *Bombax buonopozense* flowers only once a year (Fl.f. = 1) while *Bombax sessilis* flowers twice a year (Fl.f. = 2). *Albizia zygia* flowers once a year (Fl.f. = 1) while *Albizia lebbeck* flowers twice a year (Fl.f. = 2). Again *Terminalia catappa* flowers four times a year (Fl.f. = 4) while *Terminalia superba* flowers only once (Fl.f. = 1).

Flowering frequency may also be of use in determining the age of a tree. Thus a tree which flowers twice a year may be expected to show two seasonal rings per annum if the species is sufficiently sensitive to the seasonal changes in the weather and hence grows in between flowering periods. One of the reasons why annual rings have not been found to be consistent in tropical vegetation may well be the lack of knowledge about the flowering frequencies of different tropical species.

Peak (Seasonal) Periods of Flowering

In the West Indies most varieties of *Mangifera* flower between December and March in the dry time of the year; some varieties flower sporadically at other times. The fruits develop mostly during rainy periods. *Anacardium* behaves similarly.

A detailed analysis of the time of initiation of 136 flowering periods of 92 West African woody species (using the month of initiation of flowering), as reported by Ewusie (1969), is given in Table 8. The data are also represented graphically in Figure 23.

Table 8. Peak periods of flowering in trees of West Africa

	JAN.	FEB.	MAR.	APR.	MAY	JUNE	JULY	AUG.	SEP.	OCT.	NOV.	DEC.
No. of plants flowering twice a year (44 plants)												
First flowering	4	1	1	4	1	2	1	3	1	13	12	1
Second flowering	1	6	8	10	5	2	1	1	1	7	0	2
Sub-total	5	7	9	14	6	4	2	4	2	20	12	3
No. of plants flowering once a year (48 plants)	9	6	3	3	2	2	2	2	1	7	8	3
TOTAL	14	13	12	17	8	6	4	6	3	27	20	6

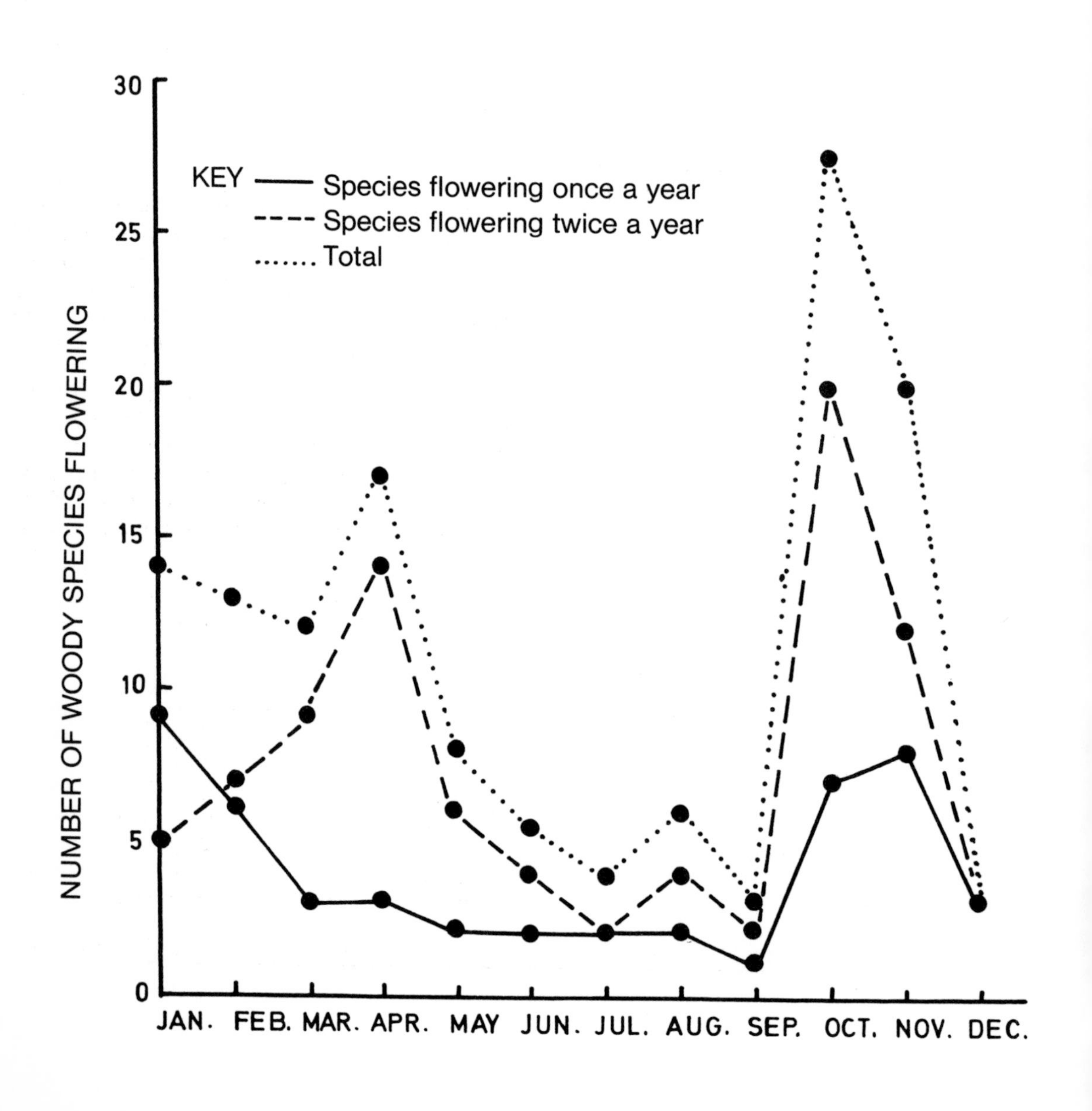

The figures in Table 8 show that there are two peak periods of flowering in woody species of West Africa.

1. **The Primary Flowering Peak Period**. This takes place before, during and after the second and smaller rainy period when the sun is little obscured by clouds and the humidity is fairly low, that is October and November with the peak of flowering in October.
2. **The Moderate Flowering Period**. This takes place during the middle and towards the end of the dry period and before the heavy rains set in, when the sky is partially clear of clouds, that is from January to April with the peak of flowering in April.

The lowest flowering period takes place, as already stated, during the wet period when the sun is obscured by heavy clouds, that is May or June to September. There seem therefore to be two peaks of flowering and two relatively inactive periods. The existence of two peak periods of flowering in tropical vegetation has also been reported in Malaya (Medway, 1972) in parts of Brazil (Davis, 1945) and has been observed in Uganda.

The report of Medway on the Malayan rain forest tree was based on seven years' observations (from 1963 to 1969) and his results were expressed as percentages of trees flowering in particular months, as shown in Table 9.

Medway found a relatively high incidence of flowering from February to July, a low incidence from August to November, and least of all in December and January.

Seasonal Flowering in a Tropical Herbaceous Community

An intensive study made in a coastal grassland in Ghana over a three and a half year period revealed growth responses and flowering phenology closely correlated with the incidence of rainfall (Adams, 1958 and unpublished correspondence with author). The 96 species studied included low shrubby phanerophytes (2), chamaephytes (12), hemicryptophytes (19), geophytes (12), therophytes (50) and a parasite (*Cassytha filiformis*).

Most of the perennial species showed vegetative activity following a single rainstorm (or several showers on consecutive or nearby days) totalling 12 mm, but very few species would then develop flowers. Following 20–50 mm of rain falling over a few days, i.e. within about a week, 10–25 per cent of the species flowered, and these totals were contributed to about equally by perennials and therophytes. Greater flowering intensity followed all heavier rains.

Maximum flowering responses occurred in March–July, with most in April–June, when in all years over 50 per cent of the species were recorded in flower on one or several days. During prolonged wet weather at this time of year high flowering frequencies were the rule, but through the period the totals were contributed to by different species in overlapping succession to one another. The greatest number of species recorded in flower on any one day was 55, contributed to by 25 perennials and 30 therophytes. At other times of the year, e.g. October, November, December or February, in various years flowering intensity reached 25–30 per cent of the species for periods of a few days,

Table 9. Distribution of flowering in a Malayan rain forest, expressed as the percentage of observed trees blooming during the months indicated (from Medway, 1972)

YEAR	JAN.	FEB.	MAR.	APR.	MAY	JUNE	JULY	AUG.	SEP.	OCT.	NOV.	DEC.
1963	–	–	–	–	–	–	24	12	7	0	0	0
1964	2	9	12	3	9	7	7	2	3	3	2	2
1965	2	14	16	12	2	3	7	0	2	0	3	0
1966	0	7	10	12	16	17	12	2	7	7	2	2
1967	0	4	7	11	9	10	16	7	0	0	2	0
1968	0	11	11	22	30	35	24	2	2	4	6	0
1969	0	4	15	4	4	6	4	–	–	–	–	–
MONTHLY AVERAGE	1	8	12	11	12	13	13	4	4	2	3	1

23 Peak periods in flowering in trees in West Africa

the totals being contributed to more by perennials than by therophytes on these occasions.

Only one species, *Jacquemontia ovalifolia*, a deep-rooted trailing chamaephyte, was seen to open flowers in every month of the year, including dry periods when all other species were inactive. At the other extreme, very few species showed calendar seasonal behaviour by flowering exclusively during one period in any year. These included the perennial geophyte *Curculigo pilosa* (February–June) and three therophytes (true annuals) *Mollugo nudicaulis* (March–June), *Crotalaria glauca* (June–August) and *Micrargeria filiformis* (August–September). It is noteworthy that the perennial geophyte could flower with the earliest rains in February, the flower buds being already differentiated underground in a probable response to a shortening day length stimulus acquired in the previous August or September when the plants were leafy; this would also account for their inability to flower at other times of the year. The therophytes took varying lengths of time to reach flowering maturity after germination. The slowest of these was *Micrargeria* (*Scrophylariaceae*) which, being a semi-parasite of grasses, may have first required to establish its dependent relationship.

In this example of phenological characteristics of herbaceous vegetation, the total annual rainfall of the location was only 720 mm on a six-year mean, ranging from 550 mm to over 750 mm in various years. In these near-desert conditions, vegetative activity and flowering could be determined very accurately between long rainless inactive periods. The main rainy periods at this site were during April–June and October–November, but in some years a third could occur at some time during December–February; August was always dry.

The Relationship Between the Water Regime and Flowering Peak Periods

A comparison of the position of the peak periods of flowering in relation to the annual rainfall distribution in West Africa and Malaya as portrayed in the data cited above is instructive. In the forest zone of West Africa the main rainy season is from April/May to July, while the second and smaller rainfall season is around October/November. The main rainy season is the period of least flowering, while the period of the second rains is the period of greatest flowering.

In the west–central Malayan states the principal rainy season is from October to January, while the second less-pronounced rainy season is in April. Here again the main rainy season is the period of least flowering, while the period of the second rains would appear to lie in the middle of the period with the greatest number of species flowering.

In both cases the dry period, and especially towards the end of it, seems to correspond with moderate flowering. Thus the water regime would appear to be a significant factor in flowering in the tropics in both tree and herb communities.

The role of the water regime in flowering is further emphasized by observations on the same species receiving rainfall at widely different times of the year, such as occurs north as opposed to south of the equator. Observations made by the author on *Jacaranda* sp., *Cassia siamea* and *Plumeria acutifolia* in West Africa (north of the equator) and Zambia (south of the equator) showed that in both places the period of flowering was related to the beginning of the rainy period. In the case of West Africa this was in April, while in Zambia it was in November.

Flowering and Fruiting Relationship

The maximum observed number of times a species sheds its fruits in a given geographical area in a year is here referred to as its 'fruiting frequency' (shortened to 'Fr.f'.). In certain species with more than one period of flowering a year it is not in respect of each flowering that fruiting results. For example, *Erythrophleum guineense* flowers in West Africa from January to February and also in May, but it is only as a result of the latter flowering that fruits are formed. *Oncoba spinosa* flowers from November to December as well as from May to July, but it is only the May to July flowering which seems to result in the production of fruits (Ewusie, 1968).

Thus while flowering (Fl.f) and fruiting (Fr.f) frequencies are often the same for a species, in some cases the fruiting frequency is lower than the flowering frequency. In the examples referred to above the flowering which does not lead to fruiting often takes place from November to February in West Africa when the weather is dry and warm.

While in the deciduous forests in which Ewusie made his observations it was evident that, with few exceptions, most of the species that flowered also fruited during the same year, it appears that the situation may be different in the rain forests.

Thus Medway (1972) reports that in the Malayan Dipterocarp rain forest, out of a sample of 45 species studied from 1966 to 1969 only six set fruit every year, even though an average of 27 species flowered annually. This means that a great majority of the rain forest trees did not fruit annually even where they might flower annually.

Defoliation and the Growth of New Leaves (Leaf Activity)

Loss of leaves and the growth of new leaves are common phenomena in flowering plants. The period during which a species loses most of its leaves varies from one species to another, not only in respect of its relation to flowering but also in duration. In *Ceiba pentandra*, for example, the old leaves are shed before flowering occurs and new leaves are formed shortly after flowering has ended, but in *Bombax* the plant remains bare for about a month after flowering before new leaves are formed (Ewusie, 1969). In most other cases the plant stays bare for some weeks before the new leaves appear, as in *Bombax* and *Hildegardia barteri*. In a case like *Terminalia catappa*, which may flower about four times a year, there is hardly any break between the shedding of the old leaves and the growth of fresh ones. However, in the rainy season new flowers may be formed without loss of the old leaves (Ewusie, unpublished).

Intensity of Defoliation and the Relationship Between Leaf Fall and Flowering

It is estimated that about 30 per cent of the trees of West African semi-deciduous forests lose an appreciable quantity of their leaves at one time or the other during the year (Ewusie, 1968). The intensity of leaf fall in tropical vegetation, however, varies among species and while it is not easy to grade this, it is at least possible to distinguish between the truly deciduous species which lose all their leaves at one time or the other in the year, and those which only lose a proportion of their leaves at various times of the year. The latter, of course, include some 'evergreens'. Species with complete leaf fall (deciduous species) in the West African forests include *Adansonia digitata*, *Bombax* spp., *Ceiba pentandra* and *Antiaris africana*. In the Malayan rain forests, these include *Parkia speciosa*, *Ficus glabella* and *Palaquium hispidum*. Among 'evergreen' species which show partial leaf fall in the West African vegetation are *Flacourtia flavescens*, *Kigelia africana*, *Fagara xanthoxyloides* and the introduced *Azadirachta indica* and *Mangifera indica*.

Holtum (1954) states that, from his observations on evergreen trees in Malaysia, flowering takes place in the dry period without being associated with an increased loss of leaves, the formation of new leaves taking place in the rainy season. In deciduous and semi-deciduous species, however, he reports that defoliation and the formation of new leaves are closely associated with flowering. Among the evergreens in West Africa, *Flacourtia flavescens* flowers during the dry season in December, but an appreciable loss of leaves and the flushing of new ones in replacement does not take place at that time. Instead this occurs with the coming of the rains in April–June. This seems to follow the observations of Holtum on the Malaysian evergreens. However some evergreens, like *Balanites*, behave like deciduous trees in this respect.

Koelmeyer (1959) studied the phenology of the tropical wet evergreen forest in Sri Lanka (Ceylon) and emphasized as a very common feature the relation between the times of leaf change and flowering. This evidently varied in different types of forests. In the West African vegetation, for example, this relationship cannot be said to be all that common, and in the Malayan forest studied by Medway this relationship was present in only a few species. He recorded only 6 out of the 45 species in which flowering and leaf activity were related. These were *Scaphium affine*, *Cynometra malaccensis*, *Parkia speciosa*, *Elateriospermum tapos*, *Erythroxylum cuneatum* and *Ficus sumatrana*. Interesting cases of absence of floral and foliar relationship included the case of *Ficus glabella* which, though annually deciduous, produced figs at any stage of the foliar cycle. In some cases fruiting seemed to suppress the renewal of leaves, and Medway cites as examples of this *Artocarpus lanceifolium* and *Koompassia malaccensis*.

The loss of leaves before flowering in tropical deciduous trees has the obvious ecological significance of facilitating pollination, as this exposes the flowers to the pollinators; the dispersal of pollen by the wind is also easily achieved. Particularly brightly-coloured flowers of trees like species of *Parkia*, *Bombax* and others attract bats and birds which move about freely in the dry period to effect pollination.

Defoliation Frequency

The number of times in a year that a plant sheds its leaves is here referred to as its 'defoliation frequency' ('D.f.'). Of the species studied by Ewusie in West Africa (1968), 70 per cent were evergreens (D.f. = 0). Of the 30 per cent of the deciduous trees 26 per cent shed leaves once a year (D.f. = 1), 1 per cent had D.f. = 2, and 1 per cent had D.f. = 4. In plants which are normally without leaves, such as *Sarcostemma*, the D.f. is ∞ (or infinity). Medway (1972) reports that patterns of loss and renewal of leaves in the Malayan Dipterocarp rain forest were diverse. In one species, *Swintonia schwenkii*, defoliation was on a two-year cycle; that is, with a defoliation frequency of $\frac{1}{2}$. In *Ficus sumatrana* leaf fall was thrice yearly (D.f. = 3). Of the 45 species which he studied, only three trees were found to produce leaves continuously throughout the year. These were *Shorea luevis*, *S. platyclodos* and *Artocarpus* sp. (D.f. = 0). In 21 species leaf growth was annually recurrent, in most of which there were two seasons of leaf growth which were apparently related to the two rainy periods.

Defoliation frequency has an interesting relationship with flowering frequency (Fl.f). These are the same in species which lose leaves at each flowering, as in *Balanites* and *Ficus sumatrana* (D.f. = 3), *Albizia lebbeck* (D.f. = 2) and *Afzelia africana* (D.f. = 1). It has been observed in a few cases that even though the species flowers twice a year the leaves are shed only once, during the dry period flowering, as in *Ficus umbellata* and *Acacia farnesiana*. Defoliation frequency does not therefore always parallel flowering frequency.

Defoliation Periods and their Peaks During the Year

Data from the work of Ewusie (1968) on leaf fall for 29 tree species are presented in Table 10.

It seems that while the majority of deciduous trees in West Africa shed their leaves during the months of October and April, other plants shed theirs at different times of the year. It is only during the heavy rainy period of May to July that leaf fall is rare. The peak months for leaf fall are October and April as they are for flowering periods. A close correlation between flowering and the shedding of leaves is here demonstrated. Bernhard and Huttel (1971) have also reported on litter fall in a rain forest in the Ivory Coast, which has the same vegetation as Ghana. They observed that litter fall was a seasonal phenomenon with two peaks.

The studies of Medway (1972) also demonstrate a distinct bimodality in the cycle of foliar activity in the Malayan Dipterocarp rain forest, which was here correlated with the two annual rainy periods. He found the monthly average percentage of species undergoing leaf renewal in the various months of the year to be as in Table 11.

Table 11. Percentage of species undergoing leaf renewal

JAN.	FEB.	MAR.	APR.	MAY	JUNE
9	9	26	19	18	13

JULY	AUG.	SEP.	OCT.	NOV.	DEC.
9	7	8	12	11	14

He found that a major peak in the percentage of trees growing new leaves occurred each year between the months of February and June; the second peak occurred between September and December. It will be recalled that the principal rainy season was from October to December or January and the second rainy period was in April which was in the middle of the first peak of foliar activity as well as flowering.

Note: in these data the middle month of the deciduous period is used, but where the period covered two months the first month, being the month in which defoliation was initiated, has been used.

Table 10. Annual distribution of defoliation in 29 tree species of West Africa (from 38 recordings)

MONTH	JAN.	FEB.	MAR.	APR.	MAY	JUNE	JULY	AUG.	SEP.	OCT.	NOV.	DEC.
NUMBERS OF SPECIES SHEDDING LEAVES	6	5	1	6	–	–	2	–	2	9	3	4

Annual Gregarious Flowering in Tropical Geophytes

The term 'gregariousness' has been used in connection with the observation of certain species apparently producing flowers suddenly. This is exhibited by a number of geophytes and some forest trees which flower rather infrequently. Special studies have been made of the geophyte *Pancratium*, a bulbous plant found in patches on the savannas of Africa. It has been noticed that three days after a rainstorm the plants would flower almost all at once over an area. Experiments showed that the flowering could also be produced by artificially watering them if no rainfall occurred for a lóng period of time, and it did not seem to matter whether the water used in the experiment was warm or cold. This established the fact that the response was due to the supply of water and not to a temperature change (Holdsworth, 1961). The response could not be due to light as the bulbs were completely underground. This finding would seem to support the suggestion about the role of water in tropical phenology.

Recent observations by the author have shown that a similar phenomenon occurs in other tropical geophytes such as *Urginea indica* and *Eulophia sordida* in coastal grasslands in Ghana, where the response is even quicker, being only two days after a shower of rain. Here it was found that the rain acted as an ultimate factor, and that the proximate factor must be an internal physiological one whereby the flower bud formation was brought up to a stage at which it would be able to respond to the water. In both *Urginea* and *Eulophia* the flowering period in Ghana falls between December and early February which is within the dry period of the year. By the beginning of this period some of the plants which are mature to flower have had their flower buds well formed during the preceding wet period. With each rainstorm that may occur during this period, the flower stalks of plants that are ready with flower buds elongate rapidly to bring the buds above the surface of the ground within about two days. It thus seems that the flower buds of some of the bulbs would just have reached the right stage to be able to respond to each shower of rain that falls during the potential flowering period which ends by the middle of February, after which there is no response and no flowering when it rains again. The flowers of these geophytes are short-lived and it appears that the gregarious flowering ensures that cross-pollination takes place among the plants that flower together in response to the same rainstorm.

Among forest trees, Medway (1972) has shown that dry weather has a stimulatory effect on gregarious flowering in Dipterocarps in the Malayan forest. He observed high incidence of flowering in 1963 and 1968, the two years in which the main dry season was most pronounced (see Table 9). This view was supported by published figures which showed that in those two years, except for only two months, evaporation exceeded rainfall in every month. Thus rainstorms during a dry season must lead to gregarions flowering in some tropical plants.

Photoperiodism

As mentioned in Chapter 4, the length of daytime does not vary as much in the tropics as it does in the temperate climates where there are long days and short nights in summer and long nights and short days in winter. Several examples are known of plants in temperate regions which flower in response to the length of day, so that while some are long-day plants, others are short-day plants. As may be evident, short-day plants require long nights and if the night is broken by even a flash of light for one minute the plant may not flower. The length of daylight which must not be exceeded in a short-day plant varies according to the species.

In spite of the rather slight variation in day length in the tropics, it has been demonstrated at Ibadan in Nigeria that some tropical plants flower in response to day length and thus show much sensitivity to day-length. Ibadan is situated on latitude 7°26′ north of the equator, and day length varies from 11 hours 41 minutes in December to 12 hours 33 minutes in June, a range of 52 minutes. The experiments conducted here (Njoku, 1958) showed that a food plant, the late okra (*Hibiscus esculentus*), flowers if grown in day lengths of $12\frac{1}{4}$ hours but does not flower if the day length is increased to $12\frac{1}{2}$ hours. Late okra is able to react to a difference of 15 minutes, and so does not flower during the long days from April to August.

Early okra, another variety, was found to flower in day length of $12\frac{1}{2}$ hours, but when this exceeded $12\frac{3}{4}$ hours it failed to flower. Since the longest day at Ibadan is 12 hours 33 minutes, it means that this variety can flower throughout the year at Ibadan, but will not flower at, say, Zaria, north of Ibadan, where the length of day exceeds $12\frac{3}{4}$ hours in June and July.

The study of a wild species of coffee, *Coffea rupestris* (Adams, 1968) in Ghana showed that flower bud initiation took place during the shortening days of October–November, a period also corresponding to the later rains of the year. During the ensuing dry period, December to February, the leaves of these small shrubs were shed and the flower buds remained dormant. Flowering took place during March and April following a similar pattern to that of *Pancratium* and the other geophytes described above. The flowering continued in bursts after each rain shower until all the buds had been used up. The initiation of flower buds in the geophytes was probably also brought about by the stimulus of shortening days while the plants still had leaves above ground. The opening of the fragrant coffee flowers took place in the evening of the third day after rain between 5 and 6 p.m. Anthesis took place in the evening irrespective of the time of day when the rain shower fell, and was accompanied by the visits of large numbers of moths. Flowering occurred at no other time of the year.

The foregoing indicates that this pattern of flowering is controlled by three independent factors: flower bud initiation as a photoperiodic response, flower enlargement up to the point of opening as a response to watering at a time between 64 and 88 hours previously, and an unknown factor probably concerned with carbohydrate hydrolysis determining the fixed time of day of anthesis. The evidence for the latter is the disappearance of starch in the anthers and other parts of the flower between the time of watering and the final opening of the corolla.

These examples have served to show that day length is a factor to be reckoned with in studying the phenology of flowering in tropical vegetation. From the observations of Ewusie (1968, 1969) reported above, where he found the peak period of flowering to occur in the dry season, which incidentally is also a period of short days, it is evident that in some of the species day length may be the more important factor at work. This emphasizes the need for more studies in this field in the tropics.

Breeding Periodicity in Tropical Animals

A number of activities of animals, just as in plants, exhibit periodicity. The best studied ones include breeding and migration, but there are other lesser events like the singing of birds which are also seasonal. For example, in Uganda the small robin-chat bird (*Cossypha heuglini*) bursts into melodious song whenever there is rain and remains quiet between the rains. Certain crickets, like the mole cricket (*Gryllotalpa*), also sing after rain but not at any other time. It is also found in many tropical areas that winged termites and ants fly in numbers shortly after a rainstorm. It appears already from these examples that rainfall has a strong influence on these periodic phenomena. In the breeding periods of birds it would appear that more species may be breeding in the wet season, and those which breed throughout the year have increased activity in the wet period of the year. Thus Chapin (1932) reports that in the lowland equatorial forest of Zaïre, most passerine birds showed increased breeding activity in the wet season of February to May and less in the drier months, which were also more favourable for birds of prey as well as doves. Also, while the black-headed herons (*Ardea melanocephala*) breed throughout the year in Nairobi, the number of them that breed in the rainy period is much greater than in the dry period.

Savanna ungulate mammals also breed in the wet season when there is tall grass for the young to feed on and also to use as shelter. The primates, however, appear generally to be non-seasonal. Examples are *Colobus* and the baboons (*Papio* sp.) as well as species of monkeys (*Cercopithecus*). Apparently one species of the latter, *C. ascanius*, shows a seasonal peak in breeding from December to April in Uganda (Haddow, 1952). The fruit bat (*Eidolon helvum*) is found to be strongly seasonal in its breeding near the equator at Kampala even though the cultivated fruits on which it feeds do not vary much in abundance throughout the year.

Lake and river fish, especially deep-water species, migrate up-river during the rainy period and breed in the swampy upper reaches of the river where the young fish obtain better feeding grounds and avoid early predation by larger deep-water fishes. Tropical frogs and toads breed when the weather is wet, and the characteristic sound of the toads and their spawn in temporary puddles during the rainy season is quite familiar to most tropical dwellers.

The Nile crocodile (*Crocodilus niloticus*), which lays its eggs in the dry season and hatches them in the rainy season, appears to exhibit two different breeding frequencies at different latitudes (Cott, 1961). In Uganda, where the crocodile occurs in the Nile, it is found that from Lake Kyoga northwards there is only one breeding season a year, namely in December to January, but from this point southwards in Lake Victoria there are two breeding seasons, one in December to January and the other in August to September. In some animals which breed throughout the year the rainfall peaks coincide not with two separate breeding seasons but rather with two peaks of intensity in a continuous breeding pattern. An example is found in the land snail (*Limicolaria martensiana*) in Uganda. Here again a role of water in annual periodicity seems evident.

Seasonal Movements

One of the well-known types of seasonal movements in animals is migration which is the regular movement of a species population to and from its breeding site. Many birds that breed in temperate Europe and Asia do not spend the northern winter months there, but migrate to the tropics for warmer weather. Other vertebrates, as well as insects, may also be migrants. In the case of birds and other vertebrates whose individual lives span a number of years, the same individuals are often involved in the movement to and from the tropics, but in the case of insects, a number of which live at most for a few months only, the individuals that may take part in the movement in one direction may not be the same individuals that move in the opposite direction.

Migration also takes place between the southern temperate regions, such as south Australia, and the tropics. Presumably the migrants receive the stimulus to migrate from the temperate climate when the winter approaches with its decrease in temperature and the reduction in day length. These weather changes are naturally associated with a decrease in the food at the time. However, these factors cannot be said to work in the reverse movement of the migrants returning from the tropics to the temperate latitudes during the northern or southern summer, for in the tropical areas, changes in temperature, day length and sometimes food during the year are small. Thus ecologists have often been at pains to find the proximal factor in this situation for the return of the migrants from the tropics. If we consider the fact that some plant species are sufficiently sensitive to little changes in day length with respect to their flowering, it is possible that these migrants are able to sense much smaller changes in the environment which make them leave the tropics at the right time. On general grounds the one significant change in the weather during the intertropical summer is the onset of the rains, and it is possible that climatic changes which herald the coming of the rains may constitute the proximal factor to which the migratory birds react by leaving the tropics at the time they do.

Apart from the long distance migrants described above, there are some birds which migrate for relatively short distances within the tropics. These are often savanna species since tropical forest birds are almost all non-migratory. An example is the weaver birds of Africa of which the savanna species migrate into the forest zone when they are not breeding, and return when it is time for breeding. Evidently large vertebrates and insects are involved more in short distance than in long distance migrations. Seasonal drought, with its attendant shortage of vegetation in tropical savannas, compels large mammals like the elephant and wildebeest to migrate in search of water and of vegetation for food.

Periodic swarming of insects is probably more commonly noticed. Most butterflies appear to swarm at certain times of the year in response to availability of food as well as to environmental conditions. The insects which feed on the fruits of the mango swarm when the mango tree fruits during a relatively moist time of the year. For example, in West Africa where the mango fruits twice a year in April and December, it is only during the April fruiting that the feeding butterflies swarm. In April the rains may have started, whereas December is the beginning of the dry period.

A number of tropical insects exhibit what is described as wet and dry season colour forms. For example, in the African butterfly *Precis octavia* (Figure 24) the wet season form is bright orange with black markings whereas the dry season form is black with blue markings. As the wet season approaches the dry season females produce the wet season type of offspring. Similarly, black grasshoppers are the predominant forms on

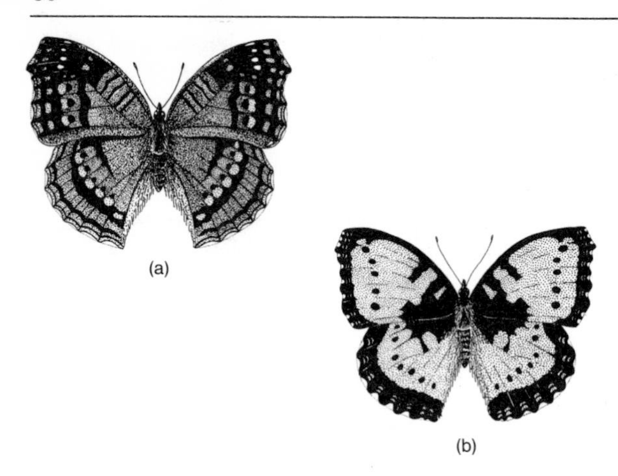

24 Wet and dry season forms of *Precis octavia*: a) wet season form, b) dry season form. The wet season form is mainly orange with black markings and very little blue. These forms are so different from each other that they can easily be mistaken for distinct species
SOURCE: D F Owen (1966)

newly-burnt savanna grasslands during the dry season. How this arises is not clear and more research work is needed to provide the necessary explanations as to the cause. Pied crows and kites in Africa and the gregarious ani (*Crotophaga ani*) in the West Indies frequently accumulate during or after grass fires to feed on escaping insects and lizards. Perhaps some protective colouration in insects is associated with these events.

The butterfly *Libythea labdaca* has been reported in West Africa (Williams, 1951) to migrate southwards in the months from February to May and northwards during the months from October to December. Evidently the southward migration starts at the peak of the dry period in the savannas, but it is not clear what factor makes it migrate back northwards when the dry season begins.

The occasional swarm of the desert locust *Schistocerca gragaria* differs from true migration in its unpredictable frequency of occurrence, and such phenomena are often referred to as **irruptions**. The desert locust is normally a greenish-coloured insect that lives as a pest in about 60 countries in tropical and subtropical Africa north of the equator and in Asia. It is normally present in very low numbers, and then its feeding activities on crops are negligible, but occasionally it increases rapidly and changes its colour to yellow and black. Swarms of about 1000 million individuals spread over neighbouring countries feeding on the leaves of crop plants that come in their way, completely defoliating large areas. The change in habit from the green solitary form to the yellow and black gregarious form seems to occur with increase in population density, but it is still not clear what causes this increase since it does not occur regularly on an annual basis.

Finally, there is seasonal vertical movement of soil fauna. In tropical areas vertical seasonal movements occur as regularly as they do in temperate regions but water rather than temperature is here the determining factor. It is found that in Trinidad soil arthropods move downwards in response to decreasing soil moisture in the dry period and move upwards in the wet season (Strickland, 1947).

Lunar and Tidal Periodicity

Periodicity induced or controlled by the moon in its orbit about the earth is practically unknown in terrestrial communities. It is slightly more prevalent in freshwater communities, and is certainly more important in marine situations.

In fresh water, lunar influence is reported to operate (Thomson, 1911) by way of a correlation between the amount of river plankton and the phases of the moon. In tropical aquatic insects it appears that their emergence shows a periodicity

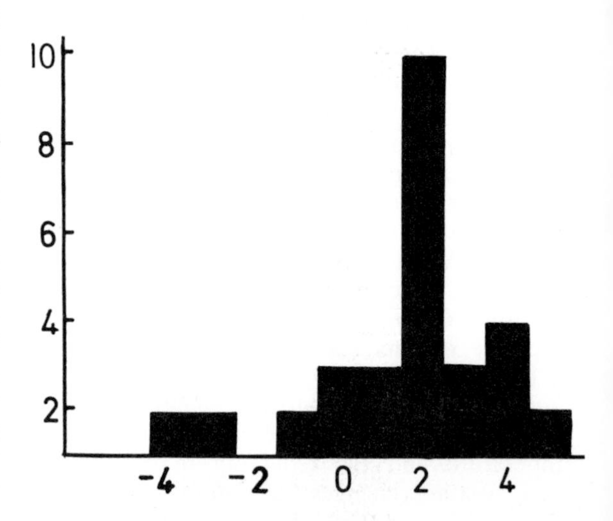

25 Lunar periodicity in the mayfly *Povilla adusta* on Lake Victoria. The number of swarms is shown on the vertical axis, and days before and after full moon on the horizontal axis
SOURCE: D F Owen (1966)

that is linked with lunar influence. It is found (Hartland-Rowe, 1955) that in Lake Victoria in East Africa the mayfly *Povilla adusta* produces a large number of adults during the period of the full moon, with a peak at the second day after the full moon (Figure 25). This occurs simultaneously over wide areas and it appears that, with the brief life of the adult of only a few hours, such swarming facilitates mating.

Lunar periodicity is also reported as being exhibited by several species of *Diptera* and *Trichoptera* in Lake Victoria (Corbet, 1958). MacDonald (1956) reports that two populations of each of two species of *Chaoborus* show emergence and swarming once a month at the time of the new moon, when they are seen like a black smoke rising from the lake.

In the marine environment the moon has the direct effect of producing the tides to which a variety of organisms shows rhythmic responses. Among these are the so-called tidal rhythms of marine animals, especially in the littoral zone. Here not only the height of tides but also the salinity, water temperature, currents, sediment and food are affected by tidal rhythms. These naturally influence the periodicities of some phenomena in the littoral zone animals. Various polychaete annelids, for example, show tidal periodicities in their reproductive behaviour.

An interesting temperate example is reported in a fish, the grunion (*Leuresthes tenuis*), in the California littoral zone (Thompson and Thompson, 1919). These fish appear exactly at high tide on the second, third and fourth nights after the spring tides. During this time the female grunions deposit their egg pods in the sand just above the high-water mark and the male grunions fertilize them immediately. In two weeks' time the eggs become ready to hatch, but they will fail to do so until the egg pods are washed from the sand by the tides at the next dark moon. The ecological advantage of this precise arrangement is provided by Pearse (1939) as follows:

> If spawning occurred just before the highest tides, when the high beach was being eroded, instead of just after, when the beach was being built up, the eggs would be washed out of the sand before they had developed for a fortnight. If spawning occurred at the very highest tide (dark of the moon), the eggs might not be exposed for a month or even two months. If grunions laid their eggs during the day, they would be exposed to the attacks of gulls and other predaceous animals.

Circadian Periodicities

Naturalists of long ago observed that some plants and animals exhibit certain activities on the basis of a daily cycle. Since some of these cycles do not have an exact period of twenty-four hours, the word 'circadian' is now preferred to 'diurnal' in describing such rhythms. A day consists of the day and the night, with their contrasting levels of environmental factors in light, temperature and humidity.

Attempts were made in the past to explain all expressions of circadian rhythmicity in terms of the environmental factor differences between the day and the night and their exogenous origin. However, recent experimental verification of these rhythms has shown that the rhythms are biological and endogenous in origin since they represent self-sustained oscillations that occur freely under constant environmental factors such as light, darkness and temperature, and that they have their own frequency which approximates to twenty-four hours. Under natural conditions, these daily activity rhythms appear to be synchronized with the period of the earth's daily rotation through the diurnal cyclic change of factors of the environment.

Circadian rhythms have been found in organisms at different levels of complexity. These include single-celled flagellates like *Euglena*, *Conyaulax*, *Paramecium*, filamentous algae like *Oedogonium*, moulds, crustaceans, insects, higher plants, birds and mammals including man. So far they do not appear to have been recorded in bacteria and blue-green algae.

In higher plants circadian rhythmicity has been studied in the leaf-movements (opening of the leaves during the day and closing at night) known as 'sleep movements' in *Mimosa pudica*, *Oxalis*, *Phaseolus*, *Canavalia* and *Acacia* species. It has also been observed in the floral mechanism (opening and closing of petals) in plants like *Kalanchoe* sp., *Bromheadia finlaysoniana* and *Epiphyllum oxypetalum* in Malaya (Holtum, 1964), in the rhythm in exudation from the root systems of severed *Helianthus* plants and in odour production in *Cestrum nocturnum* (Overland, 1960).

The mosquito, *Mansonia fuscopennata*, has been found in the Mpanga forest of Uganda to make daily vertical movements which enable it to vary its biting cycle on different animals at different levels of the forest at different hours within the twenty-four hours of day and night.

The passerine bird *Quelea quelea* commonly found in tropical Africa also shows a circadian rhythm in its feeding activity. In the Lake Chad region of Nigeria, they leave the roost in thousands soon after dawn and fly off to the feeding grounds to feed for two or three hours. After that, they return to the roost and spend the hotter part of the day there; later in the afternoon when the weather begins to get cooler, they leave the roost again to feed once more for about two hours before returning to roost for the night.

In general, studies on circadian rhythmicity have led to the impression that activity is initiated by the change in light intensity at dawn in the morning and the reversal of the activity by another change in light intensity at dusk in the evening. Thus the role of light in circadian rhythms has been deemed to be paramount. Indeed, experiments have shown that it is possible to use varying light intensities to change the phase (the stage at which the cycle may be at any particular time). Light can also be used to reduce or lengthen the period of a particular

Table 12. Circadian rhythms in petal movements

CATEGORY	CONDITION FOR FLOWER OPENING	CONDITION FOR FLOWER CLOSING
a	Low light intensity	High light intensity
b	High light intensity	Low light intensity
c	Absence of light (darkness)	High light intensity
d	Low light intensity	Absence of light (darkness) or low light intensity
e	High light intensity	High light intensity

Table 13. Plants found in the various categories of light intensity (Ewusie, 1972)

CATEGORY	SPECIES	TIME OF OPENING	TIME OF CLOSING
a	*Hewittia sublobata*	7 a.m.	1 p.m.
	Kallistroemia pubescens	8 a.m.	1 p.m.
	Merremia tridentata	7 a.m.	1 p.m.
	Luffa cylindrica	6 a.m.	3 p.m.
	Ipomoea cairica	7 a.m.	2 p.m.
	I. involucrata	7 a.m.	1 p.m.
	I. obscura	6 a.m.	1 p.m.
	Trianthema portulacastrum	8 a.m.	2 p.m.
	Passiflora foetida	8 a.m.	12 noon
	Urena lobata	7 a.m.	2 p.m.
b	*Portulaca quadrifida*	11 a.m.	5 p.m.
c	*Tradescantia* sp.	5 a.m.	12 noon
d	*Mirabilis jalapa*	4 p.m.	7 a.m.
	Operculina macrocarpa	8 a.m.	6 p.m.
	Abutilon mauritianum	8 a.m.	6 p.m.
e	*Talinum triangulare*	10 a.m.	3 p.m.
	Sida cordifolia	9 a.m.	1 p.m.
	S. stipulata	9 a.m.	3 p.m.
	Portulaca oleracea	10 a.m.	2 p.m.

a Plants whose petals open in low light intensity and close in high light intensity
b Plants whose petals open in high light intensity and close in low light intensity
c Plants whose petals open in darkness and close in high light intensity
d Plants whose petals open in low light intensity and close in darkness or low light intensity
e Plants whose petals open in high light intensity and close in high light intensity

rhythm around twenty-four hours. Temperature on the other hand appears to have little effect on period, although it can change the phasing of a rhythm.

As may be expected the extent of the effect of light varies with different species and sometimes with different individuals within the same species.

Recent observations and experiments on the floral mechanism of some tropical plants have thrown up a number of questions as to the basic mechanisms in these circadian rhythms (Ewusie, 1972, Ewusie and Quaye, 1977). The observations were made on twenty species belonging to nine different families. Unlike earlier observations, most of which were on temperate plants, these observations showed a wide range of phasing in the different species with respect to the time of day and light intensity at the time of flower opening and closing. The plants could thus be grouped into about five categories in this respect as shown in Table 12. The individual species concerned are classified in Table 13.

Thus time for flower opening in these species varies from 5 a.m. in *Tradescantia* sp. to 4 p.m. in *Mirabilis jalapa* while closing varies from 7 a.m. in *Mirabilis jalapa* to 6 p.m. in *Abuliton mauritianum* and *Operculina macrocarpa*. Evidently a wide range of light intensities must be involved in both the opening as well as the closing of the flowers. It is also worth noting that there is also much variation in the time intervals between flower opening and closing. This varies from four hours in *Sida cordifolia* and *Portulaca oleracea* to as much as thirteen hours in *Mirabilis jalapa*. This has many implications with respect to species sensitivities to light.

A few observations and experiments were conducted on some of these plants. In *Sida stipulata* a sudden drop in light intensity in nature led to the sudden closure of the petals. The ecological purpose of this may be to protect the pollen when it is about to rain. In *Mirabilis jalapa* which may be described as a 'nocturnal plant', opening its flowers throughout the night, it was observed that a sudden flash of light during the night led to closure of the flowers, a possibly protective reaction to lightning associated with rain at night. Two light intensities were used here, and it was observed that higher light intensity led to an earlier closure than lower light intensity.

Experiments involving the interaction of factors should be critical in elucidating the role of the different factors in circadian rhythmicity, but so far no such experiments appear to have been reported. The present author performed such an experiment involving light and temperature, using *Sida stipulata* whose flowers in nature open at 9 a.m. and close at 3 p.m. (Ewusie and Quaye, 1977). Under room conditions of lower light intensity than in the open, but exposed to normal light–dark alternation of twenty-four hours, the opening phase of the flower seemed to be delayed by one hour, whereas closure appeared to be advanced by one hour; thus the flower opened at 10 a.m. and closed at 2 p.m.

When such plants were exposed to continuous darkness for a few days, the opening time was advanced by two hours i.e. from 10 a.m. to 8 a.m. while the closing time was also advanced but by three hours i.e. from 2 p.m. to 11 a.m. Room temperature was about 28°C. Similar plants in continuous darkness were then placed in the lower temperature of 22°C for three days. They showed the same opening time of 8 a.m. but the closing time became 3 p.m. (instead of 2 p.m.) showing a delay in closing of one hour compared to the time in the room temperature conditions.

In another set of experiments, plants of *Sida stipulata* were placed in continuous light under room conditions for three days. They showed a delay in flower opening by one hour i.e. from 10 a.m. to 11 a.m. but an advance of closure by one hour i.e. from 2 p.m. to 1 p.m. Similar plants in continuous light were placed in the lower temperature of 22°C for three days. They showed the same opening time of 11 a.m. but the closing time became 3 p.m., showing a delay in closing of one hour as compared to the time in room temperature conditions.

The results of these experiments have confirmed that light and darkness have an effect on phase setting with respect to flower opening and closing. But it appears that opening and closing are differently affected, suggesting that they may be independent rhythms. The effect of temperature seems to be that of delaying flower closure, although the delay does not go beyond the time of closure under field conditions.

Various efforts have been made to discover the basic mechanism of the 'biological clock' responsible for circadian rhythms. Hypotheses based on biochemical, genetic and physical principles have been proposed, but none seems to have provided the required answer, although there is strong evidence that circadian rhythms are due to some type of cyclical variation in the state of the macromolecules of the cells. Meanwhile, there are suggestions of a relationship between circadian rhythms and photo-periodism. Evolution of life on earth has, of course, been taking place while the earth rotates on its own axis every twenty-four hours and revolves around the sun once every year. On this basis, both plants and animals seem to have developed devices for 'time measuring' some of their physiological activities. Thus, it is conceivable that the development of annual periodicity, including photoperiodism and its effects, is based on processes governed primarily by the daily duration of light. If there is such a relationship between circadian and annual cycles, then experiments will also be needed involving verification of the role of water or humidity in tropical biological rhythms as shown in the studies and reported in this book.

Turning to circadian rhythmicity *per se*, it may be said that since life evolved in an environment with a twenty-four hour periodicity, it appears that a number of living organisms have developed an inherent natural frequency corresponding to this. It also appears that this basic rhythmicity has been adapted to give particular advantages to different species. Thus two species may be able to make more efficient use of the same habitat by having different phase-timing in certain activities, thus decreasing inter-specific competition. Also inter-dependent species can have carefully synchronized phase-timing. Such a situation will be mutually advantageous if the rhythm of opening of nocturnal entomophilous flowers, for example, is synchronized with the time of activity of their insect pollinators (as has been reported by Adams (1968) in *Coffea rupestris*). In the same way the timing of activity in some animals must have a survival value. As the animal cannot consult the external environment before venturing out on certain activities, it is important that the onset of such activities be controlled internally in order to avoid exposure to an unsuitable environment at an unsuitable time. Thus the possession of a time rhythm makes it possible for an organism to set in motion, at the proper time, processes whose final stage can be synchronized with an appropriate phase in the rhythm of the environment.

Chapter 6 The Ecosystem Concept and Production Ecology

The Ecosystem Concept

Ecologists have now come to think of plants and animals together with their total environment as forming a system whose functioning depends on the part played by every component of it. When Jansley (1935) first put forward the term 'ecosystem' he conceived it to include not only the organism-complex, but also the whole complex of physical factors forming what we call the environment.

The ecosystem concept is now widely accepted and ecology has become largely the study of the structure and function of ecosystems. One basic feature of an ecosystem is that it is not a closed but an open system from which energy and matter continually escape and are replaced in order that the system can continue to function. The pathways of loss and replacement of energy frequently connect one ecosystem with another. It is therefore often not easy to delimit one ecosystem from another. Nevertheless, the widespread acceptance of the concept as a basic unit in ecology has greatly helped our understanding of the subject.

As far as structure is concerned ecosystems typically have three biological components. These are: producers (autotrophs) or green plants capable of fixing light energy; animals (heterotrophs) or macro-consumers which consume organic matter; and decomposers, consisting of micro-organisms which break down organic matter and release soluble nutrients. The arrangement of these biological units in vertical space is basically the same in different types of ecosystems whether they are terrestrial or aquatic. The arrangement takes the form of two strata, an upper autotrophic stratum and a lower heterotrophic stratum. The upper stratum is where light is often present and contains the photosynthetic machinery. The lower stratum may not have much light but has the consumer-nutrient regenerating machinery.

In all ecosystems, incident light energy enters mostly on a horizontal surface and this makes it reasonable to compare different ecosystems on a surface area rather than a volume basis. All ecosystems require the supply of the same vital materials such as nitrogen, phosphorus and trace elements. They are regulated and limited by the same requirements for existence as provided, for example by light and heat.

The functional aspect of the ecosystem consists of: the rate of biological energy flow through the ecosystem, that is, the rates of production and respiration of populations and communities; the rate of material or nutrient cycling (the biogeochemical cycles); and ecological regulation. The latter relates to both the regulation of organisms by the environment as well as the regulation of the environment by organisms. Thus, the functioning of the ecosystem involves the transformation, circulation, and accumulation of energy and matter. Living things and their activities play a vital role in these processes through photosynthesis, respiration, decomposition, herbivory, parasitism and symbiosis, the food-chain and other interactions of the organisms involved in them. The functioning of the ecosystem also involves the physical processes of evaporation, precipitation, erosion and deposition.

It is possible to identify a particular ecosystem by means of a number of regulatory factors which control the quantities and rates of movement of matter and energy in them and which also limit the numbers of organisms present and influence their physiology and behaviour. Among these factors are the processes of growth, reproduction, mortality and dispersal or migration.

Trophic Levels

An instructive way of studying the inter-relationships of functions of the components of the ecosystem is to study the basic means or levels of nourishment, or trophic levels, of all the organisms that can possibly be found in ecosystems. This approach also enables us to compare different kinds of ecosystems, some of which do not have all the trophic levels described below.

1. Producers

The producers in the ecosystem are the self-nourishing or autotrophic organisms. Essentially these consist of the green plants which harness light energy and use it to build up their own organic food from simple inorganic raw materials. Only a few specialized chemosynthetic bacteria also fall into this category of producers.

2. Consumers

Apart from the green plants and the chemosynthetic bacteria, all other organisms other than the decomposers are consumers or heterotrophic organisms. They require continued supplies of complex organic compounds as food, from which they derive the energy required to drive them and to provide them with their body-building materials. The food is normally derived from the production by the green plants, although in some cases plant material is first converted into animal material by some organisms before it is consumed by other animals. This leads to a subdivision of the consumers as follows:

(a) **Primary Consumers**. These consist principally of herbivores which in some respects should include man himself. Among typical herbivores are sheep, goats, squirrels, grasshoppers, mosquito larvae, etc. They feed directly on plants or plant products.

(b) **Secondary Consumers**. These are the carnivores of various levels, including some plants (such as the insectivorous plants) and again, in some respects, man. Typical examples of carnivores are lions, hawks and kingfisher birds. Large carnivores such as hawks further feed on smaller carnivores like snakes or toads and are often referred to as top carnivores.

3. Decomposers

These constitute the final major trophic group in the ecosystem. The group is made up chiefly of soil micro-organisms such as bacteria and fungi, although earthworms, termites, mites, beetles and other small arthropods also belong to it. Their function in the ecosystem is to decompose the organic compounds locked up in the bodies of producers and consumers (plants and animals) and in their waste matter, such as the droppings of animals or the fallen leaves of plants. This action is important because without it the supply of materials to the producers and the consumers cannot be maintained. By the action of the decomposers the mineral materials originally absorbed by green plants from the soil and later locked up in their bodies or in the bodies of consumers once more becomes available to the soil and thence to the green plants (producers) again.

The Importance of the Trophic Levels

It is instructive to determine whether an ecosystem must necessarily have the four trophic levels already described, the producers, the primary consumers, the secondary consumers and the decomposers, or whether any one or more of them can be dispensed with.

It appears that the producers are not always indispensable as long as fresh organic materials can be supplied continuously from outside to the ecosystem. Ecosystems which do not themselves support green plants and which therefore function without light are supplied with organic compounds to keep the consumers and decomposers going. An example is the terrestrial soil. Even though it may form part of a wider ecosystem it can in its own right function as an ecosystem without the producers. In the soil are large populations of consumers and decomposers. The leaf litter which may come from outside the habitat would serve as the source of minerals and energy for the consumers and the decomposers. Certain species of fungi and bacteria occupy a special ecological niche since they are capable of making all the enzymes needed to attack the most resistant substances. Of course, detritus feeders assist in the process by reducing the litter to smaller fragments and at the same time deriving some nourishment, although some of them feed on fungi. The droppings of some of the detritus feeders add to the litter which serves as food for other species. There are some carnivorous species of detritus feeders (organisms such as centipedes, beetles, spiders and false scorpions, and even some fungi), which feed on their fellow detritus-feeding organisms. These may also be fed upon by predaceous mites. However, a decaying piece of wood with similar myriads of consumers and decomposers cannot be regarded as a permanent ecosystem because there is only a vanishing 'capital' source of food and not a continuous supply, as in the case of the leaf litter of the soil.

The consumers appear to be indispensable even though they may be rather scarce in some situations, such as alpine tundra or parts of hot deserts. In fact fire seems to serve as a consumer in some habitats, such as the savannas, where it removes the accumulating dead vegetation and is responsible for maintaining the climax or sub-climax vegetation. In view of the importance of consumers in an ecosystem, any hazard for them may seriously affect the balance of the ecosystem. The hazard may take the form of toxic substances applied by man to control weeds or pests or of polluting waste products of industry. Some of these substances act as cumulative poisons. Thus, in some countries certain birds and other animals have virtually disappeared from some communities with dramatic effects on the ecosystem.

The decomposers are essential because of the problems of accumulation of waste and dead matter. As described above, the decomposers consist of a rather complex community which takes many years to form. Once the soil is stripped of its humus-rich top soil, the ecosystem ceases to have a decomposer level, and the vegetation suffers from reduced nutrient availability.

Trophic (Food) Chains and Trophic Webs

The different trophic levels described above link together in a sequence of events known as the food chain. A plant (primary producer) may be eaten by an animal (primary consumer), which in turn may be eaten by another animal (secondary consumer). The latter may itself be eaten by yet a third animal (tertiary consumer) and so on.

To illustrate: diatoms may be eaten by mosquito larvae in water; the mosquito larvae in turn become the food for Tilapia fish; and the latter are then devoured by kingfisher birds. The food chain in this case may be represented as follows:

Diatoms → Mosquito larvae → Tilapia fish → Kingfisher birds

Another example, this time typical of terrestial conditions, may be cited. The leaves of a herbaceous plant may be eaten by grasshoppers; the grasshoppers may themselves be eaten by toads; the toads are killed, swallowed and digested by snakes, which are often killed and swallowed by ducks. Finally man may slaughter and eat the ducks. The food chain here may be represented as follows:

Leaves → Grasshoppers → Toads → Snakes → Ducks → Man

The relationships between the trophic levels as illustrated by food chains provide only the basic elements for understanding the complexity of the food relations found in nature. In nature every trophic level has more than one food relationship. The same primary producer or plant material can serve as food for different kinds of herbivore or the same herbivore can feed on many plant species. These herbivores can in their turn be eaten by various kinds of carnivores. This means that rather than isolated 'strands' of food chains a complete network of relationships is built up. The network of food relationships found in nature is known as the food web. This may be illustrated by considering in detail the food chain:

Leaves → Grasshoppers → Toads → Snakes → Ducks → Man

Here the green plant may provide the leaves also as food for squirrels, grass-cutters and greenflies, apart from grasshoppers. The squirrels and grass-cutters may be eaten by man, the greenflies by beetles, and the grasshoppers by lizards instead of toads. Next, the beetles and lizards may be eaten by birds, then the birds by man. This is illustrated in Figure 26 (p. 88).

There are various ways of studying the food relations in nature and thus constructing the food chains and ultimately the food web. One of these techniques utilizes radioactive isotopes. Phosphorus-32 is often used to 'label' a number of individuals of a single plant species by spraying the leaves. Samples of animals living in the area of the plants are then taken at regular intervals, and the amount of P-32 in each of the animal species assayed. If radioactivity is detected in any animal, then it is reasonable to conclude that the animal (primary consumer) must have been directly or indirectly dependent on the 'labelled' plants for its initial food. In course of time, other animals

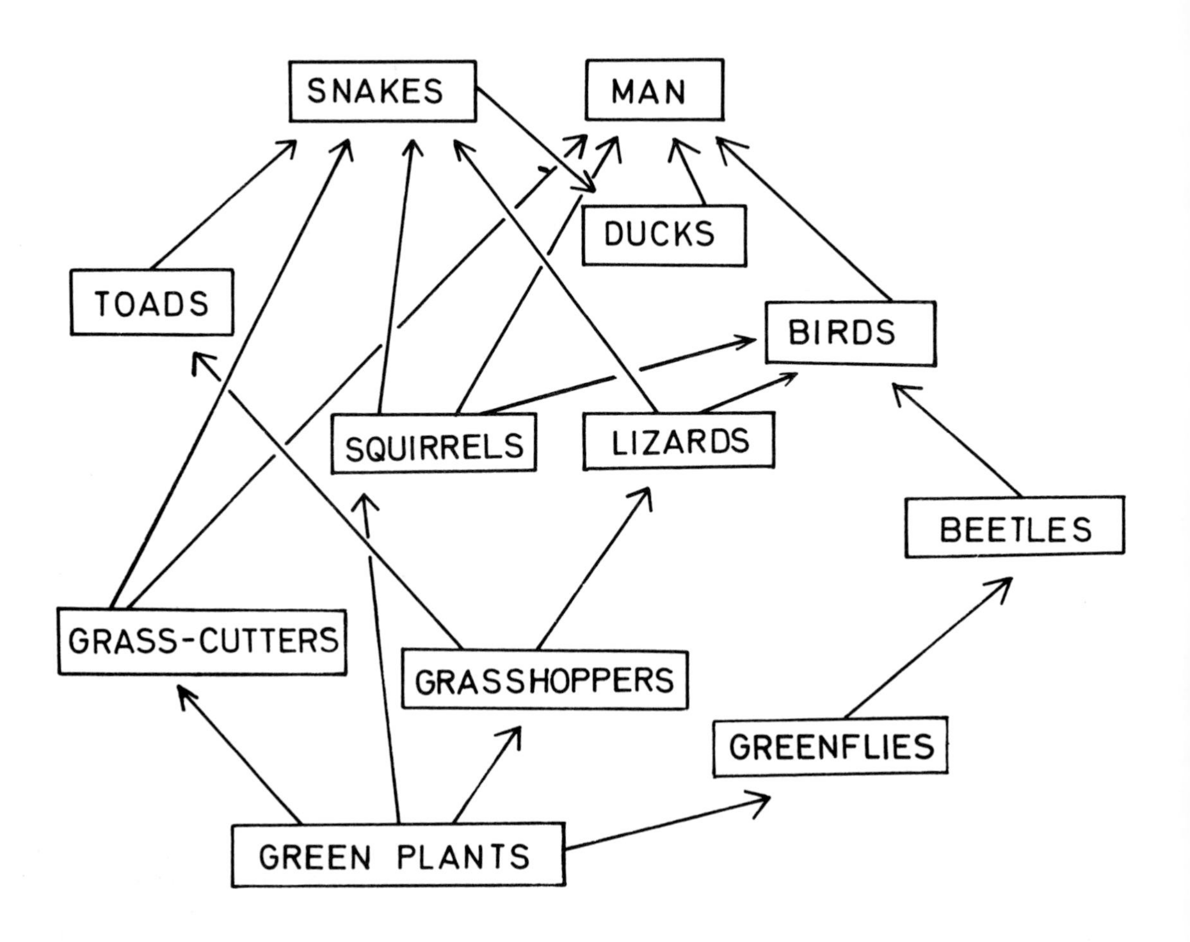

26 A simple terrestrial food web

(secondary consumers) which feed on the primary consumers are similarly detected. The process is analysed until the full picture of food relationships is revealed, and the information may be built up diagramatically into a food web like the one shown in Figure 26.

Trophic Structure (Ecological Pyramids)

It would be instructive to be able to compare the food web in one situation with that in another, but this could be very difficult in view of the complexity of the food web and also because different species may be involved at each trophic level in different situations.

Pyramid of Numbers

One method by which a common basis may be established to make comparisons of the different food webs possible has been the use of the concept of the pyramid of numbers. A comparison is made of the trophic structure of different food webs in terms of the numbers of individuals present in each trophic level of a food web, ignoring, of course, the species involved in each case. Thus all autotrophs are grouped and counted, in order to obtain the numbers per unit area (density). The same is done for the next step of primary consumers and so on. The two examples of food chains discussed above illustrate the two common types of pyramid of numbers found in nature (Figure 27).

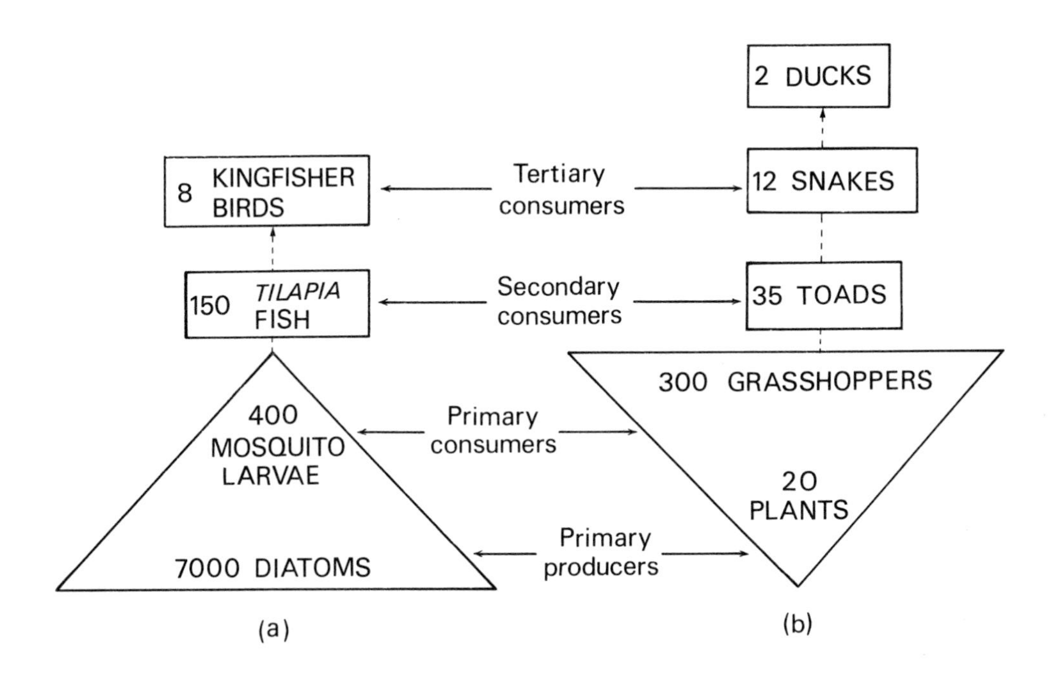

27 The two common types of pyramids of numbers: a) When primary producers are small in size but numerous b) When primary producers are large in size but few

In Figure 27(a) there are 7000 diatoms providing the food for 400 mosquito larvae which are eaten by 150 Tilapia fish, which in turn are devoured by 8 kingfisher birds. Here the primary producers are very numerous but the individuals involved are very small organisms. A true pyramid is formed by progressively decreasing numbers at the respective trophic levels. In Figure 27(b), on the other hand, the situation appears reversed. The primary producers may consist of about 20 herbaceous plants, while the grasshoppers may number about 300. The toads may number 35, the snakes 12 and the ducks 2. Here there are fewer primary producers than primary consumers. Thus the two types of pyramids differ with respect to the ratio of primary producers to primary consumers. Now the question we should ask ourselves is whether the concept of the pyramid is a reliable way of comparing different food relations. Here we are forced to count mere numbers regardless of size. A tree, a shrub, or a herbaceous plant is put on a par with a diatom. Similarly an elephant has to be equated with a grasshopper. Surely this is an unfair basis for comparison of members of trophic levels.

Pyramid of Biomass

It was in an attempt to overcome the problem of size that the concept of the pyramid of biomass was introduced. Here the weight (usually dry weight) of the organisms rather than their numbers is used in constructing the pyramid. Biomass is expressed in terms of grammes per square metre (g/m^2) or in metric tons (tonnes) per hectare (ha) of material, usually dried at 70°C in order to avoid losses of nitrogen. (Note: 1 tonne is equal to 10^6 g; 1 ha equals 10^4 m^2; and 1 tonne/ha is equal to 892 lb/acre.) If the two types of pyramid of numbers discussed above are converted to pyramids of biomass, the base of each pyramid is reversed with respect to the relationship between the primary producers and the primary consumers. This is shown in Figure 28. In Figure 28(a) let us assume that the 7000 diatoms have 10 g dry weight, whereas the dry weight of the 400 mosquito larvae is about 160 g. In Figure 28(b), in spite of the smaller number of the primary producers, their total dry weight may be very high, say about 2500 g, whereas the dry weight of the grasshoppers may be 600 g. This relationship is what is regarded as normal because it provides the normal pyramid which presumes that in no trophic level can the organisms concerned consume the whole of those in the next stage below it in a pyramid.

The situation in Figure 28(a) raises the question as to whether the pyramid of biomass solves entirely the problem posed by the pyramid of numbers, for here a smaller biomass of primary producers supports a greater biomass of primary consumers. In nature a number of interesting examples of this are known. In the famous Lake Nakuru in Kenya the microscopic blue-green alga *Spirulina platensis*, suspended in the lake, supports about 1½ million large birds, the flamingoes *Phoeniconavis minor*, which feed (as primary consumers) on the alga. Evidently the flamingoes have a far greater biomass than the blue-green alga.

The measured biomass of any organism, population or community is its dry weight per unit area at a given time, and no account is taken of the relative productivity rates of the components making up the trophic web or the duration of their life span. As the definition of biomass shows, the amount of material present at any point in time such as a particular date is known as the **standing crop**. Like biomass, this gives no indication of the period within which it was accumulated.

Nye (1959) has estimated the total biomass of two savanna grassland types in Ghana, as shown in Table 14.

Table 14. Total biomass of two savanna grassland types in Ghana (from Nye, 1959), in kilogrammes per hectare

BIOMASS	SAVANNA WITH *IMPERATA*	SAVANNA WITH *ANDROPOGON*
Aerial parts	4 200	14 000
Dead litter	600	170
Underground organs	8 400	4 400
TOTAL	13 200	18 610

Nye obtained the following results from a well-wooded savanna in Ghana:

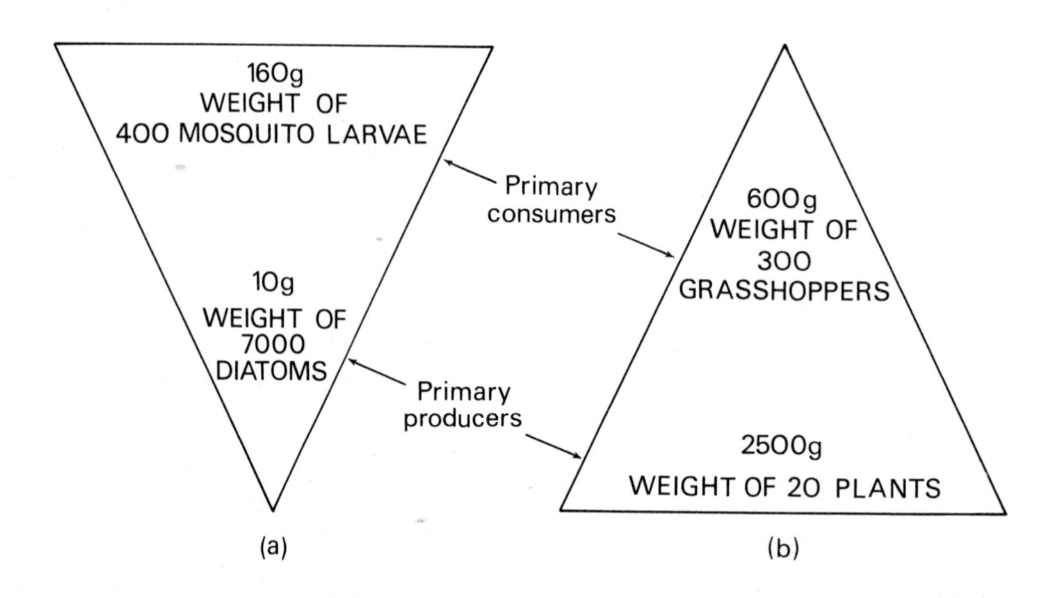

28 Base of pyramids of biomass: a) When primary producers have smaller biomass than the primary consumers b) When primary producers have greater biomass than the primary consumers

Grass stratum:	aerial parts	9 500	kg/ha
	roots	5 280	kg/ha
Tree stratum:	leaves	600	kg/ha
	wood	57 600	kg/ha

It is clear from the above data that the contribution to biomass by leafy parts is much greater for grasses than it is for trees, where a very high proportion of the biomass consists of permanent wood. Biomass for grasses can be estimated quite easily by harvesting, drying and weighing whole plants in a given sample area. For trees, this method is hardly feasible but calculations entailing height, girth and density measurements can be made. More recently, **net productivity**, that is, annual increment to biomass, has been estimated by means of determinations of carbon dioxide uptake, with infra-red gas analyser equipment, on pieces of enclosed vegetation, single trees or sample branches, or by means of leaf area counts (see p. 41). In recent years much work on estimation of biomass has been carried out in India, and some of the results obtained will be given here.

Ambasht *et al.* (1971) studied two protected grasslands of Varanasi, one dominated by *Heteropogon contortus* and the other by *Dichanthium annulatum*, and obtained the following results.

1. In July (before the rains): lowest biomass:

Community A		**Community B**	
Heteropogon	181.4 g/m^2	*Dichanthium*	252 g/m^2
Entire community	385.5 g/m^2	Entire community	691 g/m^2

2. In November (after rainy season): highest biomass:

Community A		**Community B**	
Heteropogon	2948 g/m^2	*Dichanthium*	1435 g/m^2
Entire community	3268.5 g/m^2	Entire community	2050 g/m^2

Mall and Singh (1971) have shown that, as may be expected, different species attain their highest biomass at different times of the year. They studied the seasonal variations in biomass of two principal species, *Iseilema anthephoroides* and *Indigofera linifolia* in a grassland community at Ujjain in India. They found that while *Indigofera* attained its highest biomass of 8.8 g/m^2 in July, *Iseilema* attained its highest biomass in November with a biomass of 102.2 g/m^2.

Data on biomass from aquatic habitats are rather scarce. Kaul *et al.* (1971) estimated biomass in aquatic plants of different life forms in the valley of Kashmir in India and obtained the following results.

1. **Floating Types.**
   (a) *Salvinia natans*. Maximum biomass = 320 g/m^2. It is worthy of note that there is inverse correlation between the percentage ash content and the energy value.
   (b) The lemnids *Lemna minor, L. gibba* and *L. frisulea* and *Spirodella polyrhiza*, together had a maximum biomass = 200 g/m^2.
2. **Submerged Types.**
   Myriophyllum spicatum. Maximum biomass = 384 g/m^2.
3. **Floating Leaf Types.**
   Nymphoides peltata. Maximum biomass = 185.51 g/m^2.
4. **Emergent Types.**
   (a) *Typha angustata*. Maximum biomass = 3500 g/m^2.
   (b) *Phragmites communis*. Maximum biomass for stem = 2800 g/m^2 and for leaves = 1937 g/m^2. Productivity of stem = 14.23 g/m^2 per day and of leaves = 12.80 g/m^2 per day.

It is interesting to note that in *Phragmites* an inverse correlation exists between the density of stems and biomass. In March when the biomass is at its minimum the density is highest (245 stems/m^2). In July–August when the biomass is at its maximum the density is lowest (163 stems/m^2).

The following are some biomass figures of mammalian communities in different types of vegetation in East Africa:

The Nile Basin grasslands (Mweya Peninsula in Uganda)	:	334.5 kg/ha
Serengeti plains of Uganda	:	63 kg/ha
East African ranches	:	56 kg/ha
Local farmers' grazing lands in East Africa	:	20–2800 kg/ha

Pyramid of Energy

The biomass of a tree is mostly represented by the organic material (wood) that has accumulated over a long period of time, perhaps over 50 years, whereas an estimate of the biomass of algae represents the amount of organic materials that has accumulated over a few weeks or months. It is clear, therefore, that the only reason a small biomass of algae can support a much larger biomass of primary consumers, such as flamingoes, lies in the rate at which the biomass accumulates in the algae. While a tree of about six years of age may still be growing but may not have reached the reproductive stage, the microscopic algae will in the same period have given rise to millions of its kind, and their cumulative biomass may have equalled or exceeded that of the tree. The blue-green algae in Lake Nakuru, for example, are known to have the potential of doubling themselves in a time of 24 hours, whereas the flamingoes cannot double their numbers for many years.

Biomass and productivity are analogous to capital and interest respectively. Rate of interest is always expressed in terms of time, and this stresses the importance of taking the life span of each organism in the food chain or web into account. For a primary producer that is of small size, numerous, physiologically active and rapidly reproducing, quite a small permanent biomass, sometimes referred to as standing crop, can undergo many life cycles in a year. To assess its productivity accurately it may be necessary to measure the biomass increment over a period corresponding to an average life span. The total annual productivity will be given by a single biomass increment multiplied by the number of life cycles undergone in a year. The cumulative total is what provides a sufficiently high and continuous source of food for the growth and survival of the larger species. Incidentally, such systems are fragile because any adverse circumstance, such as lake or ocean pollution, affecting the small standing crop of the primary producer, might unpredictably and suddenly cut off the food supply to the consumers with drastic consequences. Examples of such risks could be birds feeding on phytoplankton, as described above for flamingoes, or whales feeding on krill which in turn feed on plankton, or any larger animal with a specialized or restricted dependence on a particular kind of smaller organism.

It was in order to overcome the objections to the pyramid of numbers and the pyramid of biomass discussed above, that the concept of the pyramid of energy was developed. Here the pyramid is based on the amount of energy produced per unit area (often per square centimetre) by different trophic levels over a given period of time (often per annum). The rate of production of the primary producers is termed **primary productivity**, which is defined as the rate at which energy is bound or organic-material created by photosynthesis per unit area of vegetation on the earth's surface per unit time. Productivity may be expressed in terms of calories per square centimetre per year ($cal/cm^2/yr$), or dry weight per square metre per year ($kg/m^2/yr$), or other given period. There is no doubt that energy provides a unifying denominator by which to express the productivity in an ecosystem and to compare productivity in different trophic levels or ecosystems, be they deserts, tropical forests, lagoons or seas.

Energy Transformation in Ecosystems (Ecological Energetics)

In order to understand how the common denominator of energy operates at different trophic levels to give the pyramid of energy, it is necessary to understand the basis of transformation of energy in general and with respect to the different trophic levels.

As described in elementary physics, energy is the capacity to do work. There are different forms of energy, the most important of which, for living organisms, are the radiant, heat, chemical, and mechanical forms. Radiant energy is released by the sun as electromagnetic waves which consist of a rhythmic exchange between potential and kinetic energy. Heat is a very special form of energy resulting from the random movement of molecules. It is worth noting that all other forms of energy exist as a result of non-random movement of molecules. Heat is evolved in all energy transformations, that is when all other forms of energy are transformed. For example, during respiration an animal may transform the chemical (potential energy) of glucose, converting about two-thirds of it into kinetic energy, which is used for such work as locomotion and growth. The remaining one-third, of it is converted into heat and given off. Stored (chemical) energy in the form of food is a form of potential energy. This can be converted by oxidation (or combustion) into kinetic energy for work and heat. Kinetic energy which is a form of free energy resulting from motion is a form of mechanical energy, but the latter can also exist in the form of a potential or stored energy.

Every form of energy has its own units in which it is expressed. Mechanical energy, for example, is expressed in joules (J), while heat energy is expressed in calories. (The work done in raising one gramme weight to a height of one centimetre is equal to 981 ergs, and 10 million (10^7) ergs is equal to one joule. The amount of heat required to raise the temperature of 1 g of water by 1°C is the calorie.) It is necessary to reduce the different units to a common one for purposes of comparison, and in view of the special position of heat, for all forms of energy can be transformed into heat, the unit of heat is often used for the purpose. Thus, for example, 4.2×10^7 ergs are equal to 1 Kcal (Kilocalorie) which is equivalent to 1000 cal.

Since the calorie is a basic unit of energy it is usual to convert the potential energy stored in different food materials into calories of heat. Now the same mass of material from different sources produces different amounts of heat. The heat evolved from one gramme of a substance, expressed in calories, is known as the **calorific value** of the substance. The instrument used in the determination is called the bomb calorimeter. Some calorific values are as follows:

Green tops of herbs	: 4.08 cal/g dry weight
Roots	: 3.30 cal/g dry weight
Mouse tissue	: 4.65 cal/g dry weight

Thus the biomass of a substance, expressed in grammes multiplied by its calorific value, gives the energy heat in terms of calories.

The Laws Governing Energy Transformation

Many of the matters discussed so far and some of those to be discussed later derive their significance from the fact that when energy conversions occur, they do so according to specific laws of exchange known as the laws of thermodynamics. Two of these laws are given here.

The First Law of Thermodynamics is the Law of Conservation of Energy. It states that energy can be neither created nor destroyed, but can be transformed from one form into another. Thus, when the chemical energy in a given amount of food is released by oxidation, the total potential energy is equal to the kinetic energy produced for work plus the amount of heat given off. The amount of heat produced (or absorbed) in the process of energy transformation is the same whether the reaction takes place in stages or directly in one step. This is sometimes referred to as the Law of Constant Heat Sums and may be illustrated by the oxidation of glucose to carbon dioxide and water as follows:

1. Direct reaction (as in combustion):
   $C_6H_{12}O_6 + 6O_2 \rightarrow 6H_2O + 6CO_2 + 673$ Kcal of heat energy
2. Two-stage reaction (as in fermentation to alcohol, followed by combustion):
   (a) $C_6H_{12}O_6 \rightarrow 2C_2H_5OH + 2CO_2 + 18$ Kcal
   (b) $2C_2H_5OH + 6O_2 \rightarrow 6H_2O + 4CO_2 + 655$ Kcal

Thus the stages (a) and (b) produce $18 + 655 = 673$ Kcal of heat energy, which is the same as that produced in one step by direct combustion.

The Second Law of Thermodynamics states that no spontaneous transformation of energy can take place without some degradation of energy into a random or unorganized form. This in effect means that no energy transformation is expected to be 100 per cent efficient in terms of the conversion into the new energy form, because there is bound to be some loss of energy as heat. As a result of the many energy transformations in respiration and other metabolic processes, some energy is lost at each level of the ecosystem. Any energy that remains is eventually also transformed into heat at the decomposer level, so that no energy is left to re-enter the ecosystem from the decomposer level.

Nevertheless, in terms of the First Law of Thermodynamics, the total amount of heat energy that is lost in a complete trophic cycle in an ecosystem must equal the heat energies that have left the system at each of the stages.

Comparison of Mineral and Energy Cycling in the Ecosystem

The above observations lead us to study how the cycling of minerals and energy differs in nature. The comparison is shown in Figure 29. In the case of nutrients it is possible to see how the same amount of material may be used over and over again in the ecosystem. This is because eventually all plant and animal materials are decomposed to release the minerals they contain into the soil so that they may be re-absorbed by plant roots to participate in primary production.

In the case of energy, however, it cannot be recovered and re-used but it passes through the system only once and is lost. This means that, unlike materials in which a fixed quantity can maintain an ecosystem almost indefinitely, energy has to be supplied in fresh quantities all the time to enable an ecosystem to be maintained. This is so simply because, by the action of decomposers on the dead plant and animal materials and on the droppings of living organisms, the locked-up energy is converted into heat energy and released into the atmosphere.

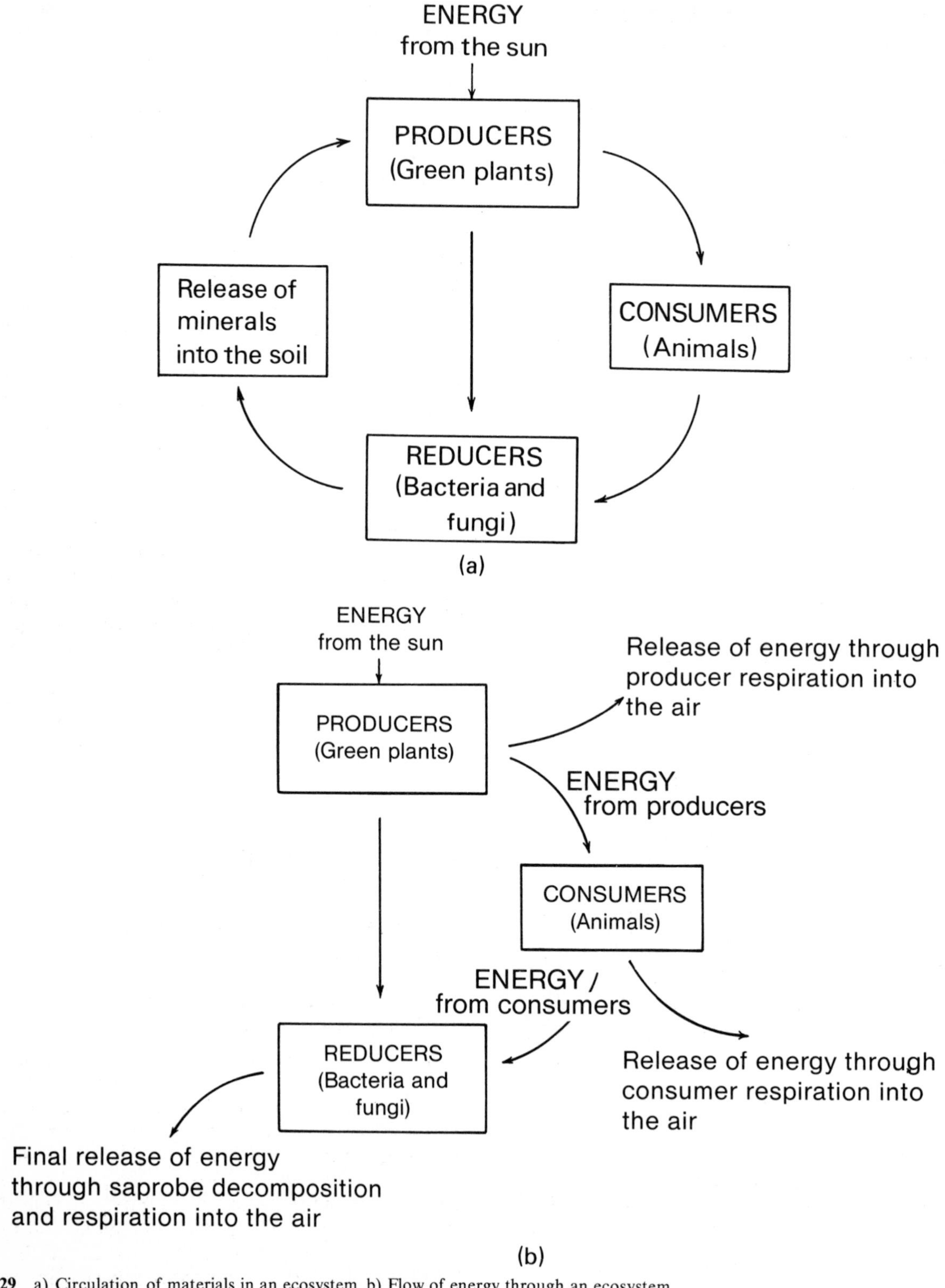

29 a) Circulation of materials in an ecosystem b) Flow of energy through an ecosystem
SOURCE: J Y Ewusie (1973)

Gross and Net Production and Production Efficiencies

The definition of primary productivity (already given), which is in terms of the rate at which organic matter is created, refers to gross production (Pg), that is, it does not take into account any loss by respiration. Let us here consider a simple trophic chain, such as grass → gazelle → lion. In nature the energy stored by the grass is not all available for the next trophic level, namely the primary consumers or herbivores. This is because some of the gross production will be oxidized through respiration into kinetic energy and heat to meet the metabolic needs of the grass itself, and some of it will be lost through death of parts of the plant. What is available after loss through respiration is referred to as net primary production. Thus,

$$P_1 = Pg - R_1$$

(where P_1 is net primary production; Pg the gross production; and R_1 the loss by respiration). Theoretically it is the net primary production only which is available to the herbivores.

Production Efficiencies

The herbivores do not assimilate all the food available from net primary production (P_1) which they consume. Some of it goes to the decomposers as leaf litter and rotting wood and thus follows a detritus channel. The rest is eaten and follows a grazing channel. About 90 per cent of this passes out of the body as faeces and urine (NA_2) without contributing to the energy intake. Of the amount assimilated to form the gross secondary production, some of it is further lost as a result of respiration (R_2) and what remains after this is the **net secondary production** (P_2). Thus

$$P_2 = P_1 - (R_2 + NA_2)$$

(Note that P_1 is the same net primary production as given above.) The same process is followed from the net secondary production of the herbivores to the **net tertiary production** of the carnivores (P_3). Thus,

$$P_3 = P_2 - (R_3 + NA_3)$$

where NA_3 is the loss from the body as faeces and urine. The only difference here is that, since the herbivore tissue is very much like that of the carnivore, the amount of food consumed which is lost in the form of faeces is much smaller, about 50–70 per cent. As a result the efficiency of production at the carnivore level is often a little higher than that at the herbivore level.

Experiments have shown that if one measured the rate of primary production, one should be able to make rough predictions of productivity at other levels in the ecosystem. It is noted that, in general, productivity at any level is approximately one tenth of that shown by the preceding trophic level. Thus if, for example, a given area of land or water produces 1000 kg dry weight of plant matter each year, then one would expect an annual production of about 100 kg dry weight of herbivore material, and 10 kg dry weight of carnivore material in the ecosystem. The factor of one-tenth is, however, only an approximation and there are some distinct departures from it. Between the primary producers and the herbivores in an ecosystems, in perhaps a forest, where a large proportion of the primary production is not used by the herbivores but is passed on directly through the detritus channel to the decomposers, the factor is much less than one-tenth. Also between the herbivore and the carnivore levels the factor is greater than one-tenth, largely because of the similarity between the herbivore and carnivore flesh.

Primary Productivity

Methods of Measuring Primary Productivity

1. The Light and Dark Bottle Method (Productivity in Water)

This method is suitable for use in aquatic situations. Productivity is here measured in terms of the balance of oxygen produced as a result of photosynthesis. Two samples taken from the standing water and containing plankton are sealed in glass bottles and suspended in the pond at the depth at which the samples were collected. One of the bottles is first wrapped in aluminium foil to keep the contents in complete darkness so that only respiration (but no photosynthesis) by the plankton takes place. In the other unwrapped bottle both respiration and photosynthesis occur during daylight. It is necessary to determine the initial amount of oxygen dissolved in the pond by taking and analysing a third sample at the time when the two experimental bottles are placed. The experimental bottles are removed from the pond after 24 hours and the temperature of the water is recorded. The oxygen content of the bottles is estimated by the Winkler method. The volume of oxygen in the light bottle over and above that in the dark bottle gives the primary productivity of the phytoplankton.

The weight of oxygen produced can be converted into weight of carbohydrate by multiplying by 30/32, this being the ratio of the molecular equivalents of carbohydrate and oxygen, as shown in the equation:

$$2H_2O + CO_2 + 114\ \text{Cal} \underset{\text{Respiration}}{\overset{\text{Photosynthesis}}{\rightleftharpoons}} CH_2O + O_2 + H_2O$$

MW of CH_2O = 30 MW of O_2 = 32

Where necessary the experiment may be repeated at different depths in the pond so that the overall productivity of the whole pond may be estimated.

2. Measurement of Primary Productivity in Land Plants

This can be measured by using a modified form of Sachs' half-leaf method. One plant is kept in the dark. From another, a set of discs about 20 mm in diameter are cut from one side of the midrib of a leaf, using a suitable corkborer. The dry weight of the discs is determined. The plant is left to photosynthesize for a given period (say 6 hours). A corresponding set of discs is then taken from the other side of the midrib of the leaf. The dry weight of the second set of discs is then determined and compared. The difference in weight is added to the loss in weight (due to respiration) from corresponding areas from the plant kept in the dark.

The gain in dry weight of discs from plants kept in daylight represents the net production (P_1) of carbohydrate by a known leaf area over the period of time. The carbohydrate used in respiration (R_1), together with the net production, gives the gross production (Pg) i.e. $P_1 = P_g - R_1$. An estimate of the total leaf area will make it possible to estimate the approximate productivity of the plant as a whole.

3. Measurement of Primary Productivity by the Harvest Method

Grasslands lend themselves best to studies of terrestrial productivity, for it is easy to make periodic harvests by clipping small sample areas and so obtain a direct measure of further growth of shoots at intervals during the season. Some allowance should, however, be made for the growth of the underground portions by estimating the biomass of live roots washed out from a known volume of soil each time the shoots are harvested. Respiration by a unit area of grassland could be estimated by measuring the rate of production of carbon dioxide. This is more conveniently done if the grassland samples are established in flower pots. The estimation of carbon dioxide is carried out by using the Peltenkoffer method, by which the carbon dioxide is absorbed in barium hydroxide solution and the quantity estimated by titration against a standard hydrochloric acid, or by infra-red gas analysis.

Primary Productivity in Tropical Ecosystems

Values of net primary productivities (in terms of molecules of carbon per square metre per year) in three temperature zones of the world are as follows:

Arctic region: + 0.5–1 molecules C/m^2/yr
Temperate region: + 10–30 molecules C/m^2/yr
Tropical region: + 30–40 molecules C/m^2/yr

Net production of Lake George in East Africa is estimated to be 30 molecules of carbon per square metre per year.

Estimates of net primary productivity in tropical high forests have ranged between 11 and 32 tonnes per hectare per annum. A plantation of *Eucalyptus saligna* in East Africa shows a net productivity of 45 tonnes per hectare per annum. It is estimated that forests in general have a net productivity of 1200–1500 g/m^2/yr, with tropical forests having the highest values.

The application of leaf area index or LAI (see Chapter 3) in productivity studies has been well illustrated by Heinrich Walter (1973) with respect to a comparison of the productivities of a tropical rain forest and a temperate beech forest. The luxuriance of the vegetation of the tropical rain forest would readily suggest to an observer a very high primary productivity. However, as Walter states, the preliminary estimates of about 100 tonnes per hectare per annum of dry substance proved to be too high. He drew attention to the fact that the plant material of tropical rain forests has a very high water content (75–90 per cent for the herbaceous parts) and that, although the green leaves are able to assimilate CO_2 thoughout the year, nocturnal respiration losses are especially large owing to the high temperatures. Wood productivity in tropical forest plantations can attain values of 13 tonnes/ha/yr, which is only about twice that of a good European beech forest and occurs only because the vegetative period is twice as long.

The leaf area index determined for a tropical rain forest on the Ivory Coast was found to be about the same as that for a healthy beech forest. To quote Walter (t/ha = tonnes/hectare):

This surprising result explains why light conditions on the forest floor of tropical rain forests are not much less favourable than in dense temperate deciduous forests. As was to be expected the gross production of virgin forest on the Ivory Coast proved to be very large (52.5 t/ha). However, 75 per cent of the organic substance produced is lost by respiration: respiratory losses of the leaves = 16.9 t/ha, of the axial organs = 18.5 t/ha, and of the roots (estimated) = 3.7 t/ha; in all 39.1 t/ha. Since the losses due to respiration in a beech forest amount to only 10.0 t/ha, i.e. 43 per cent of the gross production of 23.5 t/ha, it is understandable that the primary production of the tropical rain forest is no higher than that of a well-tended beech forest in Central Europe:

52.5 – 39.1 = 13.4 t/ha Tropical rain forest
23.5 – 10.0 = 13.5 t/ha Beech forest.

Vyas *et al.* (1971) have however estimated the production of the leaves of the deciduous trees *Erythrina suberosa* in Rajasthan (India) to be 68.45 g/m²/yr.

For savanna grasses generally, net productivity is around 1350 g/m²/yr. Among cultivated crops sugar cane and rice are said to have some of the highest productivities in the world. Data on productivity of tropical savanna are rather scarce. Among the data presently available are those from savannas with poor soil nitrogen in the Ivory Coast. Here savannas with *Hyparrhenia* have a productivity of 8–10 tonnes/ha/yr while savannas with *Loudetia* have one of 6–8 tonnes/ha/yr.

Ambasht *et al.* (1971) obtained a high primary productivity of 2882 g/m²/yr in a protected grassland in Varanasi (India) in which the contribution of the dominant species *Heteropogon contortus* was as much as 2766.6 g/m²/yr. In another grassland community in Ujjain, India, Mall and Singh (1971) obtained a total productivity of 6.8×10^5 Kcal/m²/yr, of which the contribution by the common species *Iseilema anthephoroides* was 3.8×10^5 Kcal/m²/yr.

There are not many data on productivity in tropical waters. Among those available is the work of Bhatnagar (1971) on the primary organic production in Kille Back waters in Porto Novo, southern India. In the shallow estuary he noticed that photosynthesis had a pronounced seasonal cycle although temperature and light conditions were more or less uniform throughout the year. He considered that salinity and turbidity changes are due to the monsoon floods which cause large variations in production and succession of phytoplankton blooms. The highest corresponding net productivity of photosynthesis was 225 mg of carbon/m²/yr. Information on marine and estuarine productivity is given in Chapter 7.

Litter Productivity in Forests

In view of the importance of wood products in national economies, the role of litter fall in the nutrient cycle of the forest, in soil development and in its relation to productivity, has received much attention. A recent comprehensive review of the literature by Bray and Gorham (1964) has shown that in equatorial forests the mean litter fall is about 10 tonnes/ha/yr, in the warm temperate forests about 5.5 tonnes/ha/yr, in the cool temperate forests about 3.0 tonnes/ha/yr and in the arctic-alpine forests about 1.0 tonnes/ha/yr. A summary of the data is given in Table 15. The data show the predominant influence of climate, with special reference to temperature, on litter productivity. Among the few studies on tropical forests is that of Bartholomew *et al.* (1953) in Zaïre (Yangambi) where a total litter fall of 12.3

Table 15. Average annual litter production in four major climatic zones (tonnes/ha/yr)

CLIMATIC ZONE	LEAVES	OTHER PLANT PARTS	TOTAL
Equatorial forests	6.7	3.5	10.2
Warm temperate forests	3.8	1.8	5.6
Cool temperate forests	2.3	0.8	3.1
Alpine forests	0.7	0.4	1.1

tonnes/ha/yr was recorded. An even higher figure of 15.3 tonnes/ha/yr was recorded in the Zaïre forests by Laudelot and Meyer (1954). Nye (1961) made studies in forests at Kade (Ghana) and estimated a litter productivity of 10.5 tonnes/ha/yr. Other studies in the Malayan forests gave litter production of 14.8 tonnes/ha/yr.

Secondary Productivity

Method of Measuring Secondary Productivity

The stock which can be supported by a given area of vegetation over a given period may be used to provide an estimate of the net secondary productivity. A relationship is often first established between a given fresh weight of an animal and its dry weight. This makes it possible to determine the dry weight of the herbivores without slaughtering them for the sole purpose of the investigation. The herbivores in a given vegetation are weighed at the beginning of the experiment. They are then left to graze on the grassland under study for a given season, usually of not less than three months. After the season they are weighed again together with any offspring that may have been produced. The increase in weight is then reduced to dry weight and this, expressed on the basis of a year's increase, constitutes the net secondary production.

In the calculation of net secondary production, it should be noted that the animals have already lost some of their gross production through respiration and excretion. If, therefore, gross production is desired, it would be necessary to determine respiration by the Peltenkoffer method and to add the value obtained to the estimated net productivity.

Secondary Productivity in Tropical Ecosystems

Here again data on tropical animals are scarce. One of the best studies on energetics in the tropics is that of Petrides and Swank (1965) on the African elephant (*Loxodonta africana*) in the African savanna. The studies were conducted on 28.5 square metres of the Queen Elizabeth National Park, Uganda. It was found that an average elephant fixed as new protoplasm 165.6×10^3 Kcal per year. The density of the elephants was 5.37 per m^2, and the net productivity was calculated to be 0.34 Kcal/m^2/yr. They calculated the amount of energy respired and, when this was added to the net production, the figure for gross productivity was found to be 25.34 Kcal/m^2/yr.

Chapter 7 Features of Tropical Aquatic Ecosystems

Aquatic ecosystems include environments ranging from cool, relatively infertile streams and lakes of high mountains to warm, highly fertile marshes, swamps, rivers and lakes (whether natural or man-made) of lowland tropical regions. They include also estuarine and near-shore coastal areas as well as the oceans. An extensive study of fresh water habitats forms a special branch of science known as **limnology**. For the present purposes we can recognize three broad categories of aquatic ecosystems: fresh water; estuarine or brackish water; and marine.

Environmental Factors

It is necessary to examine the principal environmental factors which operate in aquatic ecosystems. For one thing, some of the common environmental factors which have been discussed in Chapter 4, such as temperature and light, operate in quite different ways in the water medium. Others, like salinity and viscosity, colour transparency and currents, are essentially peculiar to the aquatic habitat. The environmental factors to be considered here are: dissolved gases; salinity and dissolved salts; density; colour and transparency; temperature; light; water currents; and surf and substrate.

1. Dissolved Gases

In the terrestrial environment the air of the atmosphere contains the normal proportions of the different gases, and it is only under polluted atmospheric conditions and special soil conditions that these proportions are significantly altered. In aquatic conditions, however, a great deal of variation exists at various depths in the concentrations of the different dissolved gases, and this becomes an important ecological factor. The percentage of oxygen in solution is far lower, being of the order of one-tenth or less, than the normal percentage in the atmosphere. The amount of oxygen in water is not as constant as in the air, but fluctuates markedly depending on the depth, temperature, wind and amount of biological activity. For example, under mats of *Pistia stratiotes* (Figure 30) in fresh water ponds, the concentration of oxygen has been found to be lower than normal. Mostly only the larvae and pupae of insects are able to live here, with adaptations suitable to these conditions. An increase in the temperature or the salinity of water results in a reduction in oxygen content. By contrast, strong winds expose more surface of the water to air and thus increase the oxygen content of the water.

Carbon dioxide (which combines with water to form carbonic acid), ammonia and hydrogen sulphide are among the gases found dissolved in water. Of these, hydrogen sulphide is rather poisonous and sometimes high concentrations of it can be harmful to water plants, and only a few plants can tolerate it.

2. Salinity and Dissolved Salts

Salinity is the total amount of dissolved salts in a given volume of water. It is expressed as parts of salts per thousand parts of water (‰). Fresh water has much lower salinity than sea water. Estuarine waters have intermediate and fluctuating salinity. The average salinity of sea water in the ocean is about 35‰.

Among the high salinity sea waters is the eastern Mediterranean which has a salinity of 37‰. Low salinity areas, with salinities of about 32‰, include the Sunda Shelf of Indonesia. In tropical estuaries and their adjoining seas, salinity may be much lower than 32‰ at certain seasons. Problems are created by such

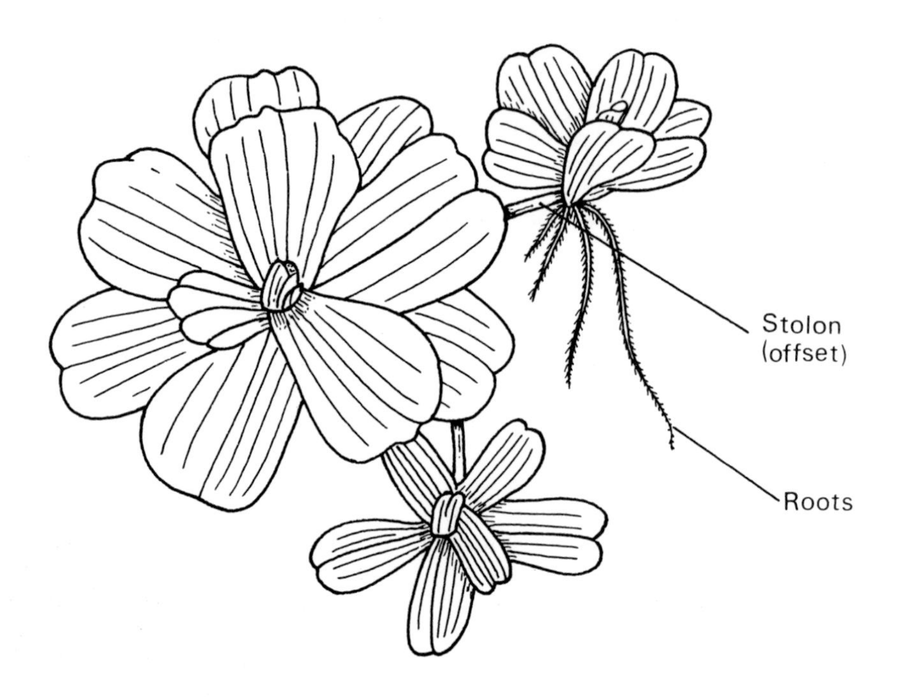

30 *Pistia stratiotes*
SOURCE: J Y Ewusie (1973)

changes for the plants and animals in the estuaries.

Some sea weeds are sensitive to salinity changes and tolerate little deviation from that of normal sea water, while others are adaptable to extremes and rapid variations. Some species of *Enteromorpha* live successfully on the hulls of ferry boats which travel daily from open sea harbours to the fresh waters of river ports. While some species tolerate low salinity others have to adapt themselves to high salinity. For example, an area that is under water at high tide but is cut off from the sea at low tide may have a high salinity as a result of evaporation as well as a high temperature. Organisms which can tolerate these conditions are the only ones normally found in each situation. Among these are *Paspalum*, *Rimerea*, some molluscs, crabs and cat fish (see also Figures 42, 58 and 59).

Salts concentrations are naturally quite low in fresh water and rather high in marine water. Table 16 shows the results of an analysis of samples of fresh water from the Sokoto river in northern Nigeria and sea water at Lagos (Olaniyan, 1968), and it illustrates the differences in salt composition between river water and sea water in a tropical environment.

Table 16. A comparison of the mineral constituents of fresh water and sea water in Nigeria, in parts per million

MINERALS	Na	K	Mg	Ca	P_2O_5	Cl	SO_4	HCO_3	NO_3
									LAGOS
SEA WATER	10 530	380	1280	400	trace	18 980	2650	140	trace
									SOKOTO
RIVER WATER	8.7	2.8	7.0	32.0	20.0	6.0	trace	77.8	0.44

Sea water contains at least traces of nearly all known chemical elements, either as salts or gases. The chemicals can be present in various states, such as ions, true solution, or particles in suspension. Certain salts, such as sodium chloride, are present in large quantities. At least 44 chemicals are present in various concentrations in the sea. Most of these are in unvarying proportional concentrations, so that, unlike terrestrial plants, marine plants are immersed in a relatively constant nutritive medium. A few of the essential elements, however, occur in variable amounts and these concentrations may be limiting to the growth of sea plants. Among these are phosphates (which frequently constitute a limiting factor in the productivity of the sea) and nitrogen, which exists in the form of nitrites, nitrates and ammonia. The relationship of the mineral constituents of sea water is upset in regions where land drainage brings down to the sea large quantities of some compounds such as calcium and magnesium salts.

Water plants are, however, selective in the absorption of minerals. Recent investigations in the Volta Lake in Ghana show that *Pistia* and *Ceratophyllum* have larger quantities of sodium potassium and calcium ions per gramme weight of ash than the lake water itself.

3. Density

Natural waters vary greatly in their viscosity. This in effect means different densities of the water. Fresh water naturally has lower viscosity than sea water because viscosity is related to salinity and temperature. Water with a high temperature and a low salinity has the lowest viscosity, and water with a low temperature and a high salinity has the highest viscosity. This means that, generally speaking, tropical waters tend to have a low viscosity because of their high temperatures and average salinity.

The viscosity of the water affects the life of floating organisms as well as the movement of gametes and spores in the water. It is because of the generally low viscosity of tropical waters that planktonic organisms have greater difficulty in floating in them.

4. Colour and Transparency

The colour of natural water determines its transparency and hence the extent to which light can penetrate. If the water is disturbed and has soil particles suspended in it, transparency is reduced and submerged plants are unable to receive enough light for photosynthesis. This results in the reduction of food for animals. Heavy rainfall, especially coming after a long spell of drought, often results in the washing of soil, debris and nutrients into rivers, ponds and the sea. The transparency of the water is reduced temporarily, but the fresh nutrients that are brought in result in increased productivity as soon as the soil particles settle down. Plankton blooms are often noticed at such periods.

5. Temperature and the Thermocline

Temperature is an important factor in aquatic ecosystems. Temperatures are comparatively more stable in large water bodies than they are on land or in the air. It is only in small water bodies that temperature fluctuations are similar to those on the land. Tropical waters are naturally warmer than temperate waters and they rarely freeze at any time of the year. Surface waters of the tropics are usually at about 25–28°C. In shallow waters temperatures are about 28–32°C. Brackish waters that are periodically cut off from the sea can reach much higher temperatures than 32°C. Higher temperature appears to lead to a decrease in the amount of dissolved oxygen available for respiration, especially at night. Lower temperatures generally favour the richer development of sea plants, and this may explain the more abundant sea vegetation on cool as compared to tropical shores.

As described in general terms in Chapter 4, the change in temperature with depth in large bodies of water is not proportional throughout. The temperature drops slowly and steadily downwards from the surface of the water. At about 120 m the drop in temperature is very rapid until a depth of about 175 m is reached, and from that point the temperature begins to drop slowly and steadily again. Thus between 120 and 175 m, that is, over an interval of 55 m, where the very rapid drop in temperature occurs, there is a layer which is discontinuous in the normal trend of temperature change with depth. This discontinuous layer is called a **thermocline**. The thermocline can be smaller or larger depending on whether the rapid drop in temperature occurs in a narrower or a broader band than 55 m. A typical thermocline is illustrated diagrammatically in Figure 31 by an example from the western Indian Ocean (p. 102).

Thermoclines are found in all tropical oceans, in nearly all large lakes and even in shallow water

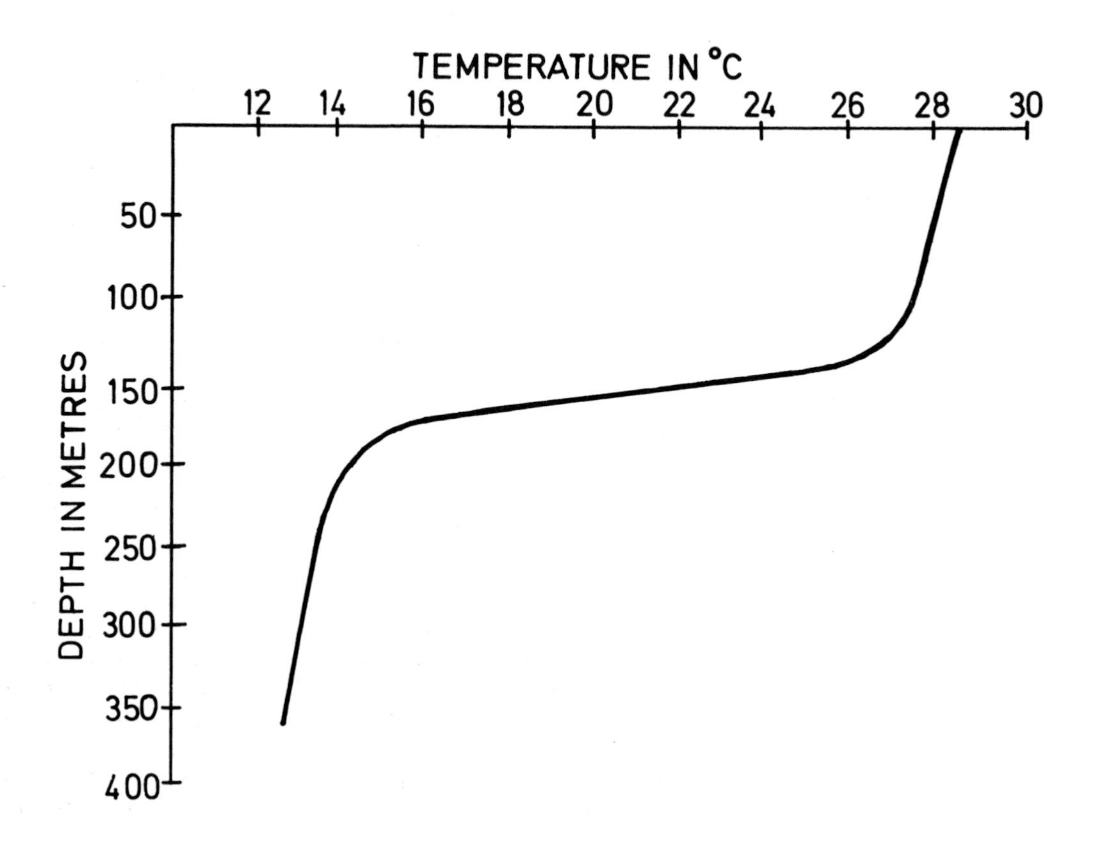

31 A diagrammatic representation of a thermocline from the western Indian Ocean
SOURCE: J H Wickstead (1965)

of about 50–100 m deep. In the latter case the rapid drop in temperature is within a much smaller range, such as 1–3°C with the depth involved being only 2–5 m. This often happens in fairly shallow water in a season with much sun and little wind. In all cases a thermocline is formed in calm conditions. When the season changes and winds become strong, turbulence breaks down the thermocline. The thermocline may also break down during cold nights when, as a result of convection currents, the temperature of the surface layer of water becomes lower than that of the thermocline. With this the surface water attains a higher density than the water of the thermocline and sinks. The role of currents and the ecological significance of the thermocline in relation to nutrition are discussed in Section 7 below.

6. Light

Light is important for chlorophyll-containing plants but as light does not penetrate to any great depth green plants cannot live in very deep water. When sunlight falls on water it penetrates to varying depths depending on the intensity of radiation, the amount of surface reflection and the transparency of the water at the time. Even in transparent water only about 36 per cent of the sunlight which enters the water penetrates to about one metre. The light then fades away quickly with increasing depth, and below 50 m only a fraction of one per cent remains. Below about 100 m there is often total darkness. Before this depth is reached there is a point at which the amount of carbohydrate used up by a particular species in respiration just balances the amount made by photosynthesis. This depth is called the **compensation depth**, or point, of the species (see also Chapter 4). The actual depth of the compensation point in any given body of water depends on a number of factors, such as the species, the time of the year, the cloudiness of the sky, and the transparency of the water in relation to whether it is a muddy pond, a coastal water after rainfall, a clear lake or oceanic water. Some plants like fucoid algae have shallow compensation depths, while *Rhodophyceae* survive at far greater depths (Figure 32). Compensation depth in tropical deep waters, including the oceans, often lies between 25 and 50 m. In the sea off West Africa, for example, it lies at about 35 m, for organisms which can survive at greater depths.

Light is selectively absorbed by water, so that the longer red, orange and yellow wavelengths are removed first as well as the violet, leaving only the shorter blue and green wavelengths to penetrate deeply. Only those autotrophic plants with pigments capable of utilizing the available short wavelengths of blue and green light can survive at the lower depths.

Considering the fact that the average depth of the sea is about 3800 m, it becomes evident that life in large bodies of water like the sea depends very much on what goes on in the thin top layer consisting of only 3 per cent of the entire depth of the water. While nutrients in the top 100 m of the water are being actively used by plants, the nutrients below this depth are virtually locked up. The food manufactured in the upper layers sinks to feed the animals and decomposers that are able to live in the dark below. The water below 100 m is therefore richer in nutrients than the water above it. Because of the discontinuity layer, the thermocline, at a depth of about 150 m in most tropical waters, the water above 150 m, which is steadily being depleted of nutrients, does not normally mix with the water below this depth, which is rich in nutrients. However in seasons of strong winds when storms and gales may occur, the turbulence that results breaks down the discontinuity layer (the thermocline) and this permits the mixing up of the deeper nutrient-rich waters and the upper nutrient-deficient waters in which photosynthesis is occurring. The supply of fresh nutrients results in increased photosynthesis and rapid growth of plants. This in turn affects the growth and increase in the animal communities and explains why it is often observed that the period of the year in which both primary and secondary productivity are highest on the continental shelf is during the time when turbulent waves occur. Temporary breakdown in the thermocline may occur also in the open sea or in the centre of heavy storms. The mixing of lower and upper waters that occurs for short periods results in a short burst of growth of phytoplankton.

7. Water Currents (and Upwelling)

Water currents in the oceans and large lakes constitute one of the important factors in the ecology of aquatic ecosystems. Ocean currents are of special significance here. The most conspicuous feature of the surface currents in the two major ocean basins, the Pacific and the Atlantic, is that both show clockwise circulation in the northern hemisphere and anti-clockwise circulation in the southern hemisphere. This has the effect of bringing widespread warm water, and consequently a warm tropical climate, to much of their western shores. At the same time, the influence of cold water along the eastern shores provides more extensive temperate marine conditions.

An influence that further characterizes the two is brought about by the prevailing winds that tend to parallel the cool currents along the continental west coasts. These winds cause a coastal upwelling of sub-surface waters, from about 200–300 m, that helps to maintain the cooling effects far into the tropics and also breaks down the thermocline and brings richer (and colder) waters from lower depths to the surface, resulting in high productivities.

One of the best-known areas for upwelling is off the Peruvian coast of South America (Humboldt current) where a temperate ocean climate is produced, which is not the case in New Guinea which lies at the same latitude as Peru. The cool waters of Peru abound in plant life that supports multitudes of invertebrates, fish, birds, sea lions and whales. Other areas where upwelling occurs include the Somali coast of East Africa, the west coast of North Africa, the Guinea coast of West Africa, and the west coast of South Africa (the Benguela current).

Upwelling is a very important phenomenon in the cycling of nutrients in the waters of some tropical areas. It is therefore important to appreciate just how it arises. Strong and persistent winds from the land move surface waters away from the shores out to sea, and these are replaced by deeper water which moves towards the shores. A continuous movement of water is thus set in motion. When the deeper, upwelling, water also reaches the surface it is in turn blown out to sea by the wind. Thus there is formed a long stretch of surface sea water which is parallel and

FUCOID ALGAE

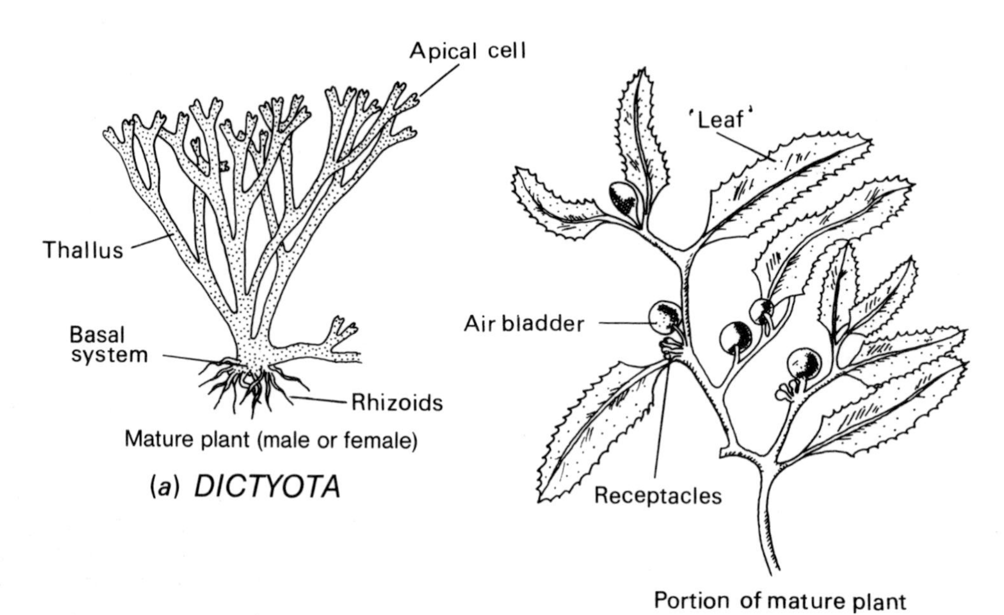

Mature plant (male or female)

(*a*) *DICTYOTA*

Portion of mature plant

(*b*) *SARGASSUM*

RED ALGAE

Habit drawing of thallus

(*c*) *GRACILLARIA*

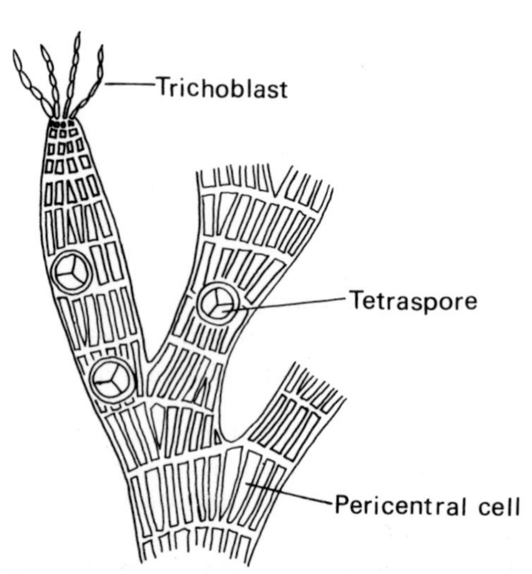

External features of part of thallus

(*d*) *POLYSIPHONIA*

close to the coast and is continually being replaced from below by water rich in nutrients.

8. Surf and Substrate

Despite favourable temperatures, larger plants in the sea cannot survive easily the agitation of surf and the movement of current and surge without secure attachment to the substrate. Thus, except in calm situations, marine vegetation is confined to permanent rocky outcrops and is absent from shores of shifting sand and gravel. At depths below the effect of surf and surge, however, a rich vegetation may be found, even along the most exposed shores, on any substrate.

In tropical waters it is common to find fairly extensive floating patches of *Sargassum* (gulfweed). Both plankton and larger animals are associated with it, some creeping on them while others swim underneath them. Some of these animals appear to be found only in these microhabitats created by the floating *Sargassum*. The largest of these mats are found in the Sargasso Sea and in the Gulf of Thailand, where large masses of *Sargassum* are found drifting more or less permanently. Among the animals associated with the *Sargassum* here are small polyps, barnacles, moss animals (bryozoans), several kinds of crustaceans with a variety of colours, small molluscs and several kinds of fish known collectively as *Sargassum* fish.

Freshwater Habitats

Freshwater habitats are mostly inland waters. The composition and concentration of dissolved salts are relatively low or negligible. They are divided into two types on the basis of their mobility; these are the **lotic** fresh waters and the **lentic** fresh waters. The lotic fresh waters consist of running waters flowing continuously in a definite direction, including all rivers and streams of all sizes. Running or riparian waters originate inland and eventually empty into the sea. The lentic fresh waters consist of standing waters like lakes, ponds and swamps.

32 (a) and (b) Fucoid algae. (c) and (d) Red algae (*Rhodophyceae*)
SOURCE: J Y Ewusie (1973)

Running (Lotic) Waters

Running water has certain features which clearly distinguish it from standing water even though both are aquatic habitats. These differences naturally affect the nature and life of the plants and animals inhabiting them. One fundamental difference between a lake and a river is that a lake exists because its basin is there and the water fills it, but it may at any time be filled by deposits to become dry land. On the other hand, a river is there because the water is there, so that it is the water which carves out and maintains the channel which persists as long as there is water to fill it.

This fundamental difference means that rivers are ancient habitats which have spread their floras and faunas among themselves over a long period of time. Their persistence has facilitated the evolution of many groups of organisms. A number of plants are thus confined to running water. They are adapted in various ways to the water current. These include species of red algae and water ferns like *Salvinia*. There are also angiosperms like *Ceratophyllum* (horn wort; Figure 33), and the *Podostemataceae* which are among the few angiosperms typical of running water which regularly reproduce by seed. The animals include freshwater snails, hydroids, the leech and blackfly larvae (*Simulium*).

The important features of the running water or the riparian habitat may be brought out in a comparison with standing water conditions as follows:

1. In running water the flow is often turbulent, but in standing water the flow, if any, is rather gentle.
2. In running water stratification is rare and unimportant in view of the turbulent flow, but this is important in standing water even though convection currents sometimes mix up the layers.
3. In running water convection currents are not important and so the relationship between density and temperature is irrelevant. In standing water, on the other hand, the special relationship between density and temperature is an important factor.
4. Deoxygenation is rare in running water. If it should occur it is often confined to stagnant pools in temporary streams or close to where a polluting effluent is entering the stream. In standing water, however, deoxygenation is

common even in the absence of pollution.
5. The accumulation of gases like carbon dioxide and hydrogen sulphide is minimal in running water as compared with standing water where, especially in the hypolimnion, they exert a negative effect on the plants.
6. Rooted plants are not as many in running water as the plants here are much subjected to wash-out. Thus there is often not much vegetation in running water, except in particularly stable areas such as spring streams or lake overflows. The flow, however, makes firm rooting essential, e.g. *Podostemataceae*. On the other hand, rooted plants are important in standing water, especially in the shallow parts, and great amounts of algae grow on solid objects.

33 *Ceratophyllum demersum*
SOURCE: J Y Ewusie (1974)

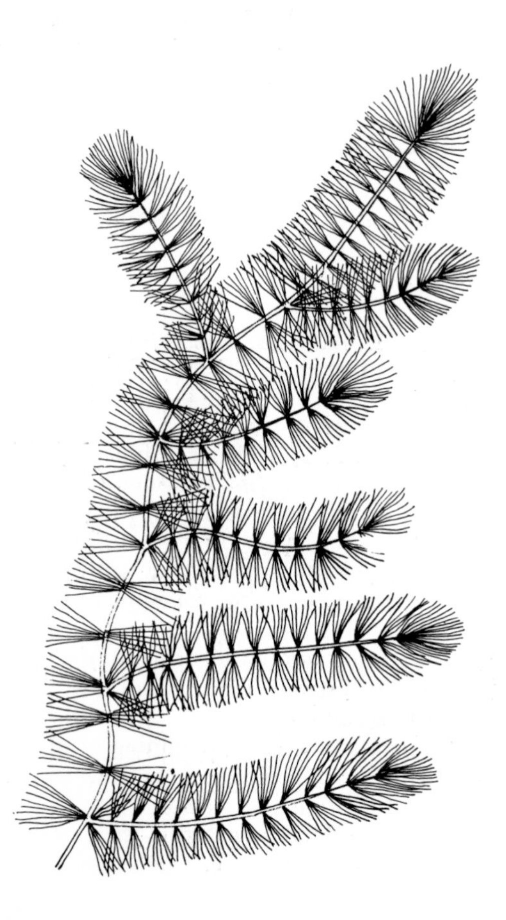

7. In running water plankton are poorly developed. This is because most of them are eliminated by floods and turbidity. Among the common plankton found are diatoms and rotifers. The benthic flora and fauna are, however, richer in species which are adapted to the movement of the water for their respiration and for obtaining food.

Some of the animals in running water maintain their positions by clinging to some stones or floating substratum while others avoid the current. Still others maintain their positions by swimming actively. In all these devices perfect orientation of the animals is called for. A few examples may be cited. Some snails have a smooth shell which reduces the friction of the running water. Flattened bodies are typical of some of the insect nymphs in water. Special suckers are developed by some insects for attachment to firm objects in the water. Submerged aquatic plants like *Potamogeton* are firmly rooted and have long trailing stems; *Podostemataceae* are always attached to rocks in flowing streams.

Many tropical rivers are characterized by heavy seasonal flooding which brings about the inundation of definite areas along their routes. When this happens there is an increased growth of phytoplankton and a clearly defined reproductive period of some animals.

Standing (Lentic) Waters

The environmental factors affecting life in standing waters have already been discussed above. Standing waters include the large natural inland lakes such as Lakes Victoria and Tanganyika, as well as the man-made lakes like the Kariba on the Zambesi river, the Volta on the Volta river and the Kainji on the Niger, in Africa, and the Brokopondo Lake in Surinam.

Like sea water, large bodies of standing fresh water are not subject to great changes in temperature, and most fresh water animals can survive a wider range. The nature of the underlying bed of a body of standing water may be rocky, gravelly, sandy or muddy. A soft substratum is usually favoured by the higher plants that take root on the bottom of a lake. The amount and type of nutrients in the water are also important. Some may have calcareous material which renders the water hard and the pH high. Phosphates and nitrates are important biologically and may vary in amount in different seasons. The availability of some of these minerals determines the composition and luxuriance of the vegetation in the water.

Standing fresh waters are of three types in terms of nutrient status. These are: **oligotrophic**, which refers to waters that are poor in nutrients and humus; **dystrophic**, being waters that are poor in nutrients but rich in humus and which are also acidic in reaction; and **eutrophic**, being those that are rich in nutrients including combined nitrogen, phosphorus and calcium, but poor in humus.

Biotic Factors in Fresh Water

Both plants and animals have species that are adapted to the conditions in fresh water.

Plant Adaptations to Fresh Water Conditions

In the case of water plants we find adaptations to the conditions mentioned below.

1. **Reduction in Gaseous Exchange**. Since the plants are immersed in water there arises a problem of storing gases. This is overcome by the development of large air spaces separated by diaphragms (aerenchyma) (Figure 34). These spaces are intended to store the surplus oxygen produced during the day by photosynthesis; the oxygen is used for respiration at night and the resulting carbon dioxide replaces it in these spaces. These spaces are evident in *Nymphaea* and *Pistia* as illustrated in standard botany textbooks.
2. **Absence of Water Absorption**. Since water plants of submerged types are surrounded by water, there is generally no need for the absorption of water in the same way as happens in land plants. There are often no root hairs and water enters the plant by simple diffusion. This is exemplified in *Ceratophyllum* and floating types like *Salvinia* and *Lemna*. Some, in fact, even have no roots, as in *Utricularia* (bladderwort) (Figure 35). For the same reason there is often a drastic reduction in the amount of xylem tissue as compared with land plants.
3. **Virtual Absence of Transpiration**. Since there is no active absorption of water, there is virtually no need for transpiration in submerged plants. These plants have only vestigial stomata or no stomata at all. There is also no cuticle. Stomata may be found on the leaves that are exposed to the atmosphere.

34 Transverse section of *Nymphaea* (diagrammatic)
SOURCE: J Y Ewusie (1973)

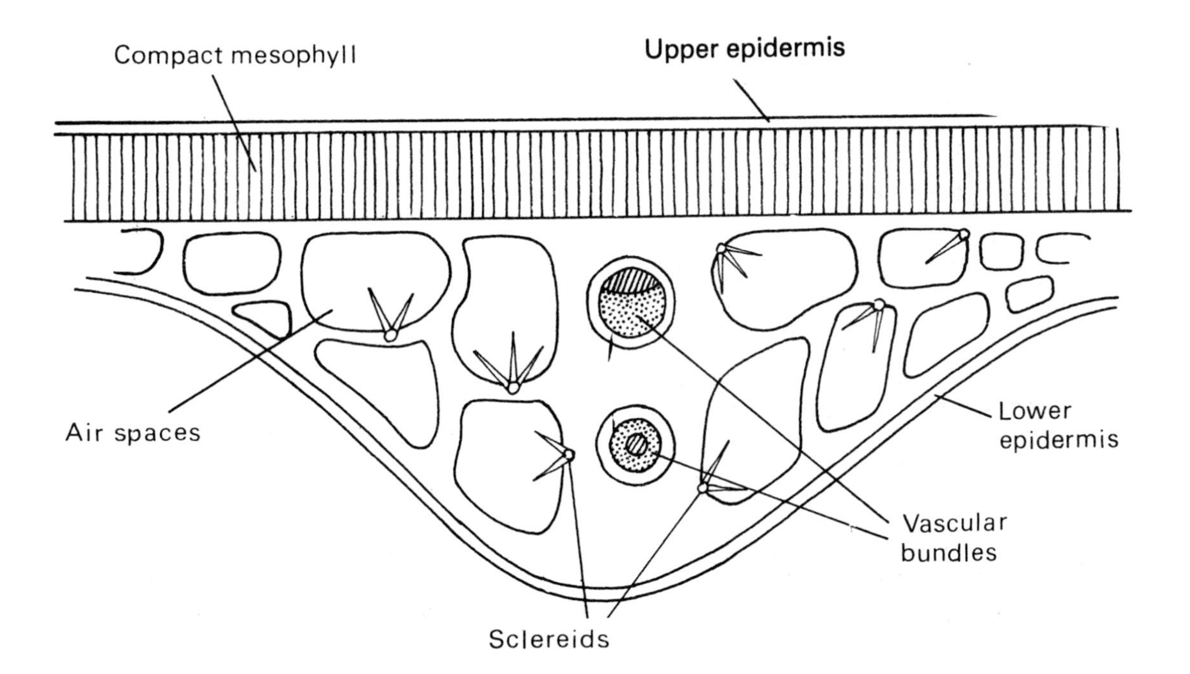

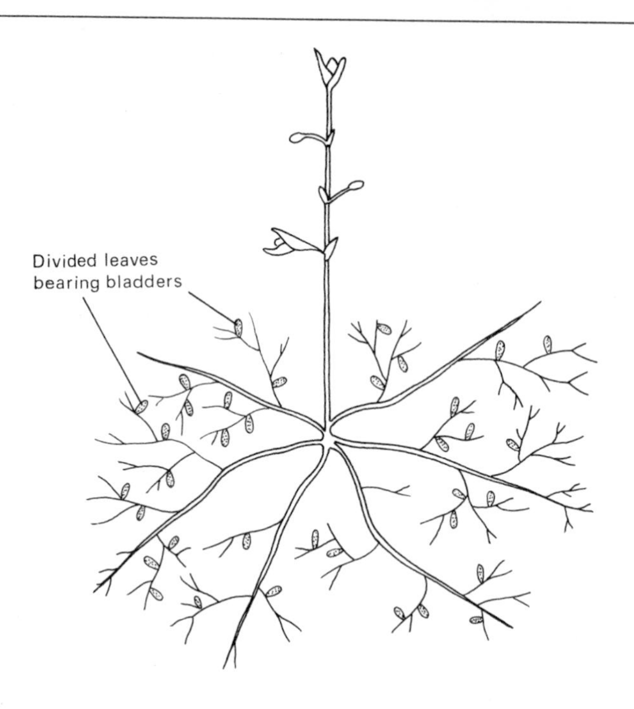

35 *Utricularia*
SOURCE: J Y Ewusie (1973)

36 Floating leaves of water plants. Lotus (*Nelumbium nelumbo*) projecting out of the water and *Pistia stratiotes* floating on the water, Laguna Bay, Philippine Islands
SOURCE: W H Brown (1935)

4. **Buoyancy**. Unlike land plants the weight of the water plant is carried by the water; there is often little or no lignified tissue like sclerenchyma in water plants. The air spaces used in gas storage also help in buoyancy.

5. **Availability of Light**. The light that reaches submerged plants is often rather weak and this may limit photosynthesis. Submerged water plants are adapted to this situation by having very thin leaves with chloroplasts in the epidermal cells, unlike most land plants. This adaptation enables the water plant to make maximum use of low light intensities.

Fresh Water Plant Forms

In fresh water, the plants are of four different forms.

1. **The Floating Plant Type**. This includes the microscopic floating algae (phytoplankton) *Lemna, Wolfia, Salvinia*, the water lettuce (*Pistia stratiotes*) and the water hyacinth (*Eichhornia*). In the floating plants the upper side of the leaves is mesophytic in structure, with cuticle, while the lower part has the typical aquatic structure described above. In *Eichhornia* the spongy swollen bases of the leaf stalks help in floating.
2. **The Floating Leaf Type**. In these the plants are rooted, but the leaves are carried by long petioles to the surface of the water. Examples are the water lilies like *Nymphaea* and *Nymphoides* and lotus (*Nelumbium nelumbo*) (Figure 36). The petioles have large air spaces, while the leaves are structured like the floating types.

3. **The Emergent Types**. These are also rooted, but part of the stem projects above the water. Among these are species of *Typha* and *Phragmites*.
4. **The Submerged Types**. These are the most typical of water plants and show most of the adaptive features described above. Among these are *Ceratophyllum demersum* (see Figure 33), species of *Myriophyllum* as well as species of *Chara*, water crinums (*Crinum natans*) and the large types of green algae. They often have elongated leaves or sometimes dissected leaves (as in *Ceratophyllum*), which facilitate the flow of water through them without the risk of their being torn off (Figure 37). This group also includes the diatoms and some of the species of *Utricularia*.

Animal Adaptations to Fresh Water Conditions

Some water animals, too, spend their lives completely in fresh water. These include protozoans, rotifers (Figure 38), annelids, platyhelminths, snails, crabs and other crustaceans, fish and turtles.

37 Aquatic floating weeds on the River Densu, Ghana. Light-coloured patches are *Pistia stratiotes. Ipomoea aquatica*, with pointed triangular leaves is seen in the right front. Small weed covering the largest area is *Azolla africana* with some *Lemna*. In the foreground are half-submerged plants of *Ceratophyllum demersum* and *Utricularia*
SOURCE: G W Lawson (1966)

It is worth noting the devices against low oxygen, such as an oxygen-carrying pigment in *Chiro nomus* larvae or a long siphon in *Eristalis* larvae, in Figure 38. The manatee is also found in fresh water.

Animals face a number of problems in fresh water and develop differing adaptations to it. These are some of them:

1. **Buoyancy**. By virtue of the high temperatures and sometimes low salinity, tropical waters are less dense and so plankton have much greater difficulty in staying afloat than do temperate plankton and larger animals. As a result it is noticed that tropical plankton are much smaller in size than their counterparts in the cold waters of the higher latitudes. Other animals like fresh water molluscs have smaller shells than the temperate types. The small size of the body increases the surface area of the body relative to the volume, and thus improves its flotation. The phenomenon is the same in both fresh and marine waters generally in the tropics. Other ways of achieving buoyancy are described under marine habitats below.
2. **Osmo-regulation**. The problem of osmo-regulation is greater than in the marine habitat. The animals may have impermeable membranes, or may produce hypotonic solutions. The protozoans tend to have contractile vacuoles which help in osmo-regulation. It is worth noting that marine protozoans do not have these vacuoles.
3. **Reproduction**. In view of the very low salinity and viscosity of fresh waters, there is a greater danger of loss of reproductive structures like gametes and eggs. Freshwater organisms tend to produce more eggs of smaller size than their marine counterparts.
4. **Disperal**. The stable conditions in fresh water lakes permit confinement and isolation which often result in the development of endemic species. An example may be cited from Lake Victoria where a number of endemic fish species are present.

Man-made Lakes and Water Weeds

In recent years a number of large rivers in the tropics have been dammed to create lakes for various purposes including the generation of electricity, water transport, fish production, irrigation and agriculture. The creation of a lake in a

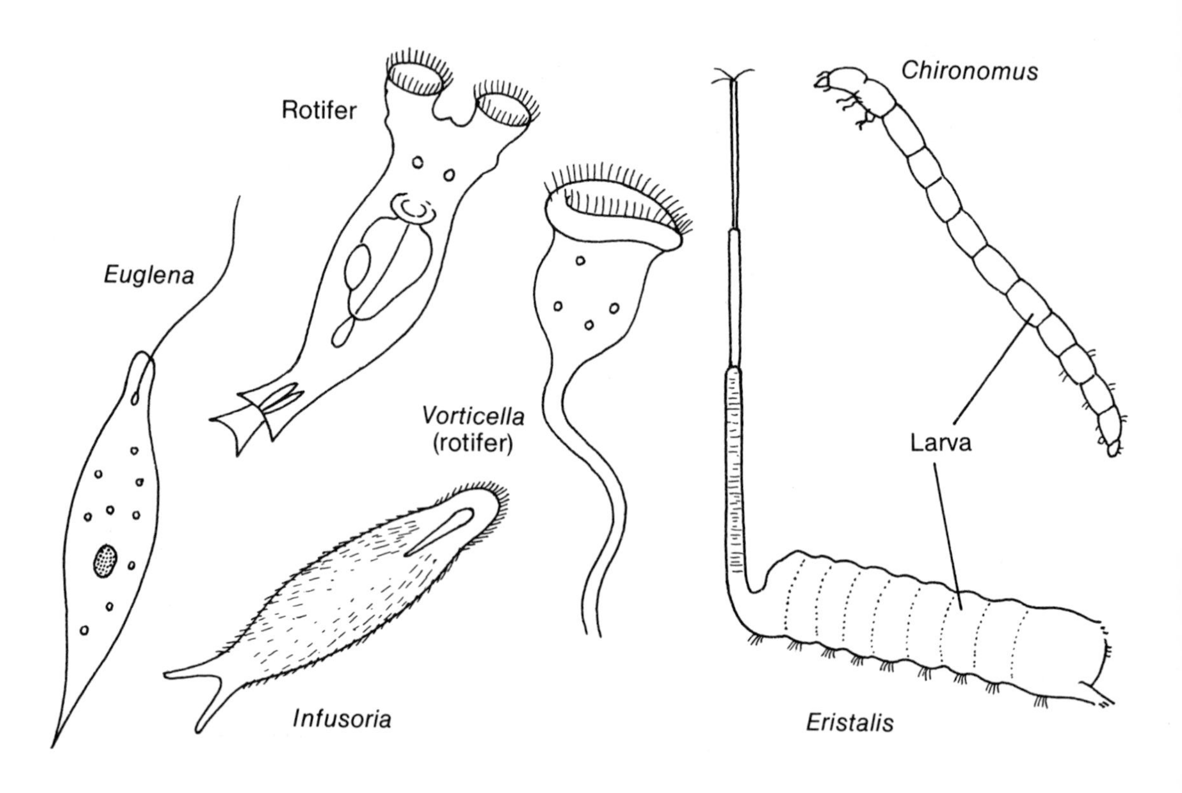

38 Animals at the bottom of the pond
SOURCE: J Y Ewusie (1974)

river valley sets up changes in biological succession and affects competition between species as well as altering their trophic and energy relationships. A comparison of the changes in the Volta with those in Lake Kariba found that in both cases the proportion of cichlids and plankton-feeding *Tilapia* rose sharply, while more riverine species decreased sharply in numbers.

One important problem which arises with the formation of a large lake is the rapid growth of certain plant species to constitute masses of weeds. A notable example is the small floating fern *Salvinia auriculata* which has grown on Lake Kariba to form a thick mat over 520 km² of the surface of the lake. As light cannot penetrate, carbon dioxide accumulates below the mat and as a result many fish are unable to survive. Sudd forms where grasses and other weeds grow on the *Salivinia* mats. These weeds can interfere with the electricity generating machinery in the dam, or cause serious damage by breaking booms that are placed in the lake to control the movement of the mats. Other species that can cause a similar trouble include the water hyacinth *Eichhornia crassipes*, already a serious problem on the Zaïre and Nile rivers and *Pistia stratiotes*, *Ceratophyllum demersum* and *Azolla* spp. in tropical America.

Removal of these mats by mechanical means is impossible in terms of cost and labour. The use of herbicides poses other problems of polluting the water and contaminating the fish which is valuable food. It will be some time before we know enough about these weeds to enable us to take precautions to prevent their explosive growth.

The Marine Habitat

The environmental factors operating in the marine habitat have already been discussed above. In order to deal systematically with the adaptations of plants and animals to the marine habitat, it is

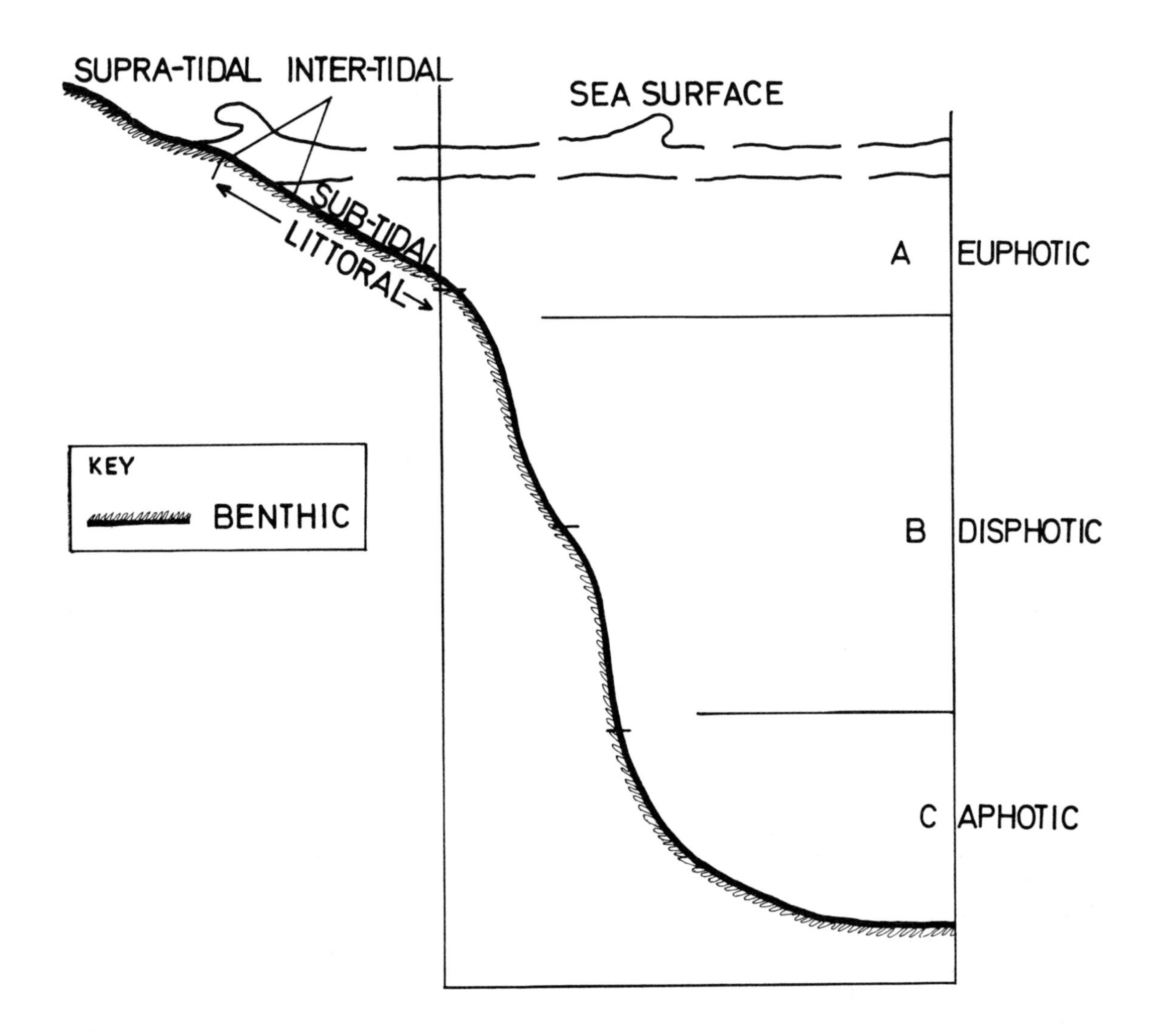

39 Zones of the marine habitat
SOURCE: D A Maxwell (1971)

necessary, in view of the size and depth of the sea, to delimit the major parts of the marine habitat. These are as shown in Figure 39. The four main zones are the euphotic zone, the disphotic zone (or the nektonic layer) the aphotic zone and the benthic zone. These will now be described.

The Euphotic Zone

The most important organisms in this zone are the plankton. The plankton consist of all the small-sized plants and animals which live in the water but drift about at the mercy of the wind and currents. However, many plankton animals are strong swimmers and are able particularly to swim in a vertical direction. Especially in the tropics, the plankton consist of a wide range of plants and animals. Some of the animals are carnivores. Plankton in fact include larval stages of non-plankton worms, starfishes and crabs.

The plankton are not distributed at random in the water but exist in distinct communities associated with particular environmental features, such as warm or cold currents; sometimes warm and cold waters may adjoin each other, but the temperature discontinuity may enable each community to support different species.

One way of overcoming the lower viscosity of tropical seas is the possession of body fluids less dense than the surrounding sea water. An example is *Noctiluca*, a dino-flagellate, which contains a solution of ammonium chloride within its cell walls. This solution is found to have a lower specific gravity than that of sea water, and so helps to provide the buoyancy (Figure 40). Other animals achieve buoyancy by means of a gas filled sac. A good example is the purple, gas filled float of the Portuguese man-of-war. Many other siphonophores have similar gas sacs. Some soft-bodied animals such as salps, heteropods and ctenophores possibly reduce the density of the sap of their vacuoles by secreting out the heavy ions like potassium and calcium and replacing them with lighter ions such as sodium.

A further adaptation is the relatively small size of tropical plankton as compared with temperate ones, since the surface area of the body relative to the volume is thereby increased. Finally, the commonest device to achieve buoyancy is the extension of the body or skeleton into numerous spines and projections as found in the *Sergestid* protozoans (Figure 40). Such extensions increase the surface area with only a slight increase in weight.

The Disphotic (or Nektonic) Zone

This zone has only dim light and the organisms are described as nektonic, or actively-moving. Among these are most marine invertebrates, squids, fishes including large ones like sharks, and sea mammals like whales. These animals often swarm in very large numbers or shoals. Since very little light enters this zone there is very little growth of plants.

The Aphotic Zone

There is not enough light here for photosynthesis and living organisms depend in the main on the 'rain of detritus' from the surface. This zone is therefore much less densely inhabited than either the euphotic or the disphotic.

The Benthic Zone

The benthic zone contains all forms of life which remain fixed to the seafloor, such as red algae,

40 *Noctiluca* and *Sergestid* protozoans
SOURCE: J H Wickstead (1965)

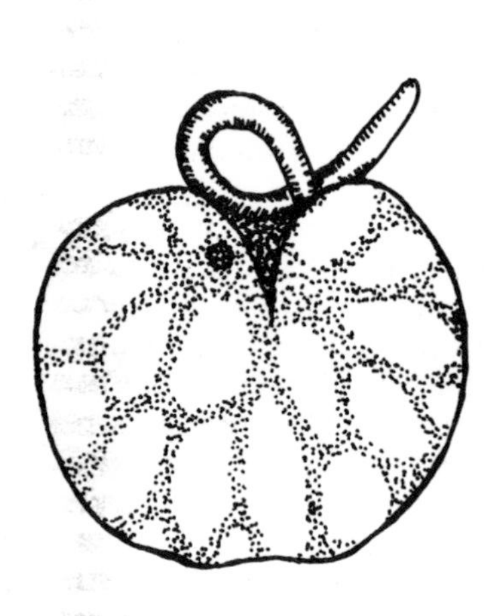

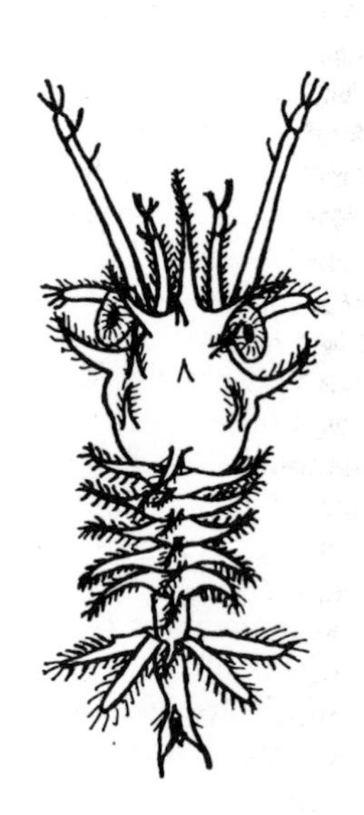

sponges, oyster, corals and some polychaetes. It also contains animals which burrow into or creep along the bed of the sea. These include crabs, lobsters, echinoderms, bivalve molluscs, crustaceans and some worms.

The littoral forms part of the benthic zone, and consists of two unlike environments, namely the sub-tidal and inter-tidal provinces. The sub-tidal province is strictly aquatic and more uniform than the inter-tidal, although it has marked variables of substrate, temperature, light, salinity and nutrients. The inter-tidal province is subject at intervals to desiccation, direct solar heat, rain, freshwater run-off and wind. The nature of the substratum and the action of predators are also important factors, which affect the life of the organisms found here.

In the tropics, blue-green, green and brown algae are frequently found in the littoral. The blue-green algae here form conspicuous filamentous or spongy tufts. Among the cosmopolitan green algae are species of *Enteromorpha* and *Ulva*. The best evolutionary development of the green algae in the tropics is found in the Indian Ocean, the Caribbean Sea and the western Pacific Ocean. Tropical regions have only a few brown algae, the commonest of them being *Sargassum*. Rocky sub-tidal areas may be covered by tufts of red algae. In some tropical bays and sub-tidal areas are some sea angiosperms (referred to as sea grasses). They are common in the Caribbean Sea, East Africa and the tropical western Pacific. The west coast of Africa and the coasts of tropical South America are markedly deficient in sea grasses, because of the absence of coral reefs and the large tidal amplitudes.

The animals show a range of adaptations to life in the littoral. Animals that are not physiologically adapted to desiccation, avoid it by burrowing or hiding in areas with cool water. Other burrowers do so to escape predators, such as sea gulls and other birds. The rocky substratum is often inhabited by a few animals such as barnacles which possess strong adhesives for attachment to the rocks.

Productivity in Tropical Seas

Productivity in nearly all areas of open ocean in the tropics is relatively low compared with that of the temperate seas. It is only in limited areas with regular or occasional upwelling water movements or around coral reefs that higher productivity is found. This would appear surprising in view of the fact that, with the greater transparency of tropical seas, light wavelengths necessary for photosynthesis can penetrate much deeper, and indeed to as much as twice the depth that they do in temperate open seas. But this does not help since, with a few exceptions, an almost permanent thermocline divides this deep photosynthetic zone. Below the thermocline the nutrients are locked up while above in the surface waters they become exhausted very quickly. It is found that at 20° latitude north or south of the equator the concentration of total nitrogen in the surface waters of the Atlantic Ocean is only about $\frac{1}{100}$ of the winter concentrations in temperate seas, while phosphate concentrations are even less than $\frac{1}{100}$ of temperate seas. No significant annual changes seem to occur in the phytoplankton and so primary production remains low throughout the year. Very little seasonal fluctuation occurs in the zooplankton in tropical seas. The high nutrient content of temperate seas makes them more productive.

It should, however, be pointed out that despite the low productivity of life in tropical seas the variety of species in the zooplankton is much greater than in temperate oceans. For example, while northern temperate seas have about 5 species of ctenophores, tropical seas have as many as 65 species; similarly there are 5 species of chaetognaths in temperate seas as opposed to 19 species in tropical seas; and 21 species of copepods as opposed to over 500 in tropical oceans. However, none of the species is abundant, unlike the position in temperate seas where one often comes across a pure population of, say, *Calanus*. Shoals of pelagic plankton-feeding fish are rare in the tropical seas.

The level of productivity expressed as grammes of carbon per square metre of sea surface per year in different climatic regions of the world, based on available research data, has been given by Ryther (1963) as follows:

1. In tropical open oceans, productivity is 18–50 g C/m²/yr, as in the Sargasso Sea. Where some mixing occurs, as off the Bermuda coasts, the value may reach 70 g C/m²/yr.
2. In temperate and sub-polar waters, productivity is about 70–120 g C/m²/yr.
3. In the Arctic region, productivity is very low, possibly less than 1 g C/m²/yr.
4. In the Antarctic region it is about 100 g C/m²/yr.

Exceptions to the low productivity in tropical seas are found in regions of upwelling and coral reefs. Before this is further considered, it would be worth mentioning the situation in the Indian Ocean which also seems to be among the exceptions. Here it is the changes in weather conditions brought about by the monsoon rains which result in a seasonal pattern of productivity.

The upwelling which occurs in the seas in certain regions of the western coasts of the continents seems to result from the world's pattern of prevailing winds and their effects on the surface currents of the oceans. Thus, the upwelling that is found off the Atlantic coast of Morocco and the southern California coast is due to the fact that north of the equator a prevailing wind from the north-northwest blows nearly parallel to the west coast of the continents of Africa and America. This causes an offshore movement of surface waters away to the southwest, to be replaced by an upwelling of colder, denser water, often rich in nutrients, which destroys the thermocline in the process. South of the equator corresponding examples of upwelling are found along the coasts of southwestern Africa and Peru. These regions then experience increased primary and secondary productivity leading to important commercial fisheries. Mention should also be made of the upwelling of the Humboldt current which supports vast populations of fish-eating sea birds. The droppings from these birds accumulate along the shores to form large deposits of guano which is a very rich source of fertilizers.

Other areas of high productivity in the tropics are centred around coral reefs. Much of the productivity here appears to depend on efficient and very local recirculation of nutrients. On and around the reef, little of each year's primary production of organic matter is dispersed and lost in the deep water. Also, it appears that symbiosis occurs between the various producer and consumer components on the reef. On and around the reef fish are in local abundance, and most of the decomposers in the ecosystem are also located in the shallow waters which the growth of the corals help to create.

It has been shown that productivity and standing crop are much higher in inshore waters than in offshore or open ocean in the same latitudes in both temperate and tropical regions (Anderson, 1964 and Teixeira, 1963). In inshore waters there is a rich fauna usually consisting of benthic invertebrates and demersal fishes which feed upon them. The benthic animals consist mainly of bivalve molluscs, crustaceans and similar organisms with a high biomass. These create a fast local cycle of nutrient salts and a rapid turnover of detritus, as in the case of coral reefs.

Estuarine or Brackish Habitats

Estuarine or brackish waters are found in depressions of the coast, where sea water meets and invades the mouth of a river. The mixture of saline and fresh water is referred to as brackish. The estuarine water is subject to river currents and tides which are often slower at the edges and faster in the centre. There is also a vertical salinity gradient of the water in the estuary since sea water is denser than freshwater. Large areas of muddy land may be exposed at low tide.

Adaptations of Organisms to Estuarine Conditions

Among the factors which affect the life of organisms in brackish waters are fluctuating salinity, mobility of the substrate, turbidity, and reduction of oxygen in the substrate. These factors, and how the organisms are adapted to them, are discussed below.

1. Fluctuating Salinity

Estuaries are typically tidal, receiving water of lower salinity from the incoming river and high salinity from the sea at different times each day.

A number of the vascular plants here are halophytes, being adapted to growing in high saline conditions. They often have a high osmotic pressure in their cell sap, and hence a large suction force in their root system, which is higher than that of the soil solution. This makes it relatively easier for them to absorb water even when the salinity of the estuarine water rises. High osmotic potential develops either as a result of accumulation of organic substances or as a result of mineral salts absorbed from the saline medium. *Sesuvium portulacastrum* is among the common tropical halophytes.

Some of the animals in this habitat have, like the halophytes, body fluids which are higher in concentration than their surrounding environment. They are described as euryhaline. Among these are the amphipod *Gammarus zaddaichi,*

Caramus moenas and *Nereis diversicolor*. Others have no mechanism of regulating their osmotic pressures, so that the osmotic pressure of the body fluids varies with the concentration of the environmental water. Examples of these include *Arenicola marina* and *Mytilus edulis*.

2. Mobility of the Substrate

The level of the soil in the estuary is never constant because, while the river brings down alluvium to raise the level of the mud, sea currents and tides come in later to wash it away. This situation makes it difficult for seedlings to establish themselves.

To overcome this, the seeds of some of the plants germinate before they leave the parent tree (vivipary). This is shown very markedly by *Rhizophora* (red mangrove) where the seed develops a prominent and conspicuous radicle before it falls into the water (Figure 41). Because of this advanced development of the seedling a relatively short time is required for it to establish itself in the mud. The mature *Rhizophora* plant also resists being dislodged by virtue of its stilt roots.

In view of the mobility of the substrate, sessile animals are rare in the estuaries.

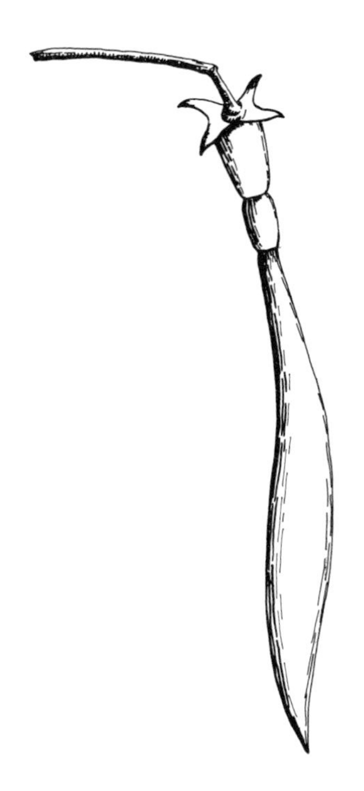

41 Germinating fruit of *Rhizophora*

3. Turbidity

During the rainy season when the river brings in large quantities of material from the adjoining land, the water in the estuary becomes quite turbid. This reduces the amount of light that can penetrate the water and so reduces the photosynthesis of the plankton. The turbid water also blocks the respiratory apparatus of some of the animals and makes breathing difficult. After the suspended matter settles down, more light enters the water. A phytoplankton bloom may then occur because of increased nutrients brought by the river.

4. Reduction of Oxygen in the Substrate

Owing to the fineness of the particles comprising the muddy substrate of estuaries, salinas and lagoons, combined with frequent inundation and high salinity, much of the micro-biological activity is anaerobic. Even if exposed surfaces are coloured brown by the presence of iron compounds in the ferric state, the immediate undersurface is often bluish-grey or black through the presence of ferrous compounds and organic matter reduced to carbon. The obvious emission of hydrogen sulphide in mangrove swamps is further evidence of reduction processes taking place.

Among the halophytes which have distinct adaptations for this situation is *Avicennia* (white mangrove) which has aerial roots known as pneumatophores (Figure 42) which grow vertically above the level of the water or mud. They have lenticels through which air enters and leaves to serve the roots and the part of the plant that is under water. Crabs bore numerous holes

42 Pneumatophores of *Avicennia*
SOURCE: G W Lawson (1966)

in the mud into which they retreat. These holes also bring more oxygen nearer to the roots of the vegetation.

Estuaries have special ecological significance by virtue of the fact that they serve as spawning grounds for some valuable fishes.

Productivity in Estuarine Waters

Estuaries can support dense animal populations most of which are benthic invertebrates. This dense animal population is possible because of the variety of food sources. These are the phytoplankton, water-borne detritus derived from crops of attached plants, and the local crops of attached diatoms and microscopic algae.

The fluctuations in salinity in estuaries occur not only over periods of hours within each tidal cycle but also over periods of months corresponding to seasonal fluctuations in the run-off of fresh water. The mixing of two kinds of waters in estuaries contributes to high productivity. When the waters of different densities meet the velocities are reduced on impact, and so suspended material becomes concentrated in the maximum turbidity that takes place in mid-estuary. Materials remain suspended for rather longer periods in an estuary than would be the case in a simple flow system like a river. In large estuaries it is possible to see from aerial viewing the resulting turbidity as discoloured zones. The turbidity has an important biological importance, for dissolved materials are also involved and nutrient traps are developed where much higher concentrations of inorganic salts are held in circulation than in either the open sea or the incoming river.

Although this local nutrient concentration may not always result in excessive growth of phytoplankton, the algae which live attached to surfaces and on mudbanks exposed at low tides show extremely high productivity. The amounts of suspended organic detritus more than compensate for the surprisingly low level of phytoplankton, because the total organic material per unit volume of estuarine water is about five to ten times the content of rich temperate marine waters.

The accumulation and retention of suspended detritus and dissolved nutrients by estuaries means that pollution matter, such as raw sewage, can be very persistent in estuaries. This is in contrast to rivers and sea coasts where such pollution, particularly when discharged in great dilution, can have more a fertilizing than a polluting effect.

Estuaries tend to support very few species of benthic animals as compared with the sea. These are mostly detritus-feeding molluscs, particularly bivalves such as clams and oysters. Bivalved molluscs are particularly well adapted to life in the estuaries. The ciliary collection and sorting mechanisms of bivalves, for example, are well adapted for collecting the finer detritus suspended in the water or the settled detritus from the mud–water interface. They are also adapted to living in the almost anaerobic bottom deposits and can take in better oxygenated water from above by means of ciliary apparatus. This is often seen in the form of a thin circle of lightly-coloured oxygenated deposit in the reducing mud of a tidal flat in an estuary. Finally bivalves seem to be able to close themselves off completely from the environment during unfavourable conditions. This enables new bivalves, on entering the estuaries, to be able to remain closed during phases of the tidal cycle when the salinity is unsuitable to them. A number of shrimps and the young stages of some fishes feed directly on the rich deposits of detritus in estuaries. Estuarine flatfish show very high rates of growth when the estuaries are unpolluted.

Chapter 8 The Soil in Tropical Ecology

The soil may be regarded as the natural thin layer covering the surface of the earth in which life is supported. It is not just disintegrated rock with particular chemical and physical properties. When the mineral matter has been acted upon by the life processes and the remains of plants, animals and micro-organisms, and has also been influenced by the prevailing environmental factors, soil is formed.

The soil forms the environment for the complex root systems of plants and other underground parts such as rhizomes, corms and bulbs, as well as for a number of soil organisms. It is therefore important to understand how this complex environment can vary and how such variations can affect its inhabitants. The soil provides the medium in which plant roots can grow and have anchorage. It also provides the plant with steady supplies of water and mineral salts. Anchorage is a delicate problem with trees and some of them cannot grow on certain types of soil unless they are specially adapted, as is the red mangrove (*Rhizophora*) which has stilt roots. For roots to grow healthily the amounts of water and oxygen must also be adequate. The amounts of water and oxygen available in a given soil depend on the spaces between the soil particles. This is better appreciated in the study of how the soil is formed.

The Formation of Soil

Soil is formed from rock or other parent material by processes referred to as weathering. The first action in the formation of soil parent material is mechanical weathering or disintegration of rocks and minerals. Rain and temperature change contribute to this cracking and fragmentation action. The roots of colonizing plants help to force open the fissures provided by cleavage planes. *Ficus umbellata* is a common tropical flowering plant that is involved in this action. In time, fine rock particles result from all this action and form the parent material of soil. At this stage there are few soluble compounds in the soil since very little decay of organic matter from plants has taken place, but some plants are able to live directly on the raw mineral material. For example, lichens are able to live on the bare rock surface and thrive well on the very meagre supply of nutrients produced by the direct solvent action of substances excreted by the fungal symbiont of the lichen. Drainage water carries plant remains to accumulate in any small crevices, and here the humus and mineral matter form a small pocket of soil on which hardy plants, often very xerophytic, can grow. Such hardy plants, on their death and decay, yield further humus and other plants are able to grow in the greater amount of soil formed. The roots of larger plants penetrate the parent rock and help to further its mechanical break up. Physical weathering does not provide the mineral salts required by plants but chemical weathering takes place also alongside the physical weathering.

Both physical and chemical processes of weathering are important in the disintegration of rocks and minerals. Each aids the other's action. Physical or mechanical disintegration makes smaller pieces from larger pieces of rocks and larger surface areas are exposed. This promotes chemical weathering since chemical processes involve primarily surface reactions. At the initial stages of weathering chemical processes may be slow but nonetheless they aid mechanical breakdown of rocks. For example, carbon dioxide dissolved in rainwater may corrode the surface of the rock, leaving channels along which mechanical forces can act. Carbon dioxide dissolved in rainwater from the atmosphere is supplemented by the carbon dioxide produced by the roots of plants in the soil to give carbonic acid. This, together with acids formed by the decay of dead plant material, enters into chemical reaction with the hydroxides of a number of elements. Igneous rocks, for example, consist of complex compounds of silicon and oxides of various metals, the commonest of which are aluminium, potassium, calcium, magnesium and iron. The chemical reaction which often takes place with water is called hydrolysis, in which the metals in the mineral complex are replaced by hydrogen and brought into solution as hydroxides (bases). Those elements which are most easily dissolved out of the mineral complex form strong bases. Among these are calcium, potassium and magnesium hydroxides. Those other elements which are more resistant to hydrolysis produce weak bases. Among these are hydroxides of iron and aluminium. The strong bases being more soluble often combine with acids produced by the decay of organic matter to form soluble salts. For example, soluble potassium hydroxide combines with nitric acid produced by the breakdown of protein, to form soluble potassium nitrate.

We have so far considered mechanical and chemical weathering. But these alone do not produce soil finally. The next stage in soil profile development is the addition of organic matter or humus to give a dark-coloured surface layer (A horizon). This is followed by leaching of soluble salts and bases and the translocation of clay and sesquioxides and other substances from the upper horizon to lower layers. Further horizon differentiation takes place in the subsoil (A_2 and B horizons; see below). Leaching is the process by which the soluble compounds produced by chemical weathering are washed out of the surface layers into the lower layers of the soil. Leaching takes place rapidly in tropical soils because of the heavy rainfall and the high temperature. This may deplete the soil of available bases and can leave the residual mineral complex in a highly acidic condition. The acidity of the soil influences the formation of humus from the decay of plant and animal remains. Leaching also removes some of the soluble substances produced by the decay of organic matter. It needs to be emphasized here that the removal of the end products of weathering is important in accelerating further weathering processes and in advancing soil development. A soil in which end products cannot be removed fast enough remains retarded in development.

It is apparent from the description given above that the steps in soil formation involve the accumulation of parent material and horizon differentiation. These steps overlap. Horizon differentiation is considered to arise from four kinds of changes in the soil system, namely addition, removal, transfer and transformation.

Highly weathered tropical soils as a group are known as **latosols**. The type of soil that is formed in a particular place depends on the influence of five main groups of soil-forming factors which are as follows.

1. The climate of the area including any changes that may have occurred in the past.
2. The vegetation of the area as influenced by the climate and the soil fauna.
3. The parent material from which the soil is developed.
4. The type of relief associated with the soil.
5. The length of time during which the soil has been forming under the influence of the various factors.

The description of the process of soil formation that has been given above is based on soil that is formed *in situ* from the solid crust of rock. Here the layers are closely related. However, in the case of alluvial deposits a type of soil is formed which has no immediate connection with the rock over which it lies, and the layers are therefore not related. Where the soil, originally formed *in situ*, has been much disturbed by activities of man as in agriculture or drainage, the relations between the original layers may be greatly altered.

Temperature and rainfall play very important roles in determining soil types in temperate and tropical regions, and in dry and wet parts of these regions. The primary difference between tropical and temperate soils is said to arise from the type of weathering to which the soils are subjected. In temperate soils aluminium and iron are much more rapidly lost than silicon. In tropical soils a reverse condition obtains, whereby silicon is more rapidly lost and iron is least leached. This results in various iron oxides, which help to bind the soil particles together, remaining in the upper layer, and helping to improve structure (aggregate formation), thereby reducing erosion.

In a tropical climate where rainfall is low, the balance between evaporation and rainfall is all in favour of evaporation (that is, high evapo-transpiration). Where leaching is slight, as in the savannas, the soil characteristics will differ markedly from those of, say, the forest areas, where there is higher rainfall, reduced evapo-transpiration and increased leaching. In general, this results in more humus in tropical forests than in tropical savannas. But, as will be explained later, the reverse obtains in the comparison of temperate forests and grassland.

The Soil Profile

A soil profile is a vertical section through the soil showing the various horizontal layers (called **horizons**) from the surface to the least weathered parent material below. Each horizon differs in some features from those above and below it. A soil showing at least the upper horizons may be observed in road cuttings, deep ditches or quarries. Special pits dug for the purpose of studying the soil profile are known as soil profile pits.

A good example of a soil profile is one which is taken of soil that has not been disturbed and where the whole profile is the product of soil formation in the same position (sedentary or residual soil). A typical example from West Africa illustrates the common features of the profiles of tropical soils. Owing to the intensity of leaching the horizons are not always as sharply separated from each other as in temperate climates. However, they are sufficiently distinguishable in the field.

A typical tropical soil profile may be said to be made up of three main horizons: the top soil, the subsoil and the weathered substratum. The top soil is quite thin, often about 15 cm deep. The subsoil seems to be the thickest of the horizons, often extending from 15 cm to about 180 cm, and the weathered substratum begins from this depth. The thickness of each of these horizons varies in different situations but their relative thicknesses are fairly uniform. For a typical tropical soil profile see Figure 43 (p. 120).

The Top Soil (The A Horizon)

The top half of this horizon (A_1 horizon) is very dark greyish-brown in colour due to the high proportion of humus, whereas the lower half is not so dark since the humus content is slightly lower. On the surface of the soil is the leaf litter layer consisting of fresh and partly decomposed leaves and other plant parts. These are continually being attacked by bacteria, fungi and termites; they may also be eaten by worms, insects and other animals. The products of these activities enter the soil and, together with roots and bodies of soil micro-organisms which die and decompose within the soil, form the humus. The humus makes the soil loamy, granular or crumbly, porous and hence well aerated and drained. The top half of this horizon is more crumbly than the lower half. Owing to the effect of the sun there are fewer micro-organisms at the surface of the soil. Below the surface the top soil is the richest layer in micro-organisms.

Owing to the faster rate of decomposition of leaf litter in the tropics, as a result of the high temperature throughout the year which speeds up bacterial activity, the litter layer is usually thinner in tropical soils. This is so even though the amount of leaf litter deposited per year is greater in tropical than in temperate vegetation (see Chapter 6).

Much leaching by soluble constituents to lower layers takes place from this horizon.

The Subsoil (The B Horizon)

The subsoil, which is the horizon below the top soil, has very little humus, so that the mineral particles are not darkly coated, but are variously coloured red, brown, brownish-yellow, yellow or grey. Owing to the reduced amount of humus, this layer has many fewer micro-organisms. The number decreases with further depth in the soil profile.

The presence of iron compounds is as typical of this horizon as it is of tropical soils generally, and the colour of the soil depends partly on the amount of iron present and partly on the form in which it occurs. When the iron is oxidized to reddish ferric oxide, it is usually a sign of good drainage and aeration. Thus red soils are commonly found on summits and upper slopes of hills where drainage is facilitated. Yellow and brown soil colours are attributed to the presence of hydrated forms of iron oxide such as limonite. Soils of these colours occur on lower slopes which are not as well drained as the soils nearer the hilltops. Finally, grey coloured soils occur on valley bottoms and in depressions where poor drainage conditions lead to the formation of reduced iron compounds. The reddish-brown soils have a fine sub-angular blocky structure with numerous pores between structural aggregates and larger cracks and channels. These facilitate the circulation in the soil of both air and water.

ON THE SURFACE

	Leaf litter layer

TOPSOIL

0 to 8 cm	Very dark, greyish-brown humous loam, crumbly and loose.
8 to 15 cm	Dark, greyish-brown loam, less humous than above, crumbly.

SUBSOIL

15 to 120 cm	Reddish-brown light clay, weak, fine subangular blocky structure, containing frequent to abundant ironstone concretions and some quartz gravel and stones.
120 to 180 cm	Reddish-brown light clay to heavy loam, transitional to the horizon below, usually coarsely mottled and sometimes indurated.

WEATHERED SUBSTRATUM

Below 180 cm	Brown or red silty loam, usually slightly mottled, with traces of soft weathered parent rock (phyllite or schist).

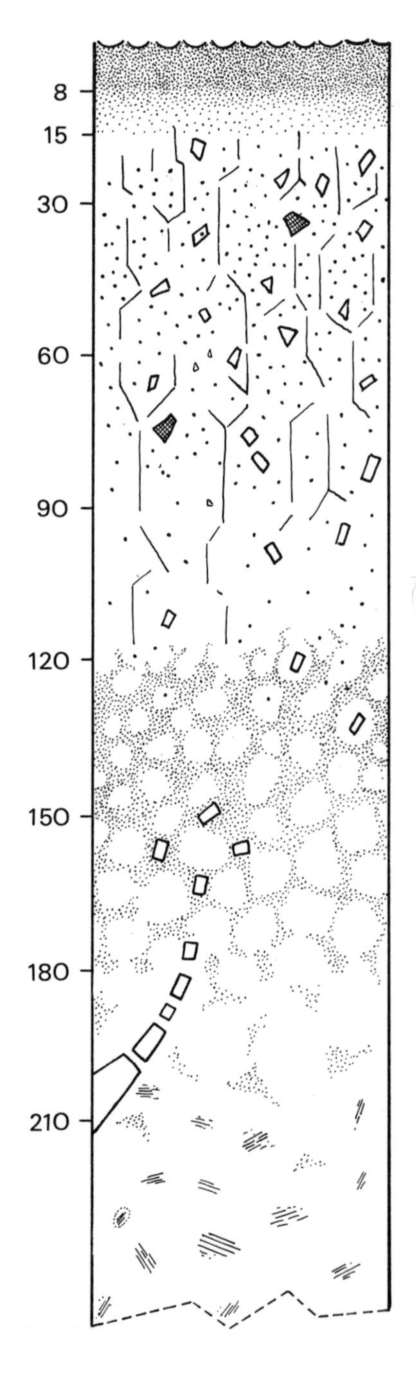

The subsoil often contains pea gravel in varying quantities. The gravel may consist of ironstone concretions or fragments of iron pan as well as fragments of quartz and stones. This horizon may also have a high clay content.

The subsoil is more compact than the top soil with fewer roots, pores and channels. It is often between 1 and 2 m thick, but in soils on steep slopes it may be much less.

The iron oxides of this zone are said to be responsible for reducing erosion in tropical soils. The lower part of the subsoil is usually coarsely mottled and sometimes indurated. This part is transitional to the next horizon (the weathered substratum).

The Weathered Substratum (The C Horizon)

The transition from the subsoil to the weathered substratum is not as distinct as that from the top soil to the subsoil. The weathered substratum is the zone where the parent material has not been much changed. It is of much lighter colour and varies according to the presence of traces or fragments of soft weathered material. It is often less compact than the subsoil, but it often does not contain gravel or stones as found in the subsoil.

The nature of the weathered substratum varies considerably depending on the nature of the parent material. The weathered substratum finally merges with the fresh rock which may be quite close to, 70 m or more below, the surface.

Soil Colour

Tropical soils vary enormously in colour. Among these are black, grey, red, brown in all its different shades, yellow, slaty-blue, grey and occasionally white. The most important factors which determine soil colour are the proportions of silicon, iron and humus which a particular soil contains. In the tropics red and brown soils are more prevalent, but black soils (the black earths) are also widely encountered in low lying situations. As already indicated, the red and brown soils result from the presence of iron and the high degree of oxidation. On less well drained sites on lower slopes the iron often combines with water, and this hydrated form gives a brown or yellow colour. Finally, in poorly drained situations, the shortage of oxygen results in the iron taking on grey colours which are given to the soils. Leaching and deposition of soluble compounds also contribute to colour changes within the above stated colours in successive horizons.

The black soils are formed in anaerobic conditions where drainage is impeded. Such soils are more basic than the red soils which lose bases through leaching. Contrary to our normal impressions the red soils have a higher humus content than the black soils. The A and B horizons in the black earths are often mixed into only one zone, followed by the C horizon. This mixing up occurs because these soils expand greatly when wet and shrink when dry. The result is that in the dry season big cracks are formed linking the A and B horizons and facilitating the mixture of soils from these horizons into a uniform one in the course of time. The red soils, on the other hand, exhibit the normal three horizons. It is considered that it is the difference in the clay fraction of these two types of soils which is responsible for this difference. The type of clay in the black earth is called **montmorillonite** while that in the red soil is called **kaolinite**. The fundamental properties of these types of clay will be discussed later.

Laterite

The term 'laterite' is presently used to cover various aspects of the tropical red soils. Originally the term (derived from Latin, *latus* = brick) was used by Buchanan in 1807 to describe a soil specimen from the Malabar coast of India which was soft and could be cut into blocks which hardened on exposure to air to become bricks. Since then the name has been applied to such a variety of red soils that it is now difficult to define it precisely. For example, the name has been applied to such hardened soil materials as ironstone sheets and boulders. It has also been used to describe pea gravel and the reddish and mottled subsoils which are hardened to some extent. Finally almost any red soil has been loosely referred to as laterite or lateritic soil.

A hardened soil horizon is referred to as a pan, and the term 'iron pan' or 'ironstone' is more appropriate to describe the hard, rock-like sheets of

43 Diagram of a typical well-drained upland sedentary soil developed over phyllite or schist. A simple description of the soil is given next to the profile. The figures on the left refer to the depth
SOURCE: Peter M Ahn (1970)

indurated iron-rich material. 'Ironstone concretion' appears also to be more appropriate a term to cover the iron-rich concretions in the soil. Expressions like 'slightly' or 'moderately' indurated soils may equally well describe the mottled, iron-rich soil horizons, but these soils are not laterite at all.

What seems to be a common feature of all these red earths is the richness in oxides of iron and aluminium and paucity in silicon. But, as mentioned above, this is no basis for regarding all the red earths as true laterite because of the process of laterite formation. Laterite is normally found as a surface layer, often on high ground. However, one explanation of this is that it is, in fact, first formed at a greater depth and later thrust up by geological movements. An alternative explanation is that the iron pan developed first on a low lying peneplain and that by geological processes, such as differential erosion or denudation of adjacent areas, it gradually came to occupy a relatively higher position. The dissolved iron and aluminium are carried down by seepage to the level of the water table where the oxygen supply is very inadequate, and so the dissolved iron and aluminium may be in the reduced condition. With the alternation of the wet and dry seasons and the consequent rise and fall of the water table, the salts forming are subjected to conditions of oxidation and reduction, precipitation and solution. Progressive precipitation results and this is aided by the presence of minute particles which form the centres of the concretions that result from these actions. The concretions collectively form the laterite hard pan.

As erosion proceeds the laterite gradually becomes exposed, and it persists because it is not so easily eroded. Exposed laterite may have shallow soil and support poor grass vegetation since root ramification is made difficult. Often the hard-pan is fully exposed and becomes a hindrance to the spread of normal vegetation and a nuisance to agriculture. However, some crops are more adapted to it than others, and *Hevea*, for example, is a crop which is adapted to laterite, except that where the hard-pan exists the trees are stunted and maturity is delayed.

Podsols

As we have seen above, the process of laterization, typical of hot and wet climates, essentially consists of the removal of silicon to leave behind iron and aluminium. Podsolization, typical of the temperate climates, by contrast removes the iron and the clays, to leave behind a silicon-rich soil. The silicon-rich soil forms a bleached sandy horizon (A_2 horizon) below the humic upper top soil and above a dark B horizon. In temperate climates podsolization is best developed in cool, wet areas, often under pine forests, and because of this it was thought that the process could not occur in the tropics. This led to the impression that it was the opposite of laterization.

However, it is now known that there exist in the tropics soils similar to temperate podsols. These are confined to very sandy soils or horizons. They are found in many parts of the tropics, but are not very widespread. They occur mainly on very sandy parent materials on tropical coasts where rainfall is very high, as typical of high forest areas. In Ghana these soils are known as regosolic ground water podsols.

Soil Texture

Any given soil consists of a mixture of particles of different sizes. It is now agreed that the soil particles are of three size classes, namely sand, silt and clay. Soil texture therefore refers to the relative percentages of sand, silt or clay in the soil. It is the texture of the soil which gives it such characteristics as its feel and physical handling properties. Russell (1961) describes methods of analysing soil to show the proportions of the size classes. Soil particles that are more than 2 mm in diameter are referred to as gravel; particles between 2.0 mm and 0.2 mm form coarse sand; particles between 0.2 mm and 0.02 mm form fine sand; particles between 0.02 mm and 0.002 mm form silt; and finally particles with a diameter of less than 0.002 mm form clay. Table 17 shows the percentages of the sand, silt and clay fractions at various depths

Table 17. Particles size fraction in Kikuyu soils, Kenya

PARTICLE SIZE FRACTION	PERCENTAGE OF SOIL FRACTION AT DIFFERENT DEPTHS (IN CM)			
	0–30	30–120	120–210	210–300
Sand (2.0 to 0.02 mm diameter)	12.7	12.9	14.4	8.9
Silt (0.02 to 0.002 mm diameter)	13.3	8.0	8.4	7.5
Clay (less than 0.002 mm diameter)	70.2	74.1	73.0	81.2

in the soil on a coffee plantation in Kenya (Pereira, 1957).

From the table it is obvious that the percentages of sand and silt decrease with depth, while the clay fraction increases with depth. When a given soil has a high proportion of clay it is described as a clayey soil, and when dominated by sand it is a sandy soil. When it has virtually equal proportions of sand and clay it is called a loam or a loamy soil. There are, of course, several grades of clayey, sandy or loamy soils. Starting from clay one of the simple classifications in the tropics gives grades of heavy loam, loam, light loam, loamy sand and sand.

It is very instructive for the ecologist to be able to determine these classes of soil with the minimum of apparatus and as readily as possible (Russell, 1961). One of the simplest methods described by Ahn (1970) is quoted below:

1. Sieve the soil and take approximately 1 tablespoonful of fine earth, enough to form a ball about 25 mm in diameter.
2. Slowly add water, drop by drop, to the soil ball until the sticky point is reached: this occurs when the soil sticks to itself and can be shaped, but does not stick to the hand. (If it does, too much water has been added.)
3. On a flat wooden board try to roll out the 25 mm ball of soil 150 mm long. If it will bend without breaking, form it into a circle.

The extent to which the soil can be shaped is an indication of the texture. If the soil is a sand, it cannot be shaped or worked at all and the most that can be done is to heap it up into pyramid or cone (Figure 44), but if the sand contains sufficient finer material to enable it to be shaped into a ball it is a loamy sand. If the sample can be rolled into a cylinder, but breaks when bent further, it is a loam (a light loam forms a short, fat cylinder, an ordinary loam one which is full length). If the cylinder can be bent into a U-shape but no further, the sample is a heavy loam. If it makes a full circle without breaking it is a light clay or clay: a light clay forms a circle with cracks in it: a clay one without cracks.

One peculiarity of tropical soils in terms of texture is that the clay particles (which are the smallest in size) often adhere rather firmly to each other. In some cases these are cemented together by oxides of either iron or aluminium or both. As a result aggregates of clay are formed with the sizes of silt particles. This fact may give an impression of a low clay content in some tropical soils as determined by the simple method described above. In actual fact tropical soils are highly weathered in view of their age, and so contain a high proportion of clay, while there is a relatively lower proportion of silt as compared with temperate soils.

The Clay Fraction

In view of the importance of clay it is necessary for the ecologist to appreciate its properties. Clays exhibit colloidal properties because of their very small particle size (less than 0.002 mm). This is because the smaller a particle of matter the greater is its surface area in relation to its volume and the greater the proportion of its molecules which are at the surface. In view of the enormous surface area of a clay particle, its surface molecules are not electrically satisfied and tend to have a high proportion of negative charges which attract positively charged ions (cations) in the soil. Thus these negative charges give the clays their cation exchange capacity.

Differences in mineralogical organization (lattice structure) distinguish the clay mineral types. The principal building elements of the clay minerals are sheets (two-dimensional arrays) of silicon–oxygen tetrahedra and sheets of aluminium or magnesium–oxygen–hydroxyl octahedra. In most clay minerals the sheets of tetrahedra and octahedra are superimposed in different fashions. The kind of arrangement of the sheets is characteristic of each type of clay mineral.

Kaolinite *(Two Layer Clays)*

This is a non-expanding two layer mineral type. The crystal unit is made up of a silica sheet and an alumina sheet. Thus kaolinite is said to have **1:1 lattice structure**. The crystal units are arranged with a fixed distance between them and the layers of silica and alumina are alternating. The units are held together by oxygen-hydroxyl linkages. The cation exchange capacity of kaolinite mineral is low and usually amounts to 3–15 m.e./100 g (m.e. = milli-equivalents. Cation exchange capacity is 1/1000 times the equivalent weight per 100 g of the substance.)

Montmorillonite *(Expanding Three Layer Clays)*

The crystal units of the montmorillonite group of clays are made up of two silica sheets with an alumina sheet tenaciously held between them by shared oxygen atoms. Montmorillonite is therefore said to have a **2:1 lattice structure**. The distance between the crystal units is variable since

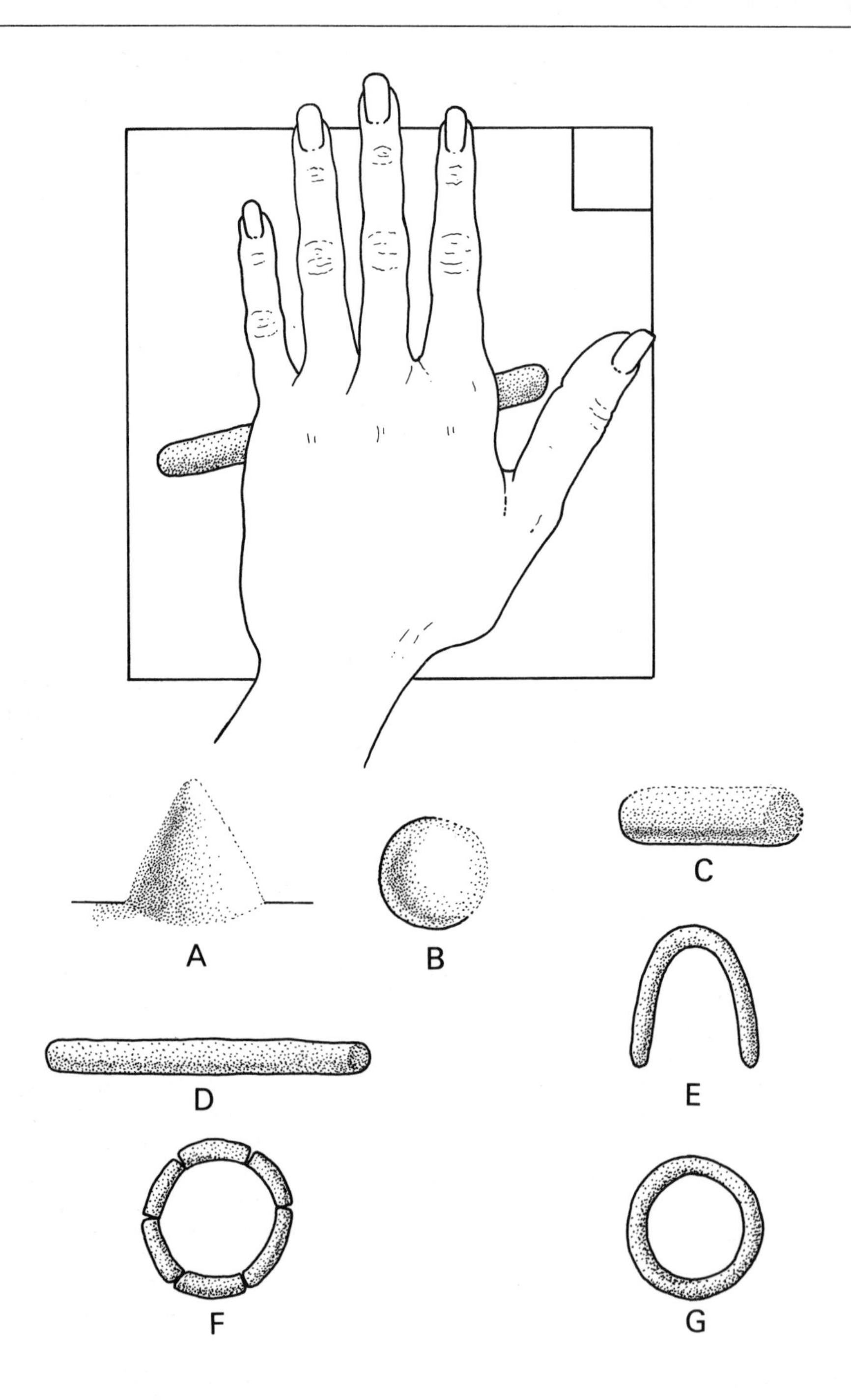

44 A manipulative test of soil texture. Enough fine earth is taken to make a ball of soil about 2.5 cm across and water is dripped on to the soil until it reaches the sticky point, the point at which the soil adheres to itself but not to the hand. The extent to which the moist soil can be worked is an indication of texture. A sand can only be heaped into a pyramid A, but a loamy sand makes a ball B. If the soil can be rolled out to a short cylinder C it is a light loam. The remaining drawings indicate a loam. D, a heavy loam E, a light clay (a circle with cracks) F, and a clay (a circle without cracks) G.
The cylinder when fully rolled out, as in D, should be 15 cm long. A board, shown below the hand, is marked with a 2.5 cm square. A line should be drawn on it about 15 cm long in order to standardize the test
SOURCE: Peter M Ahn (1970)

they are loosely held by weak oxygen–oxygen linkages. The units therefore expand and contract readily depending on the quantity of inter-layer water molecules and the type and quantity of inter-layer cations. Montmorillonite has a cation exchange capacity of about 80–100 m.e./100 g.

Illite *(Non-expanding Three Layer Clays)*

Illite has a **2:1 lattice structure** like montmorillonite but the distance between the crystal units is fixed. Hence the unit layers do not expand on addition of water. The inter-layer potassium ions which satisfy the negative changes are held tightly and are not available for exchange. Such potassium ions are referred to as fixed-potassium. Illite has a cation exchange capacity of about 15–40 m.e./100 g.

Soil Structure

Soil structure refers to the arrangement of the soil in natural aggregates. These aggregates consist of particles of sand, silt or clay which adhere together in groups or natural units. The aggregates are separated by pores and cracks. In temperate soils these aggregates (referred to as **peds**) are well-formed with certain regularity bounded by natural faces or planes of weakness. Many tropical soils also have peds in most of their horizons (e.g. latosols). However the peds here are minute and sometimes ill-defined and do not form large structural aggregates such as blocks.

Whether proper aggregates are formed or not, the important factor is the crumb structure which permits aeration of the soil and the entry of water. A good crumb structure increases the amount of pore space and allows air and water to circulate in the soil. In heavy soils the otherwise slow drainage may be accelerated if there are spaces between the structural units. The crumb structure is more clearly exhibited by loams or loamy surface soils which contain fair amounts of organic matter (humus). Soil animals including termites, worms and ants also help to produce rounded soil pellets and crumbs.

In top soils with fairly good humus content, the soil particles are loosely bound together by the glue-like humus into porous, fairly rounded aggregates. In subsoils, crumb structure is better developed in the clays, but may be masked by gravel where this is present. The weathered substratum is the poorest in crumb structure.

Factors Influencing Soils and Vegetation

We have already indicated the five main factors which influence the nature of the soil which will be formed in a broad climatic area. Of these, climate, vegetation and organisms are the active forces which influence the development of normal soils of large regions (zonal soils). Relief and parent material, the passive factors of soil formation, influence the development of local soils within a region through their effect on precipitation and temperature.

Climate

On the macro-climatic scale, rainfall is the most dominant climatic factor which affects the type of soils in the tropics. The primary effects of rainfall on the soil are weathering, leaching and soil development.

Water acts as a factor which further increases the rates of chemical weathering and change in the soil profile, even though these activities are already proceeding rapidly in the high temperatures of the tropics. Thus, the wet forests of the tropics would be expected to have much faster rates of weathering than the drier parts or the cooler latitudes of the world. During the leaching of the products of weathering, soil components are removed in a systematic sequence depending on their solubilities. The first to be removed are the chlorides and sulphates. The second are the silicates. The least soluble compounds are the sesquioxides of iron and aluminium (Fe_2O_3 and Al_2O_3). Highly leached soils with relatively high content of iron and aluminium, found in hot and wet tropical climates, form the typical latosols already referred to. The type of clay mineral that is formed in these latosols therefore depends on the degree of weathering and removal of end products through leaching and translocation. In the tropics the high degree of weathering and leaching is associated with the climate.

The influence of type of parent material, that is, the mineralogy of parent material, is also important in clay mineral formation. The kind of parent material can limit the weathering process even under high rainfall as in the tropical rain forests. A parent material of pure quartz sand will resist weathering and can hardly be weathered to form silicate clays. In the Accra Plains of

Ghana, Toje soils (latosol), developed over tertiary sands under rainfall of about 750 mm per annum, are known to have kaolinite clays. Akuse soils (black clays), developed over basic gneiss, and in close geographical association with the Toje series, are montmorillonite clays. Very acid top soils are, however, typical of areas with excessive rainfall. Soils in areas with rainfall of less than 1120 mm a year, as in the savannas, where more pronounced dry seasons occur, and where potential evapo-transpiration is greater than the total rainfall, have montmorillonite as their clay mineral. Montmorillonite clay mineral may also be formed in moderate or low rainfall areas where the weathering results from a base-rich rock. Even so, with heavier rainfall the clay mineral may eventually become a kaolin. In fairly dry savanna type climates and on base-rich parent rocks like calcareous or ferro-magnesian rocks are the well-known tropical black earths (or vertisols) which have already been described. The clay fraction here is montmorillonite and this gives these clays a heavy texture.

The less leached soils of the tropical forests experiencing moderate rainfall are richer in humus and carry more large trees than the leached acid soils with excessive rainfall.

Fire

Before considering the effect of relief, it is important to discuss the effect of fire on the soil since firing is extensive in tropical climates. The effects of burning on vegetation have already been discussed in Chapter 4 and will be again in Chapter 12. Burning robs the soil of its potential humus. Because of its action the above ground parts of the grasses and herbs merely become ash instead of turning into humus after their death and decay. In the process of burning, some of the released nutrients, especially sulphur, are lost. Further losses occur before these nutrients are washed into the soil and can be taken up by the plant. As a result of this the organic matter content is low in soils that are frequently subjected to burning, as typified by the tropical savannas. It is known that the tropical savannas are capable of supporting more wooded vegetation if fire is excluded for a number of years.

The effect of fire at the fringes of the forests has been to reduce their extent and to replace them with derived savannas, often with scattered relict forest trees. This again is accompanied by soil impoverishment, and is facilitated in the relatively dry areas where the soils are shallow or sandy. Finally, burning lays bare the soil surface and so runoff and erosion are often increased, particularly on hillsides. The increase in runoff means a reduction in the amount of percolating and subterranean stored water. With less water it becomes impossible to restore the original more wooded vegetation.

Effects of Relief on Soil Sequences (Catena)

In various parts of Africa, with special reference to East and West Africa and with the exception of deserts, a series of flat-topped hills alternate with shallow valleys, and the soils are found to change in a sequence or chain across the valley from the top of one hill to another. The tops and upper slopes have gravelly red and brown upland sedentary or residual soils. These range from deep well-drained soils formed from weathered rock to those eroded so much that little soil remains. These soils are usually more acidic in reaction, depending on the nature of the parent material, than the soils lower down the slope.

Particles of sand, silt and clay are further removed from the upper slopes to the middle slopes, where the soil is yellow-brown and of light clay. On the lower slopes are yellow-brown sandy loam and loamy sand. The eroded material settles in the valley bottoms to form water-logged soils, mostly grey to white in colour owing to the anaerobic conditions. Such soils are less acid than upland soils and are often rich in calcium and other basic elements. The sequence is seen to be repeated in the reverse on ascending the next hill. This series of change is then repeated, over and over again, producing a landscape with apparently endless monotony over huge tracts of land.

The term **catena** is used to describe these regular sequences of soil, and vegetation, types linked in their occurrence by conditions of relief, and repeated in the same relation to each other wherever the same topographic conditions occur in a landscape. The sequence is most clearly seen on a large landscape where the component soil units are zonally arranged. Each soil unit in the catena is genetically related to the one above and influences the genesis of the one below it in the series. Figure 45 illustrates the catena.

Before proceeding to give an illustrative example of soil catenas, it is important to recognize that the sequence of the soils in the catena is due to differences in parent material at different

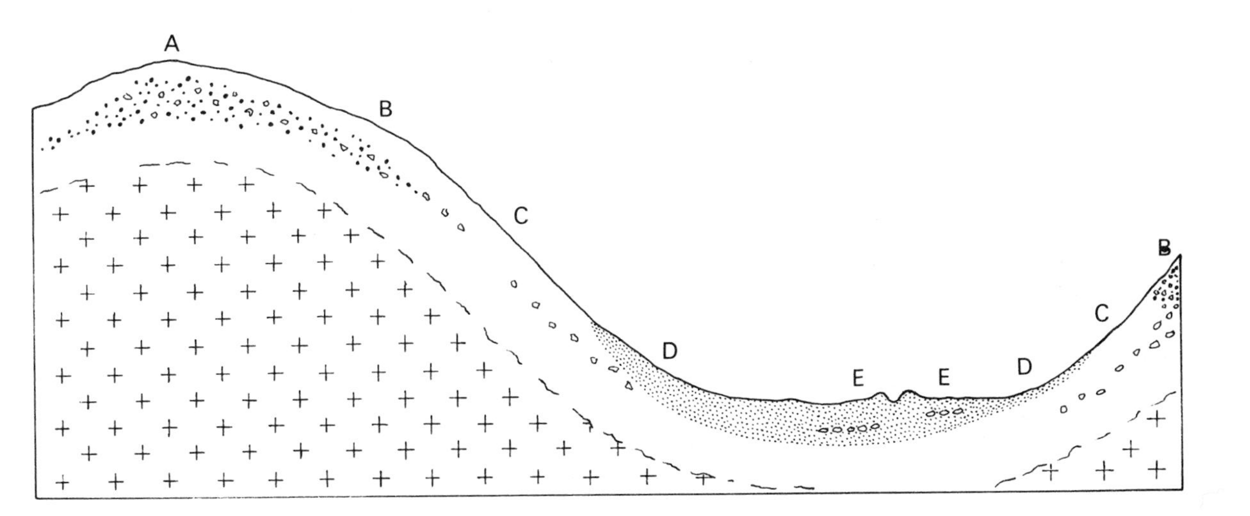

45 Diagram of a simple forest catena. Gravelly red and brown upland sedentary soils are found at A and B on summits and upper slopes. These grade downslope into yellow-brown sandy light clay soils developed in middle slope colluvium at D. The soils developed in local granite-derived alluvium at E are mostly grey to white sands with subordinate areas of gritty or sandy grey clays
SOURCE: Peter M Ahn (1970)

positions in the catena, and to differences in site, which includes slope and drainage.

Using West African catenas as an example, it is found that the weathering products of the underlying rock constitute the parent material of the summit and upper slope soils (**eluvial**) but the middle and certainly the lower slopes are often covered by parent material derived from higher up and are referred to as **colluvial**. The valley bottoms are covered by **alluvial** parent material. Thus parent materials are divided into three main complexes: eluvial (on the high level), colluvial (on the slopes) and alluvial or illuvial (at the low levels). Broadly, eluvial soils lose material, colluvium is the material moved down a slope under the influence of water and gravity, and alluvium is material transported, sorted and deposited by water. Colluvium is generally shallow, but may be extensive, with an upper area of typically brown to yellow-brown, sandy light clay, grading down the slope into yellow-brown, sandy loam and loamy sand. The alluvium on the other hand is often deeper and may contain many contrasting horizons.

The original catena concept came out of a study of African savanna soils, but it is probably of universal application where soil differences can be related to undulating topography. Whether the present vegetation is now forest or savanna, differences in floristic composition, density and cover can be attributed to the catena effect. Besides the actual long term formation of soil types, which may have become stabilized, the short term effects at different seasons are demonstrated by the movement of soluble nutrients during rainy periods and degrees of desiccation or inundation at the upper and lower parts of the topography respectively.

The Classification of Soils

From what has been discussed so far it is evident that climate, parent material and various activities of man affect the type of soil formed in any locality. Topography and age differences in land surfaces also affect the nature of soil. In order to classify types of soil it is necessary to use the criterion which best reflects the combined effects of the particular set of genetic factors playing a major part in the development of a given soil. The soil profile provides this criterion as it represents the morphology of the soil at any spot, that is, what it looks like, its colour, texture and structure, among other factors.

It is evident that soils have no definite individual boundaries, so that a soil individual is not found as a distinct entity clearly separated from others. Any recognized soil grades at its margins into other soil individuals with different properties. Nevertheless, by sinking profile pits in different areas, it is possible to recognize morphologically similar soils which, in spite of their local differences, conform to a model profile and are thus regarded as belonging to the same soil type. A group of soil types with a number of similar characteristics are grouped successively into broader categories termed soil series, families, great soil groups, suborders and orders.

Very many soil classification schemes have been put forward and efforts continue to try and move towards one that can be accepted internationally. The different schemes have arisen from the different ways in which soil profiles have been viewed. An agricultural approach, for example, would produce a scheme that was different from that produced from a purely pedological viewpoint.

Broad Classification of Tropical Soils

As discussed above, various schemes for classifying tropical soils have been proposed. However, one of them which seems to command a fair acceptability is the one put forward by Thorp and Smith (1947). This scheme recognizes three broad groups, namely zonal, intrazonal and azonal soils. Zonal soils are the soils in whose formation climate and vegetation have played a predominant role. Intrazonal soils are soils whose formation has been influenced by some local factor such as parent material, relief or drainage rather than by climate or vegetation. Finally, azonal soils are young soils whose soil forming processes have not yet been expressed and as a result of which very little profile development has taken place.

1. The Zonal Group of Soils

Among these soils are latosols and various lateritic soils. Latosols form the main zonal soils of the tropics. They are deeply-weathered, mature and free-draining upland soils. They have been described also as the tropical red earths, lateritic earths, ferrisols, ferruginous soils or ferralitic soils. They are generally formed from acidic and intermediate rocks although they are known to have been formed from some originally basic rocks after the bases have been leached out. These soils are characterized by their reddish or yellowish colour as well as their stable micro-aggregation. They are dominated by kaolinitic clay and various oxides of iron and aluminium which are hydrated to different degrees.

Latosols that are excessively leached and acidic are called **oxysols** while those with less thorough leaching are named **ochrosols**. Oxysols are formed in regions with rainfall exceeding 1800 mm per annum (the exception being only in areas over basic rocks), while ochrosols are formed in areas with less than 1650 mm of rainfall per annum. Ochrosols are peculiar in showing a progressive increase in acidity with depth, even though the top soil may be neutral or only slightly acidic.

Usually the upper top soil has pH 6.0–7.0, the lower top soil pH 5.5–6.5, while the subsoil has pH 4.5–5.0. Ochrosols are generally more prevalent, but oxysols are limited in distribution and have been described in Malaya (Owen, 1951), Sierra Leone (Martin and Doyne, 1932), Ghana (Brammer, 1962), Nigeria (Vine, 1955) and central Zaïre (Congo) (Kellog and Duval, 1949).

Latosols have rather low cation exchange capacities, so that it is the presence of organic matter which determines the amount of nutrients held. Nevertheless, they are much used for agricultural purposes because of their deep profiles, friability, good tilth and resistance to erosion.

Lateritic soils have already been discussed above.

2. Intrazonal Group of Soils

Among the intrazonal soils the more important ones are the hydromorphic soils which are of widespread occurrence in the tropics. These soils have high water tables fairly close to the surface. Among these are the humic gleys, peats and sandy podsols often found in coastal areas as in Malaya, Borneo and parts of Sri Lanka. Others are soils derived from basic rocks and referred to as basisols or krasnozems, the calcimorphic soils such as the black earths of the Deccan Plateau and the margillitic soils of Indonesia. The humic gleys refer to soils formed from material settling out of flood or tidal waters.

Differences occur between sediments deposited in salt (marine), brackish and fresh waters. In marine alluvials various stages can be identified in the development of the soil towards a mature profile. In the mature stage, the top soil is found to consist of a deep organic horizon up to 40 cm thick and the underlying soil becomes differentiated to a depth of 4.5 m with the following order of profile features.

(a) Black organic top soil.
(b) Mottled horizon, with characteristic structuring, generally finer than the underlying horizons and often streaked with charcoal.
(c) Horizon characterized by coarse blotching with oxides of iron.
(d) Horizon with heavy olive-brown or dark-green gritty iron deposition.
(e) Heavily gleyed horizon; often present are heavy deposits of crystalline gypsum and bands of sea shells.
(f) Undifferentiated clay.

In estuarine alluvials there is a deep black top soil overlying a second horizon which has prominent dark-red and extensive reddish-yellow powdery iron oxide deposits. The second horizon has a fine crumb structure and hence improved aeration. The rapid development of a crumb structure in the second horizon in these deposits in river estuaries does not take place in soils far removed from any main drainage outlet to the sea. In marine deposits crumb structure may be only weakly formed and over a much longer time.

In freshwater alluvials there is better drainage. They have organic matter in the lower horizons and the matrix colour is light grey. They can be of two types: the levée and sheet deposits. Levée soils have a lighter loamy texture and are often better drained. Sheet soils consist of finer material deposited by flood water some distance beyond the river banks. They are heavier in texture and are not so well drained.

Trough and basin deposits are soils which form on the surface of the older alluvium in depressions of old flood and drainage channels or lakes. These deposits are usually permanently waterlogged and almost structureless.

Peat. Peat or almost pure organic soil is one of the important hydromorphic soils in the tropics. Peats are in fact more extensive in the humid tropics than has been realized until recently. In Malaya alone it has been estimated by Coulter (1957) that there are about 800 000 hectares of deep peat soil. Peat is often a low-lying swamp deposit with varying amounts of mineral matter in the form of fine alluvium. Where the mineral matter forms a considerable fraction, the peat has been named 'muck soil'.

Peat soils are normally formed under waterlogged low pit conditions, and are developed over a bleached clay foundation. Drainage of shallow peat soils in coastal regions for cultivation results in the leaching of acid products of the oxidized peat and considerable shrinkage. This leads to the formation of a type of man-made podsol.

Podsols. The next group of intrazonal soils of widespread occurrence is the podsols. As already discussed, these are iron podsols developed on quartz sand (as the parent material) which is devoid of clay mineral, so that humic materials and iron oxides that are present are easily transported down the profile. Podsols show a bleached horizon (A_2) overlying a blackish brown or brown layer of sesquioxides in the B horizon. Podsols have been reported in low lying areas, generally along the sea coasts where sandy stretches prevail, in Sri Lanka, Indonesia, Thailand, Malaya and West Africa, among other areas.

3. Azonal Group of Soils

Azonal soils are free draining, relatively homogeneous recent deposits on flat or gently sloping plains. They are not differentiated into horizons. The three major groups are the regosols, alluviosols and the lithosols.

Regosols are formed from sediments other than young water-deposited materials in flood plains. Examples are windblown sands, slope colluvium and volcanic ash. The colluvium and the volcanic ash are often being constantly renewed. Volcanic ash is widely distributed in areas with active volcanoes as in Java, Sumatra and the lesser Antilles. Sandy regosols are found along the coastal strips, and are often not cultivated except for coconuts.

Alluviosols are soils developing from recent alluvium with little or no modification of the alluvial material by soil forming processes.

Lithosols are shallow skeletal soils with only A and C horizons. They are often found in situations of steep topography or on relatively flat summits where fresh rock or solid iron pan is close to the surface.

The classification system of Thorp and Smith (1947) described above was based on a geographical concept of zonality of soils. The criteria for such a system are the soil forming factors or processes of soil formation such as podsolization, laterization, etc. The system has the drawback of not using observed morphological features or measurable internal properties of the soils themselves as criteria, particularly at higher categorical levels. The more recent trend is towards the use of observable features of the soil as criteria, and the formulation of theories to seek genetic relationships between the various soil groups. Recent attempts in this direction are the United States system (the 7th approximation) and the F.A.O. system. Diagnostic soil horizons (both surface and subsurface horizons) are used as criteria for differentiating the high categories (e.g. the orders). Using such criteria soils that have been classified as zonal tropical soils (latosols) are now split into several orders. The main advantage of these recent systems is that they are more quantitative in approach and can therefore be used internationally. Moreover, the systems have taken into consideration man's influence on soils.

Effects of Soil Environment on Plants

With the general background provided above on the formation and structure of soils, the effect of the soil environment on plants can be better understood. The most important factors are: soil atmosphere; soil temperature; soil water; soil solution and reaction; and soil organisms and humus.

1. Soil Atmosphere

About one-third of the pore space is normally filled with air which is kept saturated with water vapour by the films of water covering the soil particles. Heavy rain can result in the displacement of much of the soil atmosphere by water. Fresh supplies of air will, however, be drawn into the soil as the water drains away. Unlike the aerial environment, the soil does not receive any oxygen from photosynthesis, so that the soil is comparatively low in oxygen supply even though oxygen can be replenished by rain water. Because of the respiratory activities of soil organisms, the soil contains at least seven or eight times as much carbon dioxide as the air. Seasonal changes in temperatures, between the dry and wet periods of the year, affect the carbon dioxide content of the soil, it being greater in the cooler period than in the dry and warm period of the year.

The size of the soil pores appears to be more important in soil ventilation than the total pore space. In the clay-loam the pores are often so small that diffusion is slow, and circulation of air may be blocked by water.

In the deeper parts of the soil carbon dioxide may increase to as high a concentration as about five per cent of the soil atmosphere. Such a high concentration inhibits the growth of roots, the efficient absorption of minerals and seed germination. Thus root systems in heavy soils with high carbon dioxide and low oxygen content have stunted development in the deeper layers while being richly branched near the surface. It is considered that the increase in carbon dioxide concentration with depth may be partly responsible also for the horizontal growth of some rhizomes and some roots, and for the upright pneumatophores of the white mangrove (*Avicennia*). On sand dune soils, roots are well developed because of good aeration.

2. Soil Temperature

Soil temperature seems to vary significantly between day and night only in the very top layers. Below about 50 cm marked day and night fluctuations seem to disappear. Soil temperature lags behind air temperature. In the tropica climate the high temperature of the soil by day results in rapid loss of water by evaporation from the heated surface layers. This is only partly replaced by deposition at night as dew.

The slope of the ground and the direction in which the slope faces (aspect) are among the factors which affect the absorption of radiant energy from the sun. As the declination of the sun varies from day to day so will the incident radiation. In inter-tropical latitudes, maximum radiation may strike north-facing or south-facing slopes depending on the time of year.

There are two important ways by which the water content of the soil can affect the rate of warming up of the soil. One of these concerns its high specific heat. This is the number of joules of heat required to raise the temperature of one kg of a substance by 1°C. Water has a higher specific heat than any other substance. The specific heat of soil particles is about 0.84 joules. This means that a given amount of heat will raise the temperature of a volume of dry soil through about three times the range that it would the same volume of waterlogged soil. This explains why clay is often described by farmers as 'cold soil'.

3. Soil Water (Movement of Water in the Soil)

The soil receives its supply of water from rainfall, but in soil with a rather high water table (the depth below which the soil is saturated with water) the soil may receive water from below. Soils vary in their capacity to store water in times of drought and so the behaviour of rainfall in soils will also depend on the nature of the soil.

When rain falls on porous land some of it infiltrates and percolates through the soil, while some of the remainder runs off the surface. The water which penetrates the soil is distributed in it but some of it is lost by evaporation or used by plants before it reaches the water table which may be far below the surface. The absorbed water circulates between the air spaces (of various sizes and shapes) which exist between the crumbs and their individual particle components. The soil particles have great surface areas which constitute the containing wall of the soil pores. It is estimated that one cubic centimetre of pore space has a surface area of about a million square centimetres.

Under field conditions, when a relatively dry soil is wetted by rainfall, the water front moves ahead without filling the pore spaces. It leaves behind it rather localized films of varying thicknesses held by the force of surface tension. The movement of the water in such situations is thus not directly through the pore centres but along the films. If these films are fairly thick, the movement is reasonably free, but it becomes severely hampered when the soil gets drier and the films thinner. Eventually, when the films are so thin as to be broken in places, the water can only move in the form of water vapour. The water sinks only to a certain depth after which no further movement takes place.

Soil Water Categories

It seems that there are three different levels or categories of soil water. Assume that before a fall of rain, the water content of the air-dry soil was about 10 per cent. If, two or three days after the rainfall, all free water had disappeared from the soil surface and a trench were dug, one would notice that the upper layers were moistened to a depth of about 25 cm, whereas the layers just below this would appear as dry as they were at the beginning. The wetted top portion of the soil would be found to have a water content of, say, 20 per cent. Since water surrounds the soil particles in the upper soil first, if more rain were to fall on the same soil the water content of the next layer below this one would rise to, say, 20 per cent. Although the rain water would now have reached twice the previous depth, the soil below this new depth would have remained at 10 per cent as at the beginning. This amount of soil water content is called the **field capacity** (see below). If no further rain fell and the plants were thus subjected to a long period of drought, the plants would continue to absorb water from the soil and lose most of it to the air through transpiration. This would result in a gradual decrease in the water content of the wet soil to values below 20 per cent, but this would apparently stop at 10 per cent (the original dry soil water content).

(a) **Permanent Wilting Percentage (and Hygroscopic Coefficient)**. As soon as the water content of the soil reaches the original dry soil level (here 10 per cent) the leaves of the plants will wilt. At this stage most plants cease to draw in water; a few may be specially adapted to do so. Deep-rooted plants may not wilt when surface-rooted plants do. This 'unavailable water', which is the percentage of water remaining in the soil when any plants growing in it wilt permanently, is known as the **permanent wilting percentage**. The permanent wilting percentage of the soil discussed here may be taken as 10 per cent.

Another water category that is very closely linked with permanent wilting percentage is hygroscopic coefficient. When a sample of soil is dried completely in an oven and then left exposed to normal moist air there is a limit to the percentage of moisture that it will absorb. Soil in this condition is said to be air-dry and it often contains between one and six per cent of water, depending on the nature of the soil and relative humidity of the air. The percentage of water in air-dry soil is what is referred to as **hygroscopic coefficient** and it is estimated that hygroscopic coefficient of a soil is 0.68 of its wilting percentage.

(b) **Field Capacity and Capillary Water.** In our example above, the water content of the upper layers rose from 10 per cent to about 20 per cent after the two spells of rainfall had ceased. The extra 10 per cent increase is what is referred to as capillary water.

The field capacity of the soil at this stage, however, is the total 20 per cent of the water, that is, the total of the hygroscopic water, the wilting percentage and the capillary water. It is the water which would not drain through (under influence of gravity). The water available to plants is field capacity less wilting coefficient. This is equivalent to water held at $\frac{1}{3}$ atmosphere less water held at 15 atmospheres.

(c) **Gravitational Water.** When rainfall continues for days any water above the field capacity will drain to lower layers under the influence of gravity. This water is called gravitational water. This process goes on until all the lower layers of the soil have reached field capacity too. After that, if the water is hindered in further drainage, it starts to fill the pore spaces and the soil begins to be waterlogged. This proceeds from below upwards and the upper level to which it reaches is called the **water table**, which is commonly found in many soils. In normal situations the level of the water table rises during the rainy period and falls during the dry period of the year. Near standing or running water, however, it is generally high throughout the year.

The soil water categories may be summarized as follows.

Hygroscopic coefficient
Permanent wilting percentage } Field capacity
Capillary water
Gravitational water
Water table (saturated water)

The actual values of these soil water categories depend on the properties of the soil such as the particle size and the proportion of collodial matter present. Whole clay or humus soil may have hygroscopic water of about 13 per cent and capillary water of about 7 per cent while the corresponding values for sandy soils are 3 and 13 per cent respectively. These values are expressed as percentage of weight of oven-dry soil. In natural soils, which may have different properties at different strata, the field capacity may vary at different depths. Thus different species growing side by side, but rooting at different depths, may have their roots in quite different field capacities.

Table 18 gives some of the water categories and other physical properties at different depths in East African latosols normally used for coffee cultivation (Pereira, 1957).

It has already been mentioned that in dry soil the root tips and branches grow towards the regions of the soil with a higher moisture content. This response ceases when the soil water reaches its optimum level. In general, excessive water checks the growth of roots.

4. Soil Solution

Growing plants derive from the soil solution not only water but also their entire supplies of the elements nitrogen, sulphur, phosphorus, calcium, potassium, iron and magnesium, together with the trace elements such as boron, zinc, copper and manganese. Soils vary in their capacity to supply these elements, and this results eventually in differences or changes in natural vegetation.

Colloidal clay particles, and sometimes humus particles too, behave in the soil like gigantic anions with numerous negative charges at their surface and these negative charges must be balanced by positive ones in order to attain stable conditions (Figure 46). The positive charges for this purpose are borne either by ions of hydrogen or metallic ions such as calcium or potassium present in the soil solution. In natural conditions there is always a choice of these positive ions so that equilibrium attained by any soil reflects that choice.

Clay particles are rich in aluminium and in highly acid conditions this is present in ionic form as positive charges. In such a situation the said clay becomes an alumina clay rather than a hydrogen clay. At the other end of the range it is possible to produce sodium or calcium saturation

Table 18. Some physical properties of Kikuyu soils, Kenya (from Pereira, 1957)

WATER CATEGORY/PORE SPACE	PERCENTAGE OF WATER CATEGORY OR PORE SPACE AT DIFFERENT DEPTHS (IN CM)			
	0–30	30–120	120–210	210–300
(a) Water at field capacity	37.2	40.0	42.3	44.0
(b) Water at permanent wilting percentage	24.2	26.7	29.7	34.5
(c) Available water (difference between (a) and (b) i.e. capillary water)	13.0	13.3	12.6	9.5
(d) Pore space	60.9	57.1	55.5	54.1

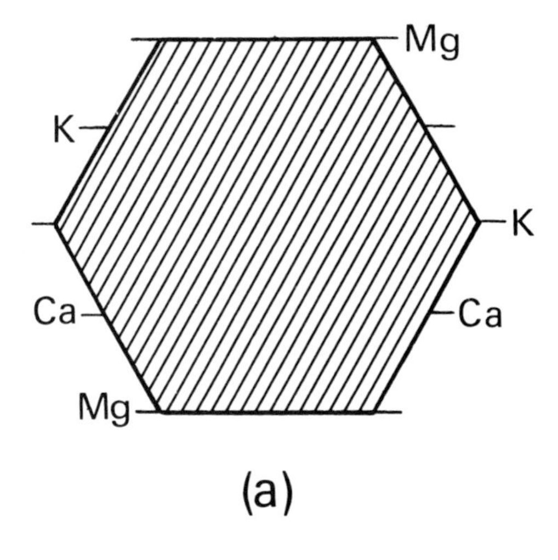

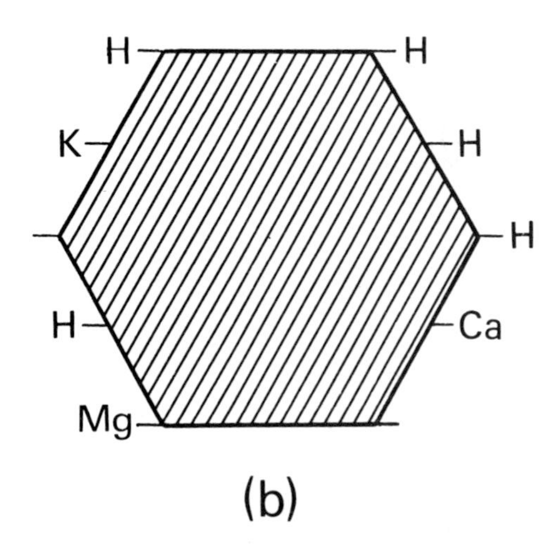

46 Ion exchange in soil particles: a) A clay particle saturated with metallic cations. The negative exchange sites are occupied by potassium, calcium and magnesium b) A clay particle showing what occurs when the place of some of the cations is taken by hydrogen ions, such as occurs during leaching
SOURCE: J Y Ewusie (1973)

by treating the clay with the appropriate solutions. If a fertile soil with calcium is flushed with ammonium sulphate solution some of the ammonium ions (NH_4^+) are retained, but a chemical equivalent of calcium ions (Ca^{++}) are displaced and will be drained off. The general picture which emerges is that of soil whose clay fraction can remove basic ions from solution and hold them in close association at the surface of the particles. However, as already described above, these tethered bases can be replaced or exchanged with others. This whole process is referred to as '**base exchange**'.

In principle soil can be saturated with any cation, but the acid radicals are not affected directly. To demonstrate this property of soil experimentally, two samples of well-packed clay or loam are used. One of them is flushed with a solution of methylene blue and the other with aqueous eosin. It will be noticed that while the eosin dye will be drained out in solution, the methylene blue will not. This is because methylene blue is methylene blue-chloride so that the methylene-blue part, which is responsible for the blue colour, is held back by the clay anions, while eosin is regarded as 'sodium eosate' and so only the sodium base is held back while the part responsible for the colour is free and appears in the drainage solution. The exchange evidently changes the chemical constitution and physical properties of a given soil. For example, calcium-saturated soils are easily worked and drained while sodium clays are sticky and less permeable.

It is thus clear that the total amount of exchangeable bases present in a soil depends upon the amount of clay it contains as well as the degree of saturation with bases, that is, the extent to which the negative charges have been balanced by positive ions. Also, the total number of charges available to be neutralized is of importance here since some clays with a greater base exchange capacity have a potentially higher level of fertility.

Base exchange can bring a change in the mechanical properties of soil. A fertile soil normally contains exchangeable calcium which helps in the production of crumb structure by the soil. If such a soil is inundated by sea water it is changed immediately into a sticky sodium soil with no crumb structure. Such land is most difficult to reclaim. The restricted plant life of mangrove swamps, salt marshes and salinas is largely due to the adverse properties of sodium soils.

Soil Reaction

Soil reaction is the term used to express quantitatively the degree of alkalinity or acidity of a soil in terms of the concentration of hydrogen ions. The greater the extent to which the ionic exchanges of clay particles are dominated by positive metallic ions to the exclusion of hydrogen, the less acid a soil becomes.

Tropical soils show great variability in hydrogen ion concentration, those at one end of the scale having a concentration ten thousand times greater than that of those at the other end. Such a scale is too cumbersome to use and so in practice it is telescoped by converting it into a logarithmic scale or scale of indices, such as $100 = 10^2$, $1000 = 10^3$, $1000\,000 = 10^6$, where 2, 3 and 6 are indices or logarithmic bases on the number 10. In actual fact the hydrogen ion concentrations are small, that of water being 10^{-7} which in ordinary terms is $\left(\frac{1}{10}\right)^7$ or $\frac{1}{10\,000\,000}$. (It is to be noted that while the ionic concentration is expressed as grammes per litre or 1000 cm^3, the units are at present immaterial.)

A highly acid soil might have a concentration of $\frac{1}{10\,000}$ or 10^{-4}, showing that its acidity is a thousand times as great as that of water. What is referred to as the pH or reaction scale is thus seen as the scale of these indices without consideration of their arithmetical signs which are always negative. Thus the pH of water is 7 and of the highly acid soil mentioned above 4. It should be noted that since no attention is paid to the negative sign, an increase in hydrogen ion concentration or acidity is shown by a decrease in the pH value and vice versa.

Among the modern precise methods for determining pH is the use of the pH-meter. In such determinations 2:5 or 1:1 soil water suspension is used. There are more rapid but only approximate methods which use organic chemical substances (indicators) which change colour at different pH values. Such indicators are shaken up with a suspension of soil and water. The range of tints produced indicates the approximate pH values of the soil.

Different plant species show their most vigorous growth of roots in soils of different reactions. By means of experiments using culture solutions that differed in their pH values it has been established that for one species maximum root development will take place at a pH different from that required for best root growth by another species. In nature each species will compete best if the soil has the optimum pH value, and is likely to be ousted by other species if the pH is changed. Since the range of tolerance of some species is rather narrow their presence in the vegetation provides an indication of the reaction of the soil.

Buffering in Soil. The change in reaction which the addition or removal of a given number of basic ions makes in a soil is not the same for all soils. For example, very little leaching of a light sandy soil by an acid solution will change its reaction markedly, while on a heavy clay or humus soil the same treatment would have relatively little effect. This resistance to change is what is described as buffering. When a given soil changes its reaction only slowly when treated with acids or alkalis it is said to be well-buffered.

The degree to which buffering occurs depends on the quantity of colloid present and its type. The importance of buffering lies in the fact that well-buffered soils are less subject to fluctuation in reaction caused by climatic and other environmental factors.

Saline and Alkaline Soils. Saline soils are soils which are highly charged with soluble salts. In the tropics such soils are typified by lagoons and mangrove swamps. Saline soils restrict the types of plants that can grow on them. In wet rain forests near the sea, toxic concentrations do not often occur in such soils because the soluble constituents are carried away in drainage water to the water table below the root range. But in other situations the high rates of evaporation and plant transpiration concentrate the salt in the surface soil, and this renders the soil unproductive.

Alkalinity is a situation found with soils that

are comparatively high in their content of both calcium and sodium. The calcium is more readily removed from the exchange complex and hence the proportion of exchangeable sodium is increased. Such sodium impregnated soils tend to deflocculate and this results in a decrease in soil permeability. This in turn impedes drainage so that salt accumulation is further increased. The soil thus becomes not merely saline but also alkaline and capable of supporting only a few kinds of plants.

It follows from the ready leaching of calcium out of the soil, that stratification will occur in humus soils with the surface layers being more acid and alkalinity increasing with depth. It is therefore sometimes observed in such situations that shallow rooted acid-loving species grow alongside deep rooting alkali-loving species.

The Role of Soil Nutrients

Water Cultures. The major elements needed for plant growth are absorbed not as free elements but combined as salts such as nitrates, sulphates and phosphates of calcium, potassium, magnesium and iron. The water in the soil contains at least small quantities of these salts and, if any of them is wholly absent, this deficiency is soon recognized by distinctive, unhealthy signs in the plant concerned. These elements are called 'essential' because the absence of any one of them prevents the plant from completing its life cycle. Each of them is directly involved in the nutrition of the plant.

In the laboratory, plants may be grown in various solutions without soil in order to find out the part played by the different elements in healthy development, and to indicate exactly what the requirements of particular plants are. This practice is known as water culture and is well described in many textbooks on plant physiology.

Nutrient Availability in Tropical Soils. Tropical soils often pose peculiar problems as far as nutrient availability is concerned. The major nutrients of importance are nitrogen, phosphorus, potassium, magnesium, calcium, sulphur, manganese and iron and the trace elements are zinc, copper, molybdenum, cobalt, chlorine and boron.

Nitrogen is closely associated with the organic matter status of soils, and in the wet tropics this has implications for agriculture. At equilibrium, the organic carbon level in forest soils of the humid tropics is about 4 per cent, while the ratio of the organic carbon level to the organic nitrogen level (C/N) is between 10:1 and 20:1. Thus Coulter (1957) analysed a tropical forest soil in Malaya (at a depth of 10 cm with pH 4.2) and found C = 4 per cent, N = 0.25 per cent and hence C/N = 16:1. In a Trinidad forest, on fine sandy soil, Duthrie *et al.* (1937) determined comparable carbon/nitrogen ratios ranging from about 8:1 to 13:1 at various depths, but the carbon component was lower, as was also the nitrogen, than in the Malayan example. Nitrogen total varied from 0.05 per cent to 0.13 per cent (Richards, 1964).

The level of phosphate in most tropical soils is low. Also, applied phosphate does not seem to be effective and the recovery rates from previously applied phosphates are not only low but fall off with time.

Except for the marine clays (humic gleys) and some black clays (vertisols), where exchangeable calcium values are high, the soils of the humid tropics are characterized by low exchangeable calcium values. Calcium carbonate is very rarely present in tropical soils except in degraded chernozems, now referred to as grumusols, and in black clays (vertisols).

Magnesium, like calcium, is present as an exchangeable cation, but owing to antagonism between magnesium and potassium in the soil there could be apparent deficiency of magnesium in certain soils.

Mineral Cycling. A plant growing in the soil absorbs nutrients needed for its growth from the bulk of the soil that is penetrated by all the roots and root hairs. When the plant dies and decays, the elements which have been used by the plant are released, but since the aerial parts of the plant fall to the surface of the ground, the elements are liberated on and into the soil. This is one of the important effects of vegetation on soil development. The liberated nutrients again become available for absorption by plants. Thus, while leaching moves the soil nutrients downwards in the soil profile, there is also an upward movement of these nutrients as a result of their uptake by roots. The decomposition of plant (and animal) remains on the surface of the soil also permits the escape into the air of some of the elements, such as oxygen, nitrogen and carbon, which originally come from the air. Thus, there is a constant recycling of nutrients (see Chapter 6). Specific examples like the carbon and nitrogen cycles are sufficiently well described in a number of textbooks and so will not be described here, but are illustrated in Figure 47.

5. Soil Micro-organisms

It is known that in the series of reactions which result in the breakdown of plant remains, each stage is brought about by particular species of fungi or bacteria. For example, in the nitrogen cycle, different fungi are responsible for the initial stages of conversion of proteins to ammonium compounds. After this, different nitrifying bacteria are responsible for the further stages of oxidation of the ammonium compounds to nitrites and nitrates. It is found that efficient working of the nitrifying bacteria in this process can be hampered by excess carbon dioxide produced by respiration and the various organic acids formed by the activities of fungi in the early stages of the process. In extreme cases, the accumulation of organic acids may completely inhibit the action of even the bacteria responsible for decay, resulting in the formation of soil poor in minerals. Incompletely decayed material is found in beds of lakes in the tropics (like peat in temperate climates). The arrest of natural decay leads to the preservation of pollen grains over thousands of years, though this takes place more easily in temperate climates than in the high temperatures of the tropics.

Among the factors necessary for the healthy growth of soil flora and fauna are aeration and adequate water supply. Over-cultivation, and its attendant increased aeration, may lead to rather accelerated bacterial activity and exhaustion of humus. On the other hand, if aeration is checked by densely matted roots or by other means, the same sort of peaty soil is produced as by high acidity.

The bacterium *Azotobacter*, which is free living in the soil, fixes extra atmospheric nitrogen in its protoplasm by oxidizing carbohydrates in the humus. The nitrogen thus fixed adds to soil fertility. It has been claimed that these organisms can add as much as 45 kg per hectare of nitrogen to the soil. Although acid and waterlogged conditions slow down the activity of *Azotobacter*, it is known that certain species of *Clostridium* are able to fix nitrogen under anaerobic conditions.

Apart from the free living saprophytic fungi and bacteria, there are a number of species of bacteria (*Rhizobium*) which are not free living but which live in symbiotic relationship with the root nodules of leguminous and a few species of other families of flowering plants. These bacteria start life as free living organisms in the soil, but later enter the roots of plants and begin to fix atmospheric nitrogen. This action proceeds with the help of traces of molybdenum which the bacteria absorb from the soil.

The nitrates built up by these organisms may, however, be reduced by the activities of denitrifying bacteria which work in the reverse direction by breaking down nitrates in the soil into free nitrogen and oxygen, the latter being utilized by them. In the tropics the high temperature accelerates the disintegration of organic matter and hence increases the rate of supply of inorganic nitrogen to the plant. Low temperatures reduce this activity. The oxidation and reduction processes of manganese compounds carried out by soil bacteria are also responsible for the availability of this trace element to higher plants.

Apart from the symbiotic relationship of *Rhizobium* in root nodules described above, other symbiotic relationships are found in the mycorrhizal or saprophytic fungi which grow in association with the roots of higher plants. There are two types of these. One is the faculative type in which either partner can exist independently. The fungi here form an external mantle covering many of the small branch roots. This suppresses completely the development of root hairs, so that absorption now becomes the function of the fungal hyphae, from which some of the absorbed ions are released to the roots. The other type of mycorrhizal fungi is the endotrophic type which is found in all species of *Ericaceae* and orchids (*Orchidaceae*). Here the flowering plants cannot survive the seedling stage without the fungus unless supplied artificially with soluble carbohydrates. The fungus grows in the intercellular spaces of the root cortex and even penetrates many of the cells.

It is worth noting here that the fungus *Penicillium*, from which penicillin is obtained, and other organisms which produce antibiotics used in medicine, are soil saprophytes. Their secretions exert some effect on competition and the balance of micro-organisms in the soil. It has been shown that these secretions can attack the roots or seedlings of arable plants.

47 The water, carbon and nitrogen cycles
SOURCE: J Y Ewusie (1973)

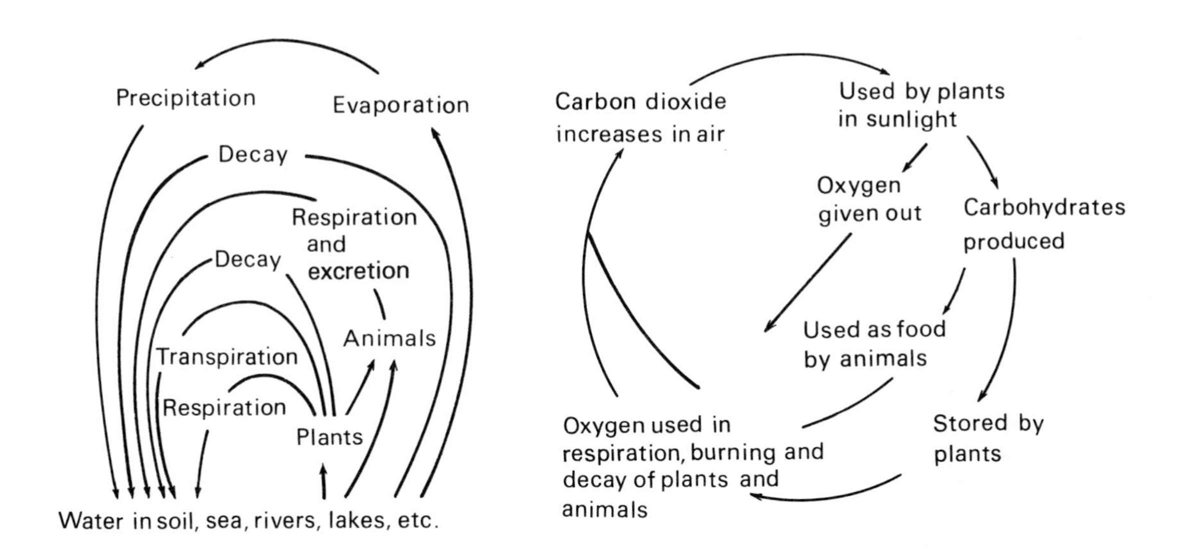

THE WATER CYCLE

THE CARBON CYCLE

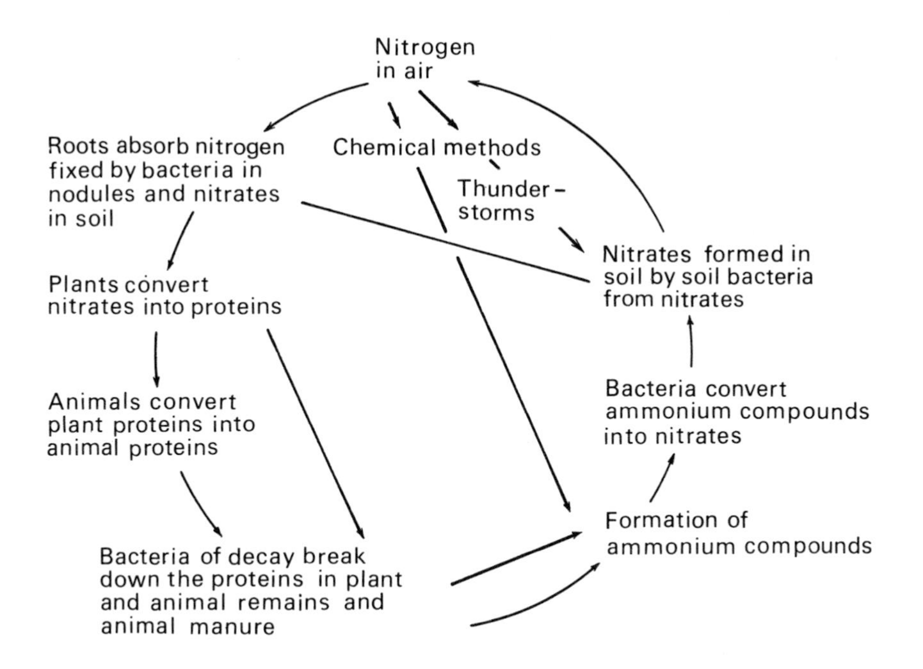

THE NITROGEN CYCLE

Part Two Tropical Ecosystem Types and Problems of Conservation

Chapter 9 The Tropical Forest Ecosystems

Since ecosystems are determined largely by vegetation, their study is best introduced through their classification. Polunin (1960) recognized five main tropical vegetation types, namely the tropical rain forest, forests with seasonal rhythm or monsoon types of forest, the savanna woodland, thorn woodland and the hot desert. These and specialized biomes within them will now be described in this and the next two chapters.

Tropical forests consist broadly of two levels: the high forests of the hot and humid regions which lack a pronounced dry season, and the forests developed in seasonally dry climates. Tropical forests form intercontinental ecological ecosystems of considerable extent and importance. They have large standing crops and exhibit rapid mineral cycling. There is much that is still unknown about the numerous biomes that are found in the tropical forests and their intensive study holds promise for the emergence of new principles and concepts in ecology.

The Tropical Rain Forest

The tropical rain forest is the most luxuriant of all types of vegetation. It is found in lowland tropical or near tropical regions of the world which receive abundant rainfall of the order of about 2000–4000 mm annually. Temperatures are high (about 25–26°C) and uniform, and the average humidity is about 80 per cent. The basic components of the forests are tall trees with an average maximum height of about 30 m. The crowns of the trees can often be recognized as comprising three layers. The trees are associated with herbs, climbers, epiphytes, stranglers, saprophytes and parasites. Flowering, fruiting and the shedding of leaves and their replacement often take place continuously throughout the year with different species involved at different times. Individual trees in the forest may, for example, be devoid of leaves at any time. Phenology in the tropics is discussed in Chapter 5. The tropical rain forest epitomizes the typical features of tropical vegetation as contrasted with temperate vegetation and, even though some of these features have been mentioned in Chapters 1 and 2, it is interesting to recount them here.

One of the prominent features is that the vast majority of the plants are woody. Only some of the epiphytes and a small proportion of the undergrowth are herbaceous. Families of plants whose members in the temperate climates are entirely herbaceous, for example, *Rubiaceae*, are, in the tropical rain forest, represented by trees.

Another feature is that the rain forests are rich in species. While in a temperate forest the common trees belong to one or a few species, it is found that in the same area of tropical rain forest there may be as many as 300 species. Such forests are called mixed forests; it is in only occasional places that forests with single strongly dominant species, referred to as single-dominant forests, occur.

In spite of the general richness in tree species, tropical rain forest trees are strikingly uniform in general appearance and form. They tend to have straight and slender trunks, with branching mostly near the top. At the base, the trunks of some of the trees have buttresses. The trees often have thin, smooth barks without conspicuous fissures or lenticels. Leaves tend to be large, leathery and dark green with entire or nearly entire margins. Typically the leaves are simple with extended 'drip tips'. It is true that some, especially the leguminous trees, have compound leaves, but even here the leaflets tend to assume the shape and size of simple leaves. Sometimes young leaves are brightly coloured, being either reddish or whitish in colour. The flowers are often small, inconspicuous, and of greenish or whitish colour. It is rare to find large and strikingly coloured flowers on tropical forest trees. The monotony of the rain forest may be broken in places where a few species with divergent forms appear. Among these are the palms and species of *Dracaena* and *Pandanus*, the monocotyledonous forms providing some structural diversity.

The trees are varied in size, but rain forest species are not the tallest on earth. The general average height of the upper stratum may be fixed at 30 m with the tallest not more than about 46–55 m. In exceptional cases heights between 60 and 90 m are attained. Thus tropical trees are usually taller than those of temperate forests where the average height is 30 m and in exceptional cases attain up to 46 m. However, they are surpassed in height by such non-tropical trees as the Californian redwoods (*Sequoia sempervirens*) and gum trees (*Eucalyptus regnans*) of Australia which can attain a height of 111 and 107 m respectively. In terms of girth, the tropical rain forests are noted for their slenderness, with girths of a metre being fairly common. It is true that larger trees attain girths of up to 17 m, but here again they are exceeded by, say, the giant trees of California (*Sequoiadendron giganteum*) and of New Zealand (*Agathis australis*), each of which attains a girth of 23 m.

The great variation in the height of the trees is reflected in the layering of their crowns into three or occasionally two layers, apart from the shrub and herb layers. This is typical of the structure of the tropical rain forests and contrasts with temperate forests. Details of the layering of the trees will be given later. Even though the undergrowth of the rain forest is made up of shrubs, herbaceous plants, saplings and seedlings of trees, the undisturbed forest itself is reasonably penetrable. Where fallen logs and branches are absent it is not difficult to walk in the mature rain forest. This is because the herbaceous ground flora is sparse, with a much lower density of ground herbs than temperate forests, and the ground is thinly covered with leaves. The canopy formed by the crowns of the trees makes the forest floor rather gloomy, with only isolated sun flecks penetrating through gaps in the canopy onto the forest-floor.

Another striking feature of the tropical rain forest is the abundance and luxuriance of climbers and epiphytes. These are found to be luxuriant in no other plant community, except some types of montane or sub-tropical rain forest. The climbers consist of a high proportion of woody varieties (lianes) whose stems can attain as great a size as that of the thigh of a man, and can be extremely long, and may either hang down in festoons on the trees or climb to the canopy. Very many species and forms of climbing plants occur. Among the epiphytes one finds high proportion of orchids and other flowering plants and a number of ferns. There are not as many species of lichens, algae, mosses and liverworts among tropical forest epiphytic vegetation as in the temperate forests. Among the tree top epiphytes are the semi-parasitic mistletoes (*Loranthaceae*) as well as the 'stranglers' most of which are species of *Ficus*. More is said about the epiphytes in the next chapter in connection with the arboreal habitat.

The preponderance of woody species and the abundance of climbers and epiphytes in tropical rain forests are best brought out by a comparison of the biological spectra of a tropical and a temperate forest (see Chapter 2).

The tropical rain forests, while representing the most elaborate of all plant communities, somehow present a first impression of sombreness and monotony by virtue of the uniformity of the trees and their foliage, and the unceasing growth and reproduction with no marked seasonal changes (see Figures 48 and 49).

Status and Significance in Ecology

The tropical rain forest is taken to represent the highest of the climax communities in the tropics according to the monoclimax theory. On this basis, and as has been discussed in Chapter 3, a number of edaphic and other factors such as fire may prevent the development of natural succession from reaching the forest climax. Among the pre-climaxes that may thus result are secondary forests, derived savannas, forest–savanna mosaics, swamp forests and mangrove swamp forests. On the other hand, the presence of other equally stable communities discussed below raises the question as to whether the polyclimax theory should apply. With the unsurpassed richness in species and complexity of the tropical rain forest, it is evident that its intensive study is likely to reveal new principles or laws governing the competition and dominance of species in mixed communities, and the relation of plant form to environment. The tropical rain forest is also seen as perhaps the oldest of vegetation types from which those of the dry tropics and the temperate regions were derived. In this connection, its closer study would also hold promise for plant evolution.

48 Ankasa Forest Reserve. Interior of well-preserved rain forest
SOURCE: Inga and Olov Hedberg (1968)

49 Mature secondary forest
SOURCE: G W Lawson (1966)

Distribution of Tropical Rain Forests

Among the regions in the world in which tropical rain forest appears to be the natural climax under present conditions are:

(a) the Amazon region of South America from which the forests range northwards into the Caribbean and the Gulf of Mexico almost to the Tropic of Cancer, southwards into Brazil south of the Tropic of Capricorn, and westwards to the Pacific Ocean coast of Columbia and Ecuador;
(b) the equatorial zone in Central and Western Africa, extending south of the Tropic of Capricorn in Eastern Africa and Madagascar;
(c) western India and Ceylon;

(d) and the Malayan region from where they range northwards to the Himalayas, northeast to Indo-China and the Philippines, and south and east through much of Indonesia and New Guinea to Fiji and adjacent archipelagos of the western Pacific, with an intermittent extension in eastern Australia well south of the Tropic of Capricorn.

In addition, there are the related subtropical rain forests which occur widely in central and southern South America and eastern China, north of the tropical rain forest in the Himalayan region and south of the tropical rain forest in East Africa, and also in Hawaii and southeastern Australasia (see Figure 1).

Stratification (Layering)

The tropical rain forest is noted for its stratification. This means that the mixed populations within it are vertically spaced out in a discontinuous fashion. In spite of some variations to be noted later, the forest typically exhibits three tree layers. These and the other layers of shrubs and herbs are described below.

1. The uppermost layer (A-storey) is made up of trees about 30–45 m high. These emergent trees rise well above the forest canopy, have wide crowns and are generally so distributed that they do not touch each other to form a continuous layer. The characteristic shapes of the crowns are often used in identifying the species within a region. The emergent trees are often shallow rooting and buttressed.
2. The second layer of trees (B-storey) below the emergents, sometimes also referred to as the upper storey, consists of trees which rise to the height of about 18–27 m. These trees are more closely placed and tend to form a continuous canopy. Their crowns are often rounded or elongated and not as broad as those of the emergents.
3. The third tree layer (C-storey), also referred to as the lower storey, is composed of trees rising to a height of about 8–14 m. The trees here often have rather variable shapes but tend to form a dense layer, particularly where the second layer does not do so.

   The three tree layers are also associated with various populations of epiphytes, climbers, and parasites, depending mainly on the light requirements of these dependent plants.
4. Apart from the tree layers, there is the shrub layer consisting of species with heights mostly below 10 m. There seem to be two forms of these shrubs: those with branching near the base and hence possessing no main axis, and those resembling small trees by having a prominent main axis, the latter often being referred to as treelets and including the saplings of the larger species of trees.
5. Finally, there is the herb layer consisting of smaller plants which are either seedlings of the larger trees of the upper layers or herbaceous species. There is less floristic diversity here than in the tree layers, and the species often belong to such families as the *Commelinaceae, Zingiberaceae, Acanthaceae, Araceae* and *Marantaceae.* Ferns and *Selaginella* are often prominent. Significantly, the ground layer is almost devoid of grasses but some broad-leaved ones like *Olyra latifolia* and *Leptaspis cochleata* are characteristically present. Likewise, there are a few broad-leaved sedges in genera like *Mapania* and *Hypolytrum.* The paucity of plants in the herb layer is due largely to the interception of the light by the upper tree and shrub layers. Luxuriant herbaceous vegetation is found only in clearings or other openings where more light reaches the ground.

The vertical stratification of the forest community seems to influence the distribution of animal populations which live in the forest. Often there are one or several bird populations which seem to be limited in their life and feeding to the canopies of the emergents; lower down are arborescent herbivorous mammals such as lemurs, squirrels and lorises which live and feed on the plants in the middle layers; finally there are the forest floor feeders, consisting of herbivores such as okapis, deer and peccaries.

It is not always easy for the untrained and inexperienced eye to recognize the various layers of trees and other storeys in the forest vegetation. In order to arrive at the presence and number of storeys or layers present in a given forest, one first has to measure the heights of all the trees and shrubs of the area under study. The data then grouped according to appropriate height classes. By plotting on a graph the number of plants with similar heights against the height above ground, the positions of maximum frequency which indicate the layers present in that forest may be found.

The stratification in the forest is usually represented diagrammatically by means of profile dia-

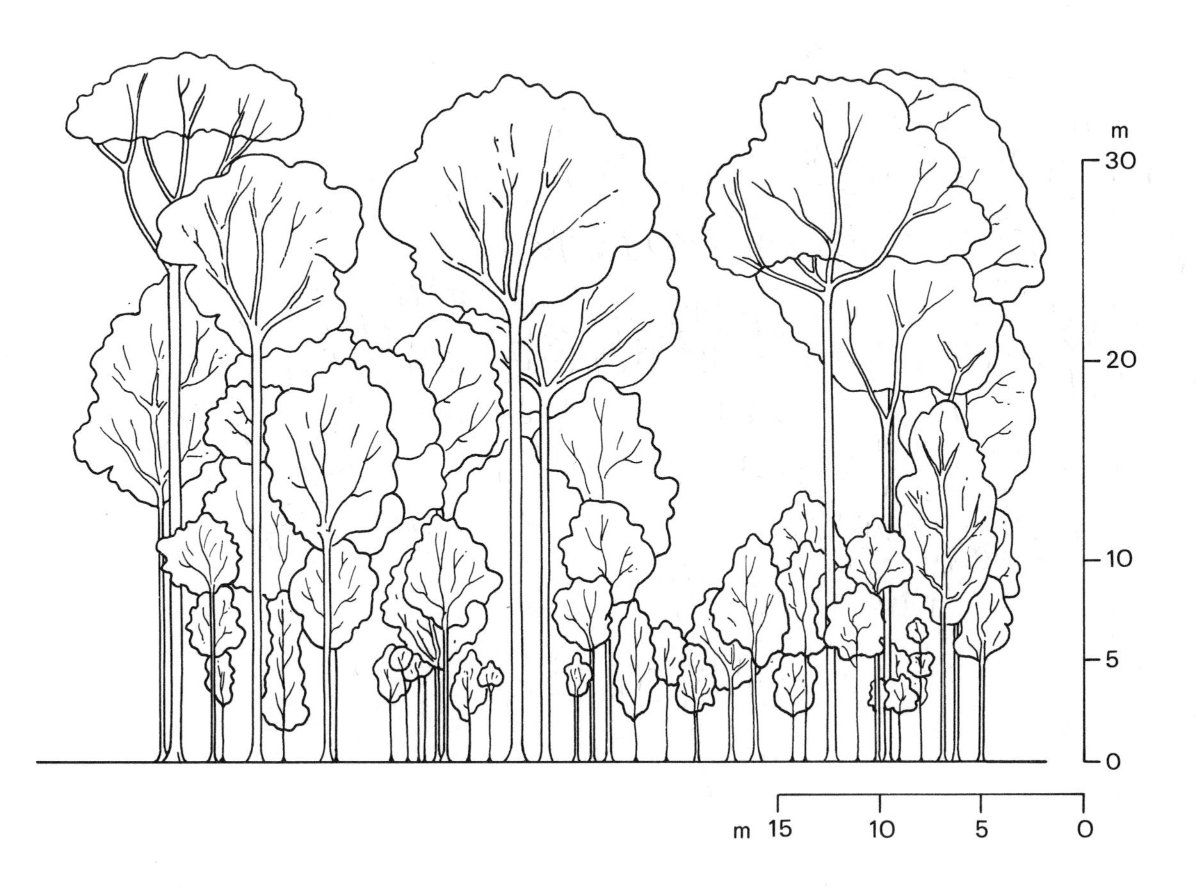

grams as illustrated from British Guiana and Nigeria in Figures 50 and 51. The profile diagram enables us to construct a scale diagram of the vegetation using accurate measurements of the position, height, height to first branch, depth and diameter of crown of all the trees on narrow sample strips of the forest. The method consists of first demarcating, by means of cords, a rectangular strip of forest. The strip is then cleared of all small undergrowth and trees which are less than a chosen lower limit of height. A map is then made of the positions of the remaining trees, whose diameters at chest height are also recorded. The data can be obtained more accurately by felling the trees in a systematic order, but the Abney level may be used to obtain fairly accurate height measurements where the trees can be clearly seen and it is not considered desirable to fell them.

50 Profile diagram of primary mixed tropical rain forest, British Guiana. The diagram shows all trees over 4.6 metres high and represents a forest strip about 45 metres long and 7.6 metres wide
SOURCE: N Polunin (1964)

Stratification in Single-Dominant Rain Forests

There are a few rain forests in which a single species of tree forms as much as 60 per cent or more of the whole stand. These communities are present in many of the principal geographical divisions of the rain forest, and of course vary in extent. As might be expected, their stratification is more regular and distinctive. The highest stratum often forms a continuous canopy, unlike in mixed forests. The formation of the canopy is also interesting in that individual crowns appear to be shaped to fit adjacent ones.

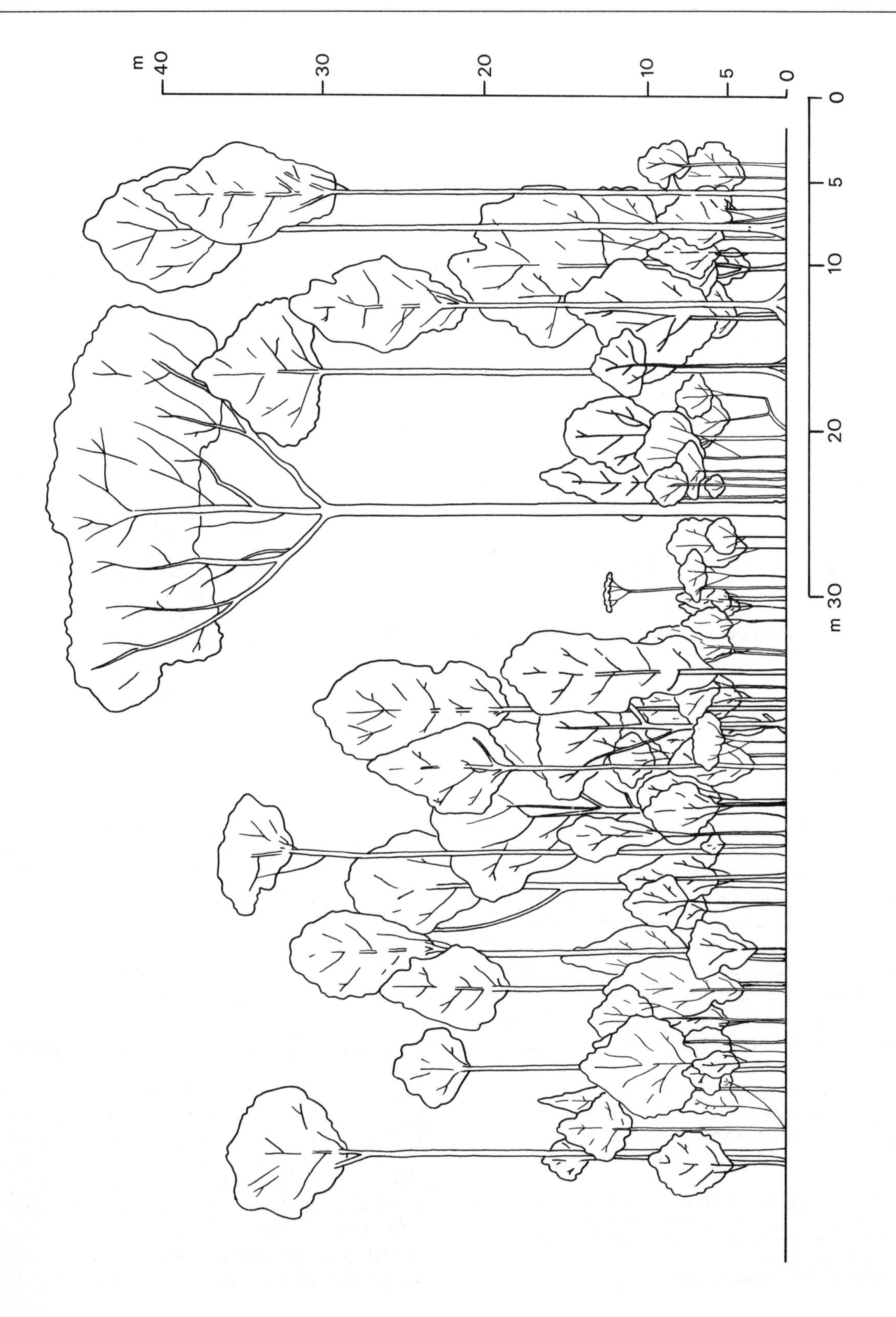
m 40
30
20
10
5
0
0
5
10
20
m 30

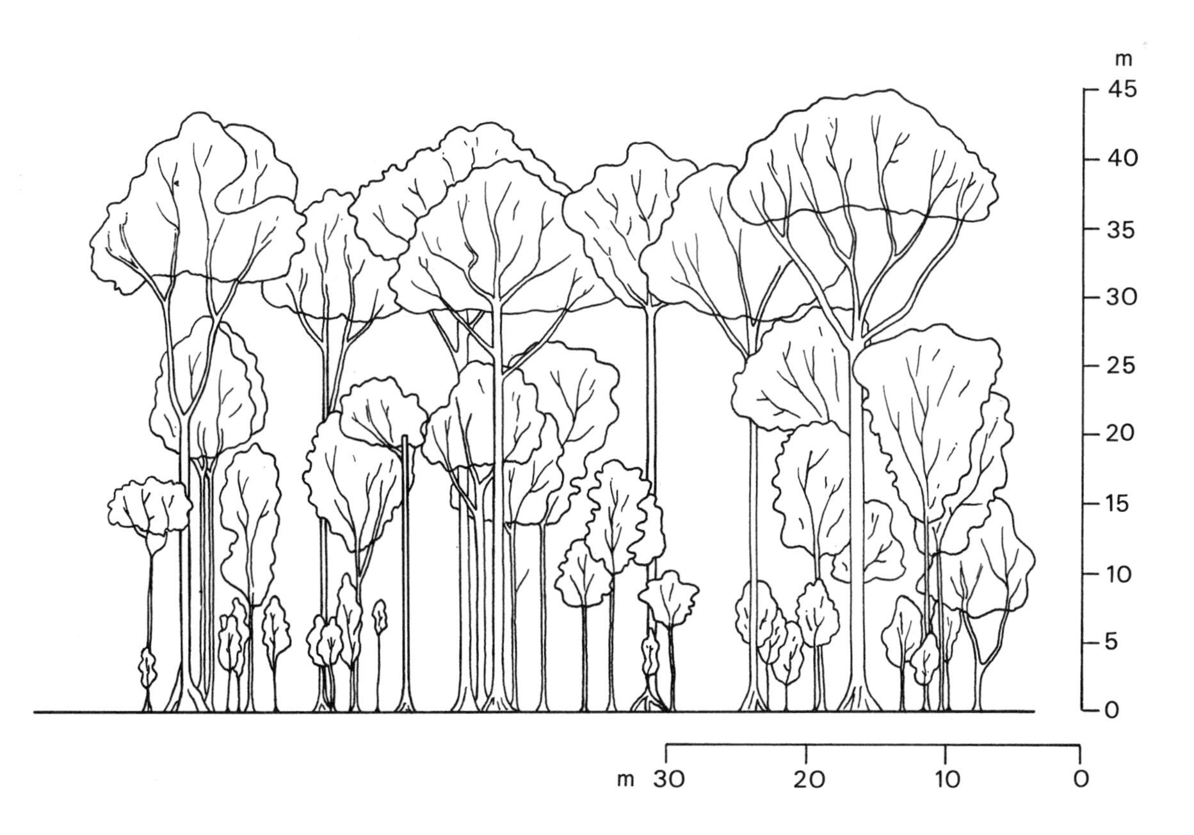

51 Profile diagram of a primary mixed forest, Shasha Forest Reserve, Nigeria. The diagram represents a strip of forest 61 metres long and 7.6 metres wide. All trees over 4.6 metres are shown
SOURCE: P W Richards (1952)

52 Profile diagram of the Mora Forest in Trinidad, British West Indies where trees grow up to about 45 metres high. The diagram represents a strip about 65 metres long and 7.6 metres wide
SOURCE: N Polunin (1964)

One example is the Mora Forest in the lowlands of Trinidad (Figure 52), in which *Mora excelsa* occupies about 85–95 per cent of the highest stratum and 62 per cent of all trees over 30 cm in diameter. Three layers are present as in the mixed forests, but the main difference here is that the A-storey is more continuous and the stratification is on the whole more noticeable. Another example is the Wallaba Forest of Guiana, dominated by *Eperua falcata* (Figure 53). In Africa there are the *Cynometra alexandri* forests of eastern Zaïre and Uganda and the *Macrolobium dewevrei* forests also of eastern Zaïre.

Localized single species dominant forests continue to be discovered. Among these is a recent discovery of a type in Ghana dominated by an endemic species, *Talbotiella gentii.*

Floristic Sub-divisions in the Rain Forests

Although most tropical rain forest communities are of mixed composition, they do not always consist of consociations with single dominants. Primary communities with a few species as dominants appear to occur in a few places. Some of these cover hundreds of square kilometres of country and may be termed floristic associations, while others are local patches in a mixed association and are regarded as societies. In some mixed forests

53 Wallaba Forest of *Eperua falcata*, Moraballi Creek, British Guiana
SOURCE: P W Richards (1952)

54 Hill Dipterocarp forest, Malay Peninsula
SOURCE: P W Richards (1952)

there is what is described as family dominance, that is, the numerical preponderance of species of the same family or of a group of related genera. The best example of this is the dominance of the *Dipterocarpaceae* in the lowland tropical forests of the Indo-Malayan region (Figure 54). There seems to be similar dominance of *Leguminosae* in some South American forests and of *Meliaceae* in some West African forests. This phenomenon of family dominance is not well understood as yet, and much research is required on it in view of its importance in elucidating the mechanism of competition in mixed communities.

The discovery of single species and family dominance in the tropical rain forests has raised the question of their status in relation to the normal mixed communities with respect to the climax theory. The climatic climax is generally taken to be the stable limit in the development of vegetation in terms of succession within the human time-scale. The two schools of thought on the development of climaxes in a given climate are the monoclimax and the polyclimax theories. According to the monoclimax theory, which was developed on the basis of temperate studies, all vegetation in a given climatic region tends to develop towards one and the same climatic climax, except where local conditions such as soil, topography or fire arrest the normal trend of this development into preclimaxes. The polyclimax theory, on the other hand, maintains that the development of vegetation does not lead to a single climatic climax but to a series of climax types, all of which are stable even though they may be adapted to different local conditions such as soil or topography. Evidently, the nature of the floristic composition of rain forest communities described above would contradict the monoclimax theory. For example, single-species communities in Guyana are all under the same climate and none of them can be considered more stable than the others. Each of them is developed on a different but equally permanent combination of soil and topography within the same climate.

The monoclimax theory seems obviously difficult to comprehend since there seems to be no way of verifying it experimentally within the lifespan of any ecologist. On the other hand, the factors that maintain the other lower biomes continue to operate in practice. It therefore seems that we might as well regard a climax vegetation as the vegetation that is best suited to the climate, soils, physiography and biotic effects that make up its habitat.

Features of the Micro-climate of the Rain Forest

Temperature

The tropical rain forest climate is marked by a high and very even temperature. The mean annual temperature in the lowlands lies between about 20 and 28°C, with the lowest in the rainy season and the highest in the dry season. In the tropics the mean temperature decreases by about 0.4–0.7°C per 100 m ascent on mountains. The small seasonal variation of temperature of the tropical zone depends partly on the small annual variation of length of day. Other important factors are the thermostatic effect of the oceans which form about three-quarters of the whole tropical zone, and of the soil which absorbs so much heat.

Studies of maximum and minimum temperatures at different heights above the ground and at different times of the year show that the undergrowth has a smaller range of temperature than the upper levels of the forest; the difference is due mainly to a lower maximum, the minimum in the undergrowth being either higher or only very slightly lower than the minimum of the upper level.

In the day the form and steepness of the temperature gradient are determined chiefly by the proportion of the sun's rays intercepted by the leaves and branches at different levels. A fairly steady decrease of temperature from the first storey downwards is observed. Temperatures in the middle of the day are some two to three degrees lower in the undergrowth than among the crowns of the first storey trees. At night the position is often reversed, although the structure of the particular forest may modify the situation. At night there is normally a decrease in temperature with increase in height from the ground level to the tree crowns. This arises because the ground and the lower levels of the forest are sheltered and at night radiation of heat takes place chiefly from the crowns of the trees. Owing to its high density the cool air surrounding the crowns sinks; the lower levels of the forest thus become cooler than the upper levels. If the space between the tree crowns and the ground were sufficiently filled up to impede the downwards sinking of the cooled air a reversed gradient would be set up; that is, the crowns would become cooler while the lower parts remained warm. The crowns would lose heat by radiation faster than the cooled air could sink to the ground. In the forests of southern Nigeria (Evans, 1939) the A (tallest) and B (next) storeys are fairly open, but the C (lower) storey is very dense and has a compact surface bound together by interlacing lianes. Thus vertical movements of air can take place with little hindrance as far down as the surface of the C-storey, but are much impeded between the top of the C-storey and the ground. This explains why the undergrowth, at 0.7 m, is always warmer at night than the B-storey, at 24 m, in this type of forest.

In conclusion it must be noted that more than one type of temperature gradient exists in the rain forest, the different types being related to differences in the forest structure. What happens locally would depend on the interaction of these factors, but, in general, the diurnal temperature range is greater among the tree crowns than it is on the forest floor.

Rainfall

The seasonal distribution of rainfall within the tropics is a function of latitude: at the equator rain falls at all seasons; in the zone ranging from 3 to 10° North or South there are two wet and two dry seasons; while still further from the equator there is a single wet and a single dry season in the year. This variation of seasonal distribution of rainfall with latitude depends chiefly on the annual passage of the overhead sun; the periods when the sun is in the zenith tend to be the periods of heavy rainfall and the intervening periods are comparatively dry, or even quite rainless.

Since the tropical rain forest is dependent on abundant rain at all times of the year and the absence of droughts, it is found principally as the climatic climax of the equatorial zone of high and evenly distributed rainfall. It may be found also in much of the zone with two wet seasons. In the zone with a single wet (but not so heavy) season it extends only where topography or other local conditions cause an increase in the rainfall. This often happens in the relatively dry areas and in some cases there are other compensating factors, as found in the case of forests along water courses (in deciduous forests and savanna) or in gallery forest where edaphic moisture compensates for the low rainfall. Such communities are regarded as post-climaxes.

In typical rain forest localities the annual rainfall is at least 2000 mm but there is a considerable range in both the total and the seasonal distribution. A few localities, such as Debundja in the Cameroons, may have totals much exceeding 4000 mm owing to special local features of topography. Differences in the length and severity of the seasonal drought result in differences in the structure and physiognomy of the vegetation. The forest of regions with a long period (4 or 5 months) of little or no rain differs little in floristic composition from the relatively non-seasonal forest (true rain forest) of regions where the rainfall is more evenly distributed. However, it is less luxuriant with a proportion of deciduous trees and tuber geophytes which have a long resting period during the dry season. This shows that the nature of the climax vegetation is related to the seasonal distribution of rainfall rather than to the total, provided that the annual rainfall exceeds a minimum of 1600 mm.

All plants intercept a proportion of the rainfall and so diminish the amount reaching the ground by a fraction, depending on the number, shape and arrangement of the leaves and branches. Since most plants absorb water only through their roots, the fraction which does not reach the ground is lost to them and has no significance except in affecting the rate of transpiration. The proportion of rain reaching the floor depends on the structure of the forest as well as on the nature of the rainfall, that is, whether it falls as thunder showers or as long continued fine rain ('land rain'). Of course, some of it runs off the surface of the ground.

Dew forms only on the crowns of the taller trees and never inside the forest. It is unlikely that the dew is ever sufficient in quantity to run down the trunks or drip off the leaves onto the soil, but it is probably important to sun epiphytes, and may be of some significance in lowering the rate of transpiration from the leaves of the trees during the early part of the day.

Atmospheric Humidity

The three chief measures of water in the air are: relative humidity, which is expressed as a percentage from readings of the wet and dry bulb thermometer or the hair hygrometer; saturation deficit, calculated from relative humidity and temperature; and rate of evaporation which is measured by means of an evaporimeter.

For the plant ecologist the most useful index of atmospheric humidity is the **saturation deficit** of the air, which is a measure of its evaporating power and therefore most closely related to transpiration. Figures for saturation deficit, however, are often not available. Since the saturation deficit is a function of the temperature as well as of the relative humidity of the air, the same relative humidity will correspond to very different saturation deficits at different temperatures. In the rain forest, however, variation in temperature is sufficiently small for it usually to be permissible to regard the relative humidity as an approximate measure of evaporating power.

In open situations in the rain forest region the relative humidity tends to be high during the greater part of the day, even in the dry season. It varies considerably during the hours of daylight and shows a seasonal variation parallel to that of rainfall. In general, even in the open, relative humidity throughout the night is always at or near saturation; during daylight on dry days it falls to the order of 65 per cent, momentarily it may touch 55 per cent or even lower.

Atmospheric humidity decreases away from the coast and this is one of the outstanding features of climatic gradient in West Africa. It has been shown that the distribution of rain forest and mixed deciduous forest in southern Nigeria seems to be determined by the distance inland reached by moist air from the sea. In the afternoon of the dry season, the sea wind may cause an increase of 60 per cent or more in the relative humidity as well as a large drop in temperature.

On the mountains in the tropics humidity rises with increasing elevation. At a certain height perpetual cloud and drizzle prevent the relative humidity from ever falling far below saturation. However, on the highest tropical mountains drier conditions are found above the zone of maximum humidity.

During the day there are numerous fluctuations in humidity so that within a few minutes the relative humidity may rise or fall by as much as 10 per cent above or below the average for an hourly period. These fluctuations are probably caused chiefly by wind and cloudiness. One of the striking features is the long period at night during which the humidity is almost at saturation point. Although this occurs both in the tree tops and in the undergrowth, it lasts longer in the undergrowth because in the tree tops the humidity begins to fall soon after sunrise, but in the undergrowth the fall begins later and is less rapid. Similarly, in the evening the maximum humidity is established much earlier in the undergrowth than in the tree tops. The duration of the maximum humidity is about the same in the dry and wet seasons in the tree tops, while it increases from about 14 hours to 18 hours in the undergrowth between the dry and wet seasons respectively. In both wet and dry seasons there is a greater daily range of humidity in the tree tops.

By plotting temperature against relative humidity for hourly intervals during the day it is possible to follow the daily march of temperature, relative humidity and saturation deficit which give much information as to changes in humidity due to transpiration and other changes like the mixing of different layers of air (see Figure 16).

Wind

Much has been said already about the effect of wind on vegetation. Wind velocities are normally lower and violent winds less frequent in tropical than in temperate regions, except locally as in the hurricane belt of the West Indies. Mean annual wind velocities in rain forest localities are commonly less than 5 km/hr and seldom exceed 12 km/hr.

In certain parts of the tropics thunderstorms are very common and are often preceded by squalls of strong wind (tornadoes) lasting for a few minutes only. In these the wind may momentarily reach very high velocities. Groups of trees, and even areas of forest several hectares in extent, felled by such squalls are common in tropical countries. On exposed sea coasts strong winds may deform the crowns of the trees and bring about the development of a strip of characteristic woodland or scrub.

One of the most noticeable features of the lower layers of the forest is the extreme stillness of the air; this contributes much to the general feeling of oppressiveness. In a typical rain forest a gentle wind blows almost continuously in the tree tops, at least during the day, so that flexible leaves are always moving; in the undergrowth the air is normally so still that smoke can rise vertically.

Light

Although the daily amount of sunshine is never less than 10 hours in any part of the tropical zone, the actual amount of bright sunshine is always much less than this, owing to the high degree of cloudiness. The neighbourhood of the equator is in fact a more or less permanent cloud belt where cloudless days are rare.

Hours of bright sunshine, as recorded by the sunshine recorder, are therefore of much ecological interest. Typically the following figures represent the hours of bright sunshine in a tropical forest belt.

Average daily sunshine:	5.5 hours
Average for sunniest month (September, dry season):	6.3 hours
Average for least sunny month (June, wet season).	4.4 hours

This shows that the amount of bright sunshine is surprisingly small. Clouds are continually passing in front of the sun even in the dry season. It is important to note that, in spite of the high altitude of the tropical sun, radiation in the tropical lowlands (at ground level) is less (rather than more) intense than in temperate regions.

An accurate knowledge of the horizontal and vertical variations of the illumination is fundamental to the understanding of the ecology of the rain forest as a whole and of the distribution of its individual components.

The measurement of light intensity in the forest poses certain problems. Light which is not reflected directly from the upper surface of the forest may either be transmitted through the leaves or it may pass between them forming sun flecks which vary continually in position, size and intensity throughout the day. In the interior of the forest light is reflected by leaves and branches, which are surfaces with a great variety of optical properties. Measurements in the undergrowth in a primary forest in southern Nigeria showed that the intensity for most of the day was about 0.5–1 per cent of the external intensity (daylight). It has been found that both when the sun is clouded and when it is unclouded the shade illumination in the undergrowth is always below 1 per cent of the outside daylight; the sun fleck intensity may be at least 100 times greater than the intensity in the shade. There is a steep gradient of intensity from the full daylight of the highest tree storey to the low intensities in the undergrowth. There is very little information as to the form of this gradient, but general observations suggest that in some types of forest there may be a fairly gradual diminution in the average light intensity from above downwards, while in other types there may be a level at which there is a relatively abrupt fall in intensity. The latter is found in forests in which the A and B tree storeys are open but where the C-storey is opaque.

Carbon Dioxide

Carbon dioxide is important ecologically because together with light it serves as a limiting factor for photosynthesis and therefore for the growth of plants. Owing to the lack of air movement in the lower storeys, carbon dioxide here has been found to be above the normal 0.03 per cent of the air and is in fact 0.04 to 0.06 per cent. In the upper storeys, with high light intensities, the rate of photosynthesis increases or decreases almost in proportion to the carbon dioxide concentration. Deviations from the normal concentrations of carbon dioxide may have considerable ecological importance. Even at the low light intensities found in the shade, a high carbon dioxide concentration would probably allow a rate of photosynthesis much higher than that at normal concentration.

The absence of a thick litter of dead leaves and the small percentage of humus in the soil show that decomposition and carbon dioxide production by micro-organisms are very rapid.

Man in the Tropical Rain Forest

The general effects of the activities of man on vegetation have already been discussed in Chapter 4. It has taken hundreds of years for the tropical rain forests to reach their present stable conditions; but with growing human populations, the destruction of these forests for farming, timber exploitation and urbanization has rapidly reduced their extent. This means that nutrients which were extracted from the soil and locked up in the trees are not returned to the soil. The fertility of the soils is decreasing. The destruction of forests is also detrimental to crops which require humid conditions. Along with the problem of increasing food production, man has had to use various insecticides and fertilizers, a number of which leave toxic residues in the natural vegetation, the crops and the soil. These in the end affect the health of man and animals through the food chain.

As forests are destroyed, the suitability of the habitat for their wild animals is also seriously jeopardized, with the result that a number of forest animals have become scarce if not extinct.

Finally mention should be made of the effect of the destruction of forests on rainfall and river resources. These matters are more fully discussed as part of conservation problems in Chapter 13. The agricultural potentialities of the tropical rain forest as well as of the forest with seasonal rhythm will be discussed in the next chapter.

Deciduous Seasonal Forest Ecosystems

In the humid tropics there are a large number of forests with one or two pronounced dry seasons. (A dry season is normally defined as a period of at least one month with less than 100 mm rainfall.) These forests are, on the whole, more extensive than the tropical rain forest and are more common in the interior of continents. Here the forest vegetation is not as luxuriant as the rain forest, being a mixture of deciduous and evergreen plants. During the dry season much of the foliage of the trees is shed, even though some trees shed their leaves at different times of the year. The degree of defoliation in the dry season depends on the severity and length of the dry season. Of course, the rainy season is the most luxuriant period of the year in terms of foliage. However, along water courses the trees are able to keep their leaves throughout the year.

A number of trees flower during the dry season, even though some species flower at different times of the year. The increased flowering in the dry season means that the vegetation is not lifeless in this season. This contrasts with the deciduous temperate forests which in the winter appear quite lifeless. Unlike the tropical rain forests, the trees of the tropical deciduous forests with a seasonal rhythm have thick barks and typically lack buttressed roots.

As far as their status in ecology is concerned, tropical forests with a seasonal rhythm are evidently the pre-climax of tropical rain forests. They are caused largely by the seasonality of the rainfall (usually monsoon rains) and the relatively lower total amount of the fall. Many of these forests come under the category of monsoon forests.

Tropical deciduous forests are, in general, found on the margins of the tropical rain forests in Africa, Madagascar, Indonesia and Central and South America, but they include also those described as monsoon forests, found in India, Burma, Indochina and northern Australia.

Habitat Factors

These forests have abundant rainfall during the wet season, which alternates with a distinct drought period which may last from about four to six months. The total amount of rainfall is generally lower than in the tropical rain forests and varies from about 1000 to 2000 mm a year.

The levels of temperature do not vary much on average from those of the rain forest areas, but there is a more marked seasonal as well as daily variation in temperature. The wet season clearly has lower temperatures than the dry season.

Unlike the still atmosphere of the tropical rain forests, the seasonal forests may be subjected to strong winds.

Structure and Phenology

The forests are much lower in height than the tropical rain forests. The A and sometimes also the B storeys are more open and may disappear, so that only one tree stratum may remain. Thus in vertical structure they do not show the numerous layers of the rain forests. They seem to consist of three layers altogether, the upper layer being the tree canopy which is often disturbed, the second layer (undergrowth) which is often dense and a ground layer of herbs.

The trunks of the trees are often rather massive but relatively short, with widely spreading crowns and rather stout and gnarled branches. The bark is is thick and often fissured, in contrast to the thin and smooth barks of trees of the tropical rain forests. The leaves of the deciduous trees are usually hygrophilous, that is, with features typical of plants of humid environments, such as being thin and large. Since they are shed during the dry season, it does not seem as if special adaptations to such an adverse period are required. The few evergreen trees, however, tend to have smaller and thicker leaves. There is a high variety of tree species in these forests.

There are fewer climbers and epiphytes than in the rain forest, and they are often smaller and more herbaceous. The undergrowth is clearly more luxuriant than in the rain forest, with shrubby thickets and more grasses. Geophytes are more common in the ground layer.

Flowering of the trees and shrubs takes place in the dry season with the initiation of the flowers at the beginning or the end of it when rains are only sporadic. The herbs flower during the rainy season and the geophytes often blossom in the dry season or when the first rains come.

Subtropical Forests

As mentioned earlier, there are regions in the world where the evergreen rain forest reaches and passes the geographical tropic and thus extends northwards or southwards far into the subtropical regions. Among the areas concerned are eastern Australia, southeast Asia including southern China and northern Burma, and the east coast of South America. In addition, there are isolated areas within the subtropics where in favourable conditions there are small outliers of broad-leaved evergreen forests, such as the hammock forests of Florida and the forests on the banks of the Rio de la Plata.

Subtropical forests are generally lower in stature than are the tropical rain forests, and they have fewer woody vines and epiphytes. Moving away from the tropics, but within the subtropical forest, the number of species (especially of trees) diminishes, and there is a gradual disappearance of the characteristic tropical physiognomic features such as buttressing and cauliflory. With this goes the tendency to single species dominance typical of non-tropical vegetation.

Agricultural Potential of the Tropical Forests

It is worth surveying the agricultural potential of tropical forests since this has ecological implications. The humid tropics support a number of crops of varying economic importance. Among those of greater economic importance are rubber, sugar, palm oil, copra, bananas, pineapples, cocoa, tea, coffee, rice, yams (*Dioscorea* spp.) and cassava (*Manihot*). Maize is an important crop in the seasonal forest regimes but the rain forest is too humid for grain ripening. Although rice is adapted to grow in water-logged situations, certain varieties will grow equally well in a wide variety of climates. Climates with about one to three months of dry season support a mixed production of the root crops like cassava and yams, even though these are essentially humid zone crops, and cereals. More arid inland regions support only cereals, among which are *Sorghum* (guinea corn) and millet.

Cocoa, although adapted to the humid tropics, does best in regions with a short dry season, since this reduces excessive leafiness and its attendant leaf diseases. Oil palm, rubber and coconuts are essentially humid zone crops, and where the dry season is longer than three months they do not do so well. It is considered that a rainfall excess over evaporation of less than 1000 mm is too dry for rubber in West Africa, unless the soil is of particularly good moisture holding capacity (Papadakis, 1965). It is also noted that mean annual temperatures below 25°C limit the production of rubber. Bananas, pineapples and sugar can thrive fairly well in relatively high latitudes of subtropical zones. Coffee is best grown in seasonal forest habitats. Tea requires cooler hill temperatures and moderately uniform rainfall.

In general it seems that at its climatic limit the tropical rain forest gives place to the seasonal, or deciduous, forest except where soil conditions are unfavourable to the growth of trees or the natural vegetation is subject to frequent fires. Further increase in dry conditions will result in the seasonal forest giving way to savanna woodland, and this in turn to thorn woodland and then to desert vegetation. The change from rain forest to seasonal or deciduous forest is quite gradual. It involves a change in floristic composition, entailing a decrease in the number of species, and a change from evergreen to a preponderance of deciduous tree species. The number of tree strata may also be reduced to only one.

Chapter 10 Specialized Ecosystems within the Tropical Forests and along the Sea Coast

It is proposed here to consider some of the specialized ecosystems within the tropical forests, such as the montane vegetation, the swamp forests and the arboreal habitat, as well as coastal ecosystems like mangrove swamps, lagoons and strand vegetation.

Montane Forest (Mountain Vegetation)

The vegetation on mountains is considerably affected by changes in the climate at different altitudes. The temperature falls steadily with increasing elevation until, on high mountains, even on the equator such as Mount Kilimanjaro in East Africa, there is perpetual snow (Figure 55). In general, rainfall is heavier on the lower slopes of mountains than on the surrounding lowlands. The reason for this is that as the warm air from the lowlands is forced up the mountainside it cools. This results in a reduction in the waterholding capacity of the air so that excess water in it forms clouds which lead to rainfall. There is an increase in rainfall on the hillside up to a certain altitude, but above that the condensation of water vapour from the air is insufficient to form much rain.

As a result of the rainfall distribution, more luxuriant vegetation often occurs at low and medium altitudes than on the neighbouring lowlands. This difference is more marked on mountains in the drier vegetation zones. Above the luxuriant zone, the vegetation becomes more sparse as the elevation increases. The zonation which results is as striking in the tropics as it is in temperate climates. In the tropics, as one proceeds upwards from the luxuriant forests at low and medium altitudes, the height of the trees decreases and the three storeyed rain forests give way to a two storeyed forest (Figure 56). This type of forest is not usually as rich in tree species as the lowland rain forest, but it is very luxuriant in epiphytes, especially mosses and liverworts, which cover the trunks and branches of the trees densely. Very few woody vines are present. This has been referred to as the submontane forest.

Above this is the true montane forest which is made up of a single storey of twisted trees of massive growth and rich branching. Here the leaves are smaller than those of the rain forest. Like the two storeyed forest, the trunks and branches of the trees are covered with a thick mat of liverworts and mosses which may hang down in festoons. This feature is called the elfin woodland

55 Mount Nimba in Guinea showing a cap of snow
SOURCE: Inga and Olov Hedberg (1968)

56 Two-storied forest on Mount Maquiling, Philippine Islands, at an elevation of 740 metres
SOURCE: W H Brown (1935)

57 Mossy elfin forest near summit of Mount Maquiling, Philippine Islands
SOURCE: W H Brown (1935)

or the mossy forest (Figure 57). A number of ferns and flowering plants grow among the liverworts and mosses. More light penetrates and the ground vegetation is heavier. Flowers of trees, shrubs and epiphytes are often numerous and conspicuous. These features make the elfin forest one of the most striking types of tropical vegetation. Stranglers are usually absent. Epiphytic mosses are numerous where mist prevails and small climbers may be found near the upper limits of the forest.

Beyond the elfin woodland, there is often a zone of dwarf shrubs, followed by a region of alpine meadow or alpine savanna grassland with short and matted plants. More and more temperate species appear as one ascends, but the total tree flora decreases even though this may not lead to single species dominance. This vegetation gives way to a region of perpetual snow at the top of the mountain.

The above account gives the general picture, but there are numerous modifications since the rate of

change of climate with height varies from one place to another depending on the topography and other factors. Sometimes, as in the high mountains of New Guinea, other zones are interpolated within the general sequence given above. Also in tropical and subtropical regions of dry climate, forest may be present only in the montane zone. It may even not occur at all, as on the western slopes of the Andes of South America. Here scrub, steppes or arid punas with large cushion plants are predominant.

A few actual examples may now be described. In Africa, much of the mountain vegetation grows on young volcanic soils and because of modification by fire, grazing and shifting cultivation it has not been easy to recognize climax types. Mount Kilimanjaro in East Africa, the highest mountain in Africa, rises to nearly 6000 m, and lies only three degrees south of the equator. At its base is the hot thorn bush savanna which is at an elevation of about 1200 m. Above this is a zone which is cultivated with coffee and bananas which extends to about 2700 m. Above this forest is a moorland occupied by giant species of heaths, groundsels and other montane species. Some typical temperate plants are also present. So also are some of the animals such as species of *Collembola* and mites. The moorland extends to about 4200 m and is followed by an alpine desert which extends to 4800 m and has little vegetation. Finally, above the height of the moorland is the permanent ice cap at the summit. As Kilimanjaro is volcanic, the zonation is subject to modification by eruptions and unfavourable soil conditions.

On Mount Cameroon, the highest mountain in West Africa (4070 m) which lies within the tropical rain forest zone, a similar pattern of zonation is found. Between sea level and a height of about 600 m the structure of the forest is typical of the three tree storeyed rain forest; but between about 600 m and 1050 m the vegetation alters its character with a reduction in the number of tree storeys from three to two. Also the general height of the trees here is reduced, and a distinct change of species occurs. Typical temperate species like *Viola abyssinica* and *Sanicula europaea* occur. Between about 1050 m and 2600 m the number of tree storeys may be further reduced to only one, except in places, and a dense layer of shrubs is developed. This type of forest is overhung by heavy clouds and the trees tend to be constantly dripping with moisture, a condition which facilitates profuse growth of mosses and lichens on the bark of the forest trees. This is a submontane forest, also known as mist forest. Here again certain species become predominant in the zone. Among these are the tree fern *Cyathea manniana* as well as other flowering plants like *Hypericum* and *Schefflera*. There are also the giant lobelias and senecios, smaller than those found on the high mountains of East Africa, *Galium, Ranunculus* and *Veronica*, typical north temperate genera. Above 2600 m the forest gives way to an open grassland which becomes very dry in the dry season when it is burnt annually by hunters. The highest part of the mountain is an alpine desert of lava, sparsely colonized by grasses, ground mosses and lichens.

In the high mountains of eastern New Guinea, an interesting zonation is present.

1. Up to 300 m: Lowland forest. The vegetation consists of typical tropical rain forest, mixed, very tall and luxuriant.
2. From 300 m to 1650 m: Foothills forest. This consists of relatively shorter and smaller trees. Species of *Ficus, Alstonia* and a few others show plank buttresses, but they may have come from the lowland forest vegetation *Quercus junghuhnii*, an oak tree, and other trees like *Cedrela toona, Elaeocarpus, Eugenia* and species of *Albizia* are common. There are few lianes and epiphytes. In the ground layer there are ferns and species of *Elatostema.*
3. From 1650 m to 2250 m: Mid-mountain forest. The forest here consists of a mixture of oaks and conifers. Trees with buttress roots are rare and lianes are much fewer. The ground layer consists of filmy ferns, mosses and *Elatostema.*
4. From 2250 m to 3000 m: Mossy forest. Here montane rain forest typical of the mountains of the Malaysian region has developed. There is only a single layer of short trees; these trees have rather crooked shapes with their trunks and branches covered by thick pads of mosses and leafy liverworts. The mossy covering can be so thick as to virtually double the diameter of the trunks or branches. The zone is characterized by permanent dampness and mist, with an average low temperature of about 10°C. In places the conifer *Podocarpus thevetiifolia* forms almost pure stands. Lianes and rattans are virtually absent, but tree ferns and bamboos are present.
5. From 3000 m to 3300 m: High mountain forest. This type of forest has taller trees than those of the mossy forest, but most of them are conifers such as *Dacrydium, Phyllocladus* and

Podocarpus. Only a few of the trees are dicotyledons. The high mountain forest exists in patches separated by grassland interspersed with shrubs and tree ferns. The zone is drier and in the absence of mist, receives more sunshine than the mossy forest.

6. Above 3300 m: Mountain savanna. This consists of grassland and bogs, and on rocky areas dwarf scrub.

An example from the West Indies is the zonation found on the mountains of Trinidad. Here the zones are the lower montane rain forest (from 240 m to 750 m), the montane rain forest (from 750 m to 870 m), and the elfin woodland (on the summit of Mt. Aripo, above 870 m). The lower montane rain forest differs from the typical lowland rain forest in the reduced height of the trees (20 m to 30 m). Lianes and epiphytes are not abundant and palms and tree ferns are not common. Relatively fewer species of trees are present as compared with those of the tropical rain forest. The montane forest is here described as two-storied with an average height that is lower than that of the lower montane forest. The upper tree storey forms a closed canopy at 15 to 19 m. Lianes and epiphytes are abundant and luxuriant. Mosses form the majority of the epiphytes. There is a layer of small palms and tree ferns underneath the lower tree layer. The characteristic dominant species of the zone are *Richeria grandis* (=*R. olivieri*) and *Eschweilera trinitensis*. Only about 28 tree species are found in this forest. The elfin woodland has *Clusia intertexta* as the dominant tree which forms a discontinuous upper layer 6.0 to 7.5 m high. Below this is a dense layer of tree ferns and small palms. Very luxuriant epiphytic vegetation exists here and consists almost exclusively of bryophytes and lichens which blanket the trunks and branches of the trees. There are only eleven woody species in this type of woodland.

Swamp Forests

In general, swamps are found in sheltered arms of lakes or gentle rivers, at the edges of standing water and in hollows. The vegetation in such a reed swamp is similar to that in temperate swamps and consists normally of erect monocotyledonous plants such as species of reed (*Phragmites*), papyrus (*Cyperus papyrus*), cattai (*Typha*) or vossia. The roots of these plants are submerged while the shoots stand well above the water. The roots and the shoots that are covered with water have air spaces to aid their aeration. In Panama *Phragmites australis* and *Typha domingensis* dominate the reed swamp.

In the tropical rain forest and other wet forest zones shrubs and trees may in later succession invade the reed swamp to form the swamp forests. These forests are poorer in species than forests on drier land, but palms, like *Raphia* in Africa and *Raystonea* in the American tropics, are characteristically present. Large trees are less common and the species differ from those of the drier land. The trees show layering; lianes as well as epiphytic orchids and ferns may be as common as in the tropical rain forest. The ground flora consists largely of members of the *Cyperaceae* and only a limited number of other plants is present.

In the tropics, unlike the temperate region, not much humus accumulates and so the majority of tropical swamp forests do not form peat, particularly where the water is fairly rich in mineral matter (eutrophic water). Silty soil is therefore formed here. In situations where the water is poor in dissolved mineral matter (oligotrophic water) and more plant remains accumulate some peat forms, resulting in the development of moor forests, regarded as equivalent to the temperate raised bogs or high moors. This means that in the tropics there are two types of hydrosere leading to two types of edaphic climax. This arises from the fact that in eutrophic waters the rise in the level of the soil with succession results from the accumulation of inorganic sediments while in oligotrophic water this is due mainly to the accumulation of plant remains. Because of the rapid breakdown of organic matter in the tropics, the soil level in both situations fails to rise beyond the highest water level, since substantial accumulation of organic matter becomes difficult at this stage. The resulting edaphic climax may form on either a silty or a peaty soil with the water table near the surface, depending on whether the water was eutrophic or oligotrophic.

An example of a peaty moor forest is that found in the rain forest region of southeastern Asia. These forests are evergreen and are dominated by dicotyledonous trees which reach about 30 m high. Some of the species peculiar to the vegetation may be gregarious and dominate. The number of species is naturally restricted. Access may be impeded due to the presence of knee-roots and other pneumatophores. Palms, screwpines (*Pandanus*) and *Podocarpus* also contribute to the impenetrability. Epiphytes are abundant. Non-peaty swamp forests are often more open with lower species density than the rain forest. Thus light-loving species become numerous in the ground layer.

Peaty moor forests are often found behind coastal mangrove swamps where the ground water is neither brackish nor salty. Peat formation was generally thought to be impossible in the high temperatures of tropical lowlands but peats have now been found in Sumatra, parts of the Malay archipelago and peninsula, tropical America and tropical Africa (in Burundi). Masses of peat are found in increasing sizes (up to a few kilometres in diameter) from the edges to the centre. The peat is reddish or greyish brown in colour, it is porridge-like and its stability is provided by the roots of the trees. Its structure shows that it has had forest vegetation throughout its development, as evidenced by the presence of tree remains throughout the horizons of the profile. The soil under the peat is often in the nature of bleached clay.

Mangrove Swamp Forests

Mangroves are the plant communities which are found covering parts of tidal lands in the tropics. The species populations making up the mangrove communities consist of evergreen trees and shrubs which are taxonomically unrelated. Nevertheless, they have a number of similarities with respect to physiognomy, physiological characteristics and structural adaptations to the habitat. They show in a marked way the phenomenon of convergent evolution by which a strong resemblance is built up between unrelated species living in a similar habitat. Mangrove vegetation is characterized by the monotonous dark green shiny foliage, the tangle of aerial roots or pneumatophores, and a more or less marked tendency to vivipary. It plays an active part in building up land from the sea. It obstructs currents and binds the soil with its roots, thus adding humus and raising the ground level seawards.

It is often found that adjoining true mangroves are communities of species with similar but less strongly marked features. The term 'semi-mangrove' has been used to describe such adjoining communities. Examples are communities of the palm *Nypa fruticans* in the mangroves of the eastern tropics, *Manicaria saccifera* in the mangroves of northeastern South America, and *Pandanus candelabrum* in the mangroves of West Africa.

Distribution

Typically mangroves are found on sheltered muddy shores where the land is rapidly encroaching on the sea in estuarine situations. They may be present also on coral reefs and on some sandy shores. They are able also to penetrate far inland along estuaries. Although mangroves are found more or less within the tropical belt, the most luxuriant of them occur in the wet tropics where the tropical rain forest is the climax vegetation. Of these the best development is found on the coasts of the Malay peninsula and the neighbouring islands.

There are, broadly speaking, two groups of mangroves: the western and eastern groups. The western group consists of the mangroves on the coasts of America, the West Indies and West Africa. The eastern group is found on the coasts of the Indian and western Pacific Oceans, and may include the Australian mangrove which has some peculiar features. The two groups differ in richness of species, the eastern group being the richer one; even though the same genera are present in both groups, the species are different in the two groups. It is only in the Fiji and Tonga islands that one finds a western species such as *Rhizophora mangle* associated with an eastern species such as *R. mucronata.*

Structure

The mangrove forest varies from a poor scrub of about 2 m high to a forest 30 m high or more. In the western group, as typified by the mangroves of West Africa, the trees are rarely more than 9 to 12 m when virgin, and 3 to 4.5 m when they have once been cut. The principal trees and shrubs are species of *Rhizophora, Avicennia* and *Conocarpus.* Another much smaller shrub is *Laguncularia.* They all have thick green leathery leaves of simple shape. The fern *Acrostichum aureum* is often present in extensive patches on the landward margin of the mangrove, as found in Puerto Rico and parts of West Africa. In relatively dry climates, the *Acrostichum* patches may be succeeded by a moist saline meadow community of *Paspalum, Kyllinga* and *Sesuvium.* In the luxuriant eastern group of mangroves, exemplified by the mangroves of the Malay peninsula, more

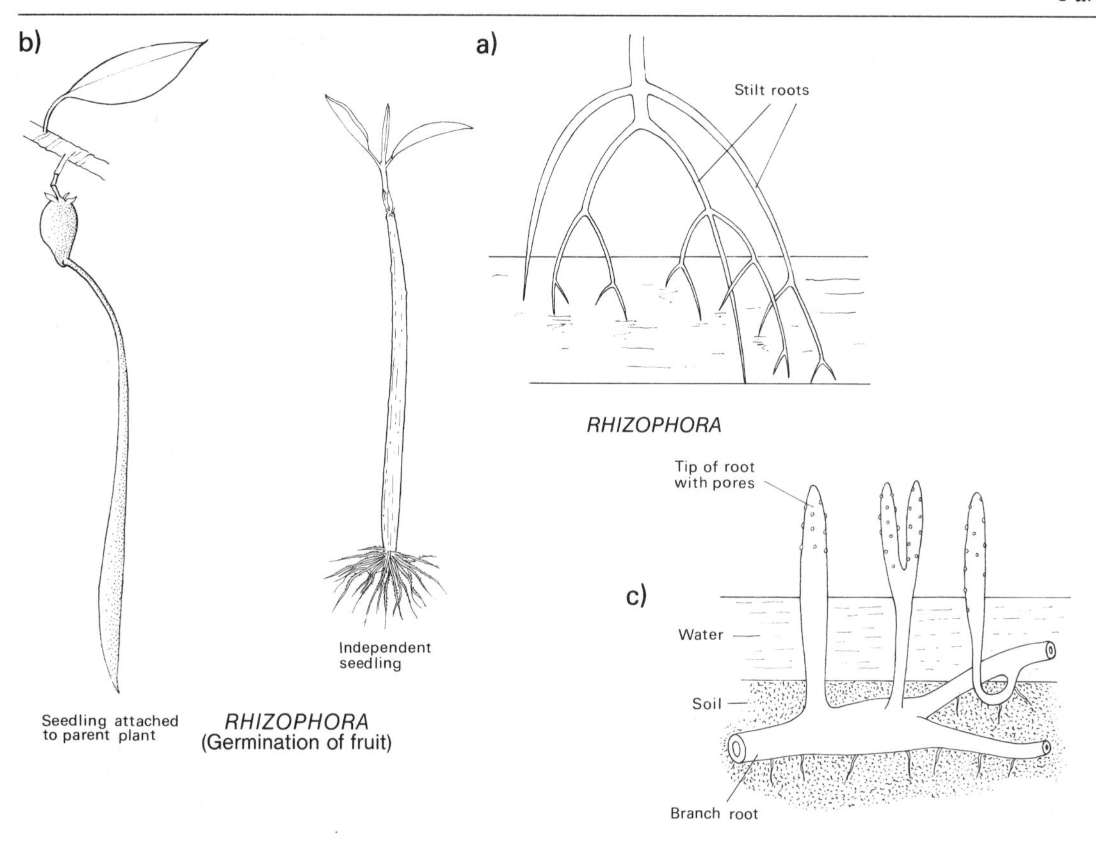
b)
a)
Stilt roots
RHIZOPHORA
Seedling attached to parent plant
Independent seedling
RHIZOPHORA (Germination of fruit)
Tip of root with pores
c)
Water
Soil
Branch root
AVICENNIA

species are found. The species of *Avicennia* are *A. alba* and *A. intermedia* and the species of *Rhizophora* are *R. conjugata* and *R. mucronata*. Among the additional species in this group are *Sonneratia griffithii, Bruguiera caryophylloides, B. gymnorrhiza* and *Xylocarpus moluccensis*. All extensive mangrove swamps support a rich and varied animal life of which many of the species are adapted to amphibious and varying salinity habitats. Vertebrates include the manatee, crocodiles and many kinds of fish, and invertebrates include crabs, shrimps, oysters and other molluscs, providing food for numerous birds.

Rhizophora, or red mangrove, is a tree whose lower part bears stilt roots which support it. At the end of each of the roots are numerous lateral roots which provide firm anchorage in the mud. The plant is also **viviparous**. A long radicle is produced by the fruit which germinates before the seedling falls into the water. The radicle is buoyant enough to keep the fruit afloat. The seedling then drifts to the bank where it becomes rooted in the mud (Figure 58; see also Figure 41).

Avicennia, or white mangrove, and *Sonneratia* are shrubs that are noted for their breathing roots which protrude from the soil into the air over a wide area surrounding the plant. Although not as strongly viviparous as *Rhizophora*, the embryo of their seeds develop to such an advanced condition that the seeds germinate rapidly when they drop from the trees (Figure 59; see also Figure 42)

Zonation and Successional Relationships

Mangrove swamps exhibit zonation of the dominant species more or less parallel to the sea shore. The zonation appears to recapitulate stages in succession which seems to take place along with changes in such factors as the relative level of land and water, the nature of the substratum, the rate of accretion and erosion, the salinity of the ground water and the frequency of tidal immersion. As is usual in most types of zonation, the community in one zone succeeds that of another as one moves away from the sea, until a transition to fresh water swamp forest or other non-maritime community is reached inland.

In the western group of mangroves, as exemplified by the Florida mangrove, the main zones corresponding to the lines of succession appear to be from the pioneer *Rhizophora* zone to the mature *Rhizophora* zone, and from this through the *Avicennia* zone and the *Conocarpus* transition zone to the freshwater marsh and eventually to the tropical forest of the region. This hammock forest is often dominated by palms. In the eastern group of mangroves, as exemplified by the mangrove forests of the Malayan coasts, the zonation seems to be from an *Avicennia* and *Sonneratia* zone to the *Bruguiera caryophylloides* zone, which is followed by a *Rhizophora* zone and a *Bruguiera gymnorrhiza* zone, through a freshwater swamp to a mixed tropical rain forest or other tropical forest of the region concerned. The pioneer species seem to differ in the eastern and the western mangrove types. It has not been easy to explain this, and the matter is further complicated by the observation that in the mangroves in Fiji (Mead, 1928) and New Caledonia (Daniker, 1929) which belong to the eastern group of mangroves, the pioneer species seems to be *Rhizophora*.

In the drier climate of the East African coast, mangrove vegetation does not seem to end in a tropical forest as the climax, but stops short at an *Avicennia* community as an edaphic climax, except at the mouths of rivers where it may be succeeded by a community of non-halophytes. The succession begins with a *Sonneratia alba* zone, representing pioneers, giving way to a *Rhizophora mucronata* zone, followed by a *Ceriope candolleana* zone and through to an *Avicennia marina* zone, which is the edaphic climax.

Information on zonation and succession is lacking in many mangrove forests, especially those of West Africa, and a great deal of further work is required.

58 *Rhizophora* and *Avicennia*: a) Stilt roots of *Rhizophora* b) Germination of fruit of *Rhizophora*: the diagram on the left shows the fruit still attached to the parent plant, the diagram on right shows the independent seedling c) *Avicennia*
SOURCE: J Y Ewusie (1974)

59 Aerating roots of *Sonneratia caseolaris* in Philippine mangrove swamp
SOURCE: W H Brown (1935)

Factors of the Habitat

1. Mobility of the Soil

The level of the soil in the mangrove swamps is not constant because the incoming rivers bring down alluvial soil to raise the level of the mud. This may be disturbed by sea currents washing the soil away, so that seedlings establish themselves with difficulty. To combat this situation a number of common species in this habitat have stilt roots which hang in the air for at least some distance before they enter the mud. The crown of tufted roots at the tips of the stilt roots help in withstanding the mobility of the soil. A well-known example is *Rhizophora* where the instability of the soil is overcome by the early development of the seed (vivipary) while still attached to the tree.

2. *FLUCTUATING SALINITY*

The mangrove swamp is essentially tidal, receiving water of lower salinity from the incoming river and of higher salinity from the sea at different times each day. In view of this most plants in the mangrove are halophytes, that is, plants adapted to growing in saline habitats. They have high osmotic pressures of their cell solutions. Among these are species of *Sesuvium*.

3. *REDUCED OXYGEN IN THE SOIL*

The soil of the mangrove swamp is muddy and saturated with water and hence contains practically no oxygen. Only a limited number of plants can live under these conditions. While the very top of the soil may be slightly oxidized and becomes brownish in colour, the rest of the soil below is bluish-grey. To combat this situation some of the plants here, such as *Avicennia nitida* and *Sonneratia caseolaris* have aerial roots known as pneumatophores which stick out upwards from the water and soil. The tips of these pneumatophores have lenticels through which air enters and leaves to serve the roots and the lower part of the plant. Crabs also bore numerous holes in the soil and these bring oxygen to some of the roots of the mangrove vegetation.

Economic Importance

Mangrove forests are of special scientific interest by virtue of their structural and physiological specializations which are adapted to the factors of the habitat. They are of economic importance in the role they play in the reclamation of land from the sea. They are used for firewood in many countries while in other parts they are used as timber. They also constitute a source of tan bark, in view of the tannins which they contain.

The mangrove habitats, by their link with the sea, provide spawning grounds for some sea fish, so that their destruction has far-reaching consequences for the future of such fish. The habitat provides also excellent areas for rice cultivation. Mangrove habitats are known to exhibit high productivity. For all these reasons it is strongly urged that mangrove habitats be preserved or, if they should be exploited, that this be done with great care in order not to destroy them.

Coastal Lagoon Vegetation

Coastal lagoons occur where a river or its tributary overflows its banks into a depression before entering the sea. The water often remains in the depression for a considerable time during the year and may even dry up during the dry season. They are found along the coastline of a number of tropical areas.

Some lagoons are completely cut off from the sea by the sand bars, and these are referred to as **closed lagoons**. These occur in relatively drier areas. Others may have daily contact with the sea; these occur in areas with heavy annual rain-

60 Top: Mouth of open lagoon, Elmina, Gold coast. Bottom: Mouth of closed lagoon, Old Ningo, Gold Coast. SOURCE: A S Boughey (1957)

fall and are called **open lagoons**. The closed lagoons may, however, gain access to the sea at certain times of the year, though not necessarily every year, especially during the rainy season when the sand barrier may be broken by the force of the rising water in the lagoon. The flooding of this type of lagoon is therefore only seasonal. As the rains subside, the sand bar may reform and cut off the sea from the lagoon. During the dry season this type of lagoon may dry out completely. Conditions for plant and animal life are therefore rather hard in this type of lagoon (see Figure 60).

Factors of the Habitat

1. Fluctuation in Salinity

It is evident that when a closed lagoon is not in contact with the sea the water will be brackish and the salinity will increase owing to increased evaporation of water from it, and before the water dries up its salinity may become greater than that of the sea, and as much as about 5 per cent of the dry weight of the soil may be salt. The plants of this type of environment are therefore subject to very great variations in salinity. Among the plants which survive under these conditions are *Sesuvium portulacastrum* and *Paspalum vaginatum*.

2. Availability of Water

It is obvious from the conditions in the lagoon that there is very great variation in the availability of water during the rainy and the dry seasons. Among the plants which may grow under these conditions are *Sesuvium*, *Enhydra* and the fern *Marsilea*.

Structure

Two clear zones can be recognized in the lagoon vegetation of West Africa. These are the herbaceous zone on the upper parts and the mangrove zone on the lower part immediately surrounding the lagoon water. These will be described in detail below, with special reference to West African lagoons.

1. The Herbaceous Zone

The plants of the herbaceous zone may sometimes be found to join up with the vegetation of the evergreen shrub zone of strand vegetation (see p. 164) and it is often distinctly zoned. At the topmost end is the dense belt of *Sporobolus virginicus* which is an ecotype differing from that of the main strand zone in being more robust and upright. Below this is a wide belt of the sedge *Fimbristylis obtusifolia*, followed by another wide belt of the grass *Paspalum vaginatum*. Patches of *Mariscus ligularis*, *Cyperus articulatus* and *Imperata cylindrica* as well as scattered plants of *Sesbania punctata* and *Hibiscus tiliaceus* are often found within these two broad belts of this herbaceous zone. The last and lowest belt consists of a pure community of *Sesuvium* or *Philoxerus*, although in certain places both species may be found together.

2. The Mangrove Zone

This zone begins with the last belt of the herbaceous zone and covers the parts of the lagoon which are flooded seasonally or daily with brackish water. The vegetation of this zone in the open and closed types of lagoons show significant differences.

(a) **Mangrove Zone of Open Lagoons**. On the mud of the lagoon bed which is exposed daily at low tides are found two species of mangrove trees such as *Rhizophora racemosa* and *R. harrisonii*. An ecologically distinct form of *Pandanus candelabrum* is often found along with the *Rhizophora* spp. The open spaces in the mangroves are inhabited by the fern *Acrostichum aureum*.

The end of the lagoon which is further away from the sea and connected with the incoming river water is not brackish but fresh, and this end is often invaded by *Raphia* and other genera to form a freshwater swamp.

(b) **Mangrove Zone of Closed Lagoons**. A fringe of the mangrove *Avicennia nitida* is found at the limits reached by the seasonal flooding of this type of lagoon, and along with this may be found the shrubby mangrove *Laguncularia racemosa* as well as the shrub *Conocarpus erectus*. The bed of the lagoon when dry in the dry season may also be covered with *Sesuvium*. On the higher ground towards the land is found a shrubby growth in which the predominant species may include *Phoenix reclinata*, *Drepanocarpus lunatus* and *Fagara xanthoxyloides*.

The end of the lagoon connected with the supplying river does not often contain much water and so the area cannot support a woody freshwater swamp community, as in the case of the open lagoons. Instead the area develops into a freshwater marsh supporting an herbaceous community. The freshwater marsh contains plants like *Hygrophila spinosa, Enhydra fluctuans* and *Marsilea diffusa*, apart from some species from the herbaceous and mangrove zones.

Successional Relationships

It appears that the lagoon bears a pioneer community that may lead eventually to normal terrestrial conditions. Thus whenever mangrove plants like *Rhizophora* establish themselves in the lagoon, succession is set into motion, for the stilt roots of the mangroves begin to trap particles of silt and dead plants. These lead to an accumulation of debris which contributes to the raising of the level of the soil. This happens particularly as the tide water recedes and more seedlings of *Rhizophora* are established on the exposed mud. In this way the lagoon may be filled up and when the old *Rhizophora* plants die out their places are often taken by other more common land plants typical of the vicinity of the lagoon.

Economic Importance

In the lagoon vegetation the mangroves are often cut for fuel while the sedges and the grasses are used for thatching. Also cultivation of paddy rice is undertaken in coastal lagoon regions in some tropical countries.

Strand Vegetation

The strand stretches from the high-water mark to where normal land conditions begin. The plants found in the strand vegetation are fairly similar throughout the tropical coasts. The vegetation is often zoned and, owing to the rather harsh factors of the environment, the plants appear to be unable to modify the habitat sufficiently to make it possible to demonstrate within a reasonable period whether the zones represent stages in succession. The observation of active succession on the fresh lava of the famous volcanic eruption of the island of Krakatau (described in Chapter 3) seems to have provided the best evidence of the successional origin of the zones of the strand. Where fresh situations for succession do not arise, the communities of the respective zones appear to be edaphic climaxes which do not encroach on each other.

Structure

Studies on tropical strand vegetation so far suggest that there are three broad zones, as in temperate strand vegetation, but with different species, even though each zone may contain recognizable sub-zones. Moving landwards from the sea the three zones which can generally be distinguished are the pioneer zone, the main strand zone and the evergreen shrub zone. These are better observed on extensive and low lying beaches, but they may be telescoped or variously modified on narrow or steep beaches. Each zone seems to have a characteristic plant community the life forms of which are also characteristic of the zone.

The Pioneer Zone

The seaward edge of all exposed beaches in the tropics is often covered by isolated clumps of pioneer species, each of which may have rather low density. Some of these plants appear to have been established from floating seeds which are carried up by the waves to the high tide mark. Among such plants are *Sporobolus virginicus, Cyperus maritima* and *Remirea maritima*. The inner part of this zone often supports perennial plants with creeping shoots such as *Ipomoea pes-caprae, I. stolonifera, Canavalia rosea, Sporobolus virginicus, Alternanthera maritima* and *Diodea maritima*. The prominent genera here are often *Ipomoea* and *Canavalia*. These have a trailing habit and send long runners over the other prostrate plants. It is reported by Richards (1964) that the only species that seems to be present in both temperate and tropical strand vegetation is *Salsola kali*, but that is far from being generally present on tropical strands. During the rainy season a number of small herbaceous weedy plants may also grow in this and other zones. Animals like ghost-crabs and spiders are often found here.

The Main Strand Zone

This zone is often most disturbed by man for various purposes. A few plants of the pioneer zone species are found here but the principal species of the undisturbed parts of the zone are perennial suffrutescent herbs and geophytes, along with a number of annuals migrating from inland places where they are also found on sandy soils. Most of them do not have the creeping or rhizomatous habit of the pioneer zone species, but prostrate or rosette habit is common. Turtles and lizards are among the common animals in this zone.

The Evergreen Shrub Zone

This zone, the last to join up with normal inland vegetation, occurs on the landward slopes of offshore bars and the final slopes of exposed beaches. The woody vegetation, consisting of a dense low evergreen scrub, is very much wind-cut and trimmed to a level with the top of the offshore bars or graded up to the vegetation further inland. Shrubs and trees become more and more frequent as one moves away from the shore.

In some parts of the tropics, such as the tropical regions of the Indian and Pacific Oceans, species of *Barringtoria* dominate to form a *Barringtoria* community which may be either dense or sparse like savanna. The height is often as much as that of the inland forests, where these exist. In the West Indies a corresponding formation is dominated by the small tree *Coccoloba uvifera*. In general, therefore, one may find among the common plants of such a community *Barringtoria specioza, Coccoloba, Calophyllum, Terminalia catappa, Pandanus tectorium, Thespesia populnea, Hippomane nancinella, Hibiscus tiliaceons, Chrysobalanus, Sophora occidentalis, Eugenia coronata* and *Phoenix reclinata*. The zone is also much disturbed especially through cutting for firewood. When this happens it is invaded by normal inland species as well as species from the main strand zone.

Factors of the Habitat

Among the prominent influences on the strand are strong winds with salt spray, high salt content in the soil, occasional flooding by the sea, rapid drainage and free mobility of the sand.

1. Strong Winds

Winds from the sea are typical of the coast. These winds increase the rate of transpiration of the plants exposed to them.

2. Salt Spray

The strong winds bring droplets of salt water from the sea to the seaward side of the strand plants and also to the sand. Since such water is saline, it is not available to the plants and it has been shown that comparatively little salt is absorbed by the roots. A far greater amount of the salt enters the shoots through mechanical abrasions and the chloride ion accumulates to injurious concentrations in the apexes of twigs and leaves. In this way terminal and seaward meristems are killed, while landward meristems develop. Asymmetrical growth takes place and this results in the irregular shapes of spray forms of shrubby plants. This is the principal factor associated with wind-cutting.

3. Salt Content of the Soil

The salt content of the soil decreases with increasing distance from the sea and this obviously plays some part in the zonation of plants, so that the more salt tolerant ones are nearer the sea and the less tolerant ones further away. Humus content increases with increasing distance from the sea. It appears that the nature of the soil and strong winds may prevent the zones from advancing seawards.

4. Occasional Flooding by Sea Water

Plants in the pioneer zone appear to be tolerant of occasional flooding by sea water as a result of wave action. Deposition of salt on the leaves left after the drying of the water increases the water stress in these plants.

5. Increased Drainage

If the sand particles are large and consequently have large spaces between them, any water, whether from the salt spray or from normal precipitation, readily drains down through the sand, which retains little or no water for the plants that grow on the strand. In view of these factors, the plants growing on a sandy shore can be said strictly to be growing in a dry environment not dissimilar to desert conditions. Many of their structural characteristics are similar to those of plants found in dry places and they can be said to be xerophytic. In the dry season they nevertheless suffer from drought and those that survive do so mainly on water from the early morning dew or they are deep rooted.

Among the plant species which are adapted to life in the sand are: *Sesuvium portulacastrum* which has thick and reduced leaves; *Sporobolus virginicus* and *Remirea maritima* which have small hard and spiny leaves; the sea lavender (*Tournefartia*) which has hairy leaves; species of *Ipomoea* which have leathery leaves; and *Canavalia obtusifolia* and *Ipomoea biloba* which have deep roots to tap water from the soil below the sand.

Rain water percolates downwards through the sand until at a depth of about five to seven metres it reaches a layer of sand saturated with salt water. The rain water here forms a saturated layer in the sand about one metre deep, without mixing much with the salt water. This reservoir of fresh water is exploited by coconuts and some other bigger coastal plants.

6. Mobility of the Substratum

In some tropical areas wave action can quite easily move the sand, so that the substratum of the plants which it forms is rather unstable. The wind rarely blows off the sand on the West African shores, probably because of the caking of the surface layers of dry sand by the deposition on it of concentrated droplets of salt spray. Wave action may either blow the sand to cover up a plant or blow off the sand to expose the stolons or runners of the plant. To meet these situations some of the plants here tend to creep on the sand and root at the nodes since such a life form makes the plants less exposed to the wind and also increases propagation by vegetative means. Examples of these plants are *Ipomoea biloba, I. pescaprae, Canavalia obtusifolia* and *C. rosea.* These plants also help to hold together the sand dunes formed by the wind. The strong fibrous roots of coconut can withstand the exposure caused by erosion through wind and water. In the beach scrub this adaptation is also exhibited by the prop roots of *Pandanus* which anchor the plant against the strong winds.

Wave action also washes away seeds, but these seeds are often adapted in structure to floating in sea water and to germinating later on arrival on a more congenial substratum. Examples are seeds of *Sesuvium,* coconut, and *Philoxerus.*

Economic Importance

The strand is a habitat which supports some economic crops like coconuts and Indian almond (*Terminalia catappa*). The sand is an invaluable ingredient in some countries for the making of cement blocks for building purposes. It is also a useful abrasive for various materials. The crabs which the strand supports are used as food.

Finally the nature of strand vegetation as described above shows that it assists very much in preventing encroachment of the sea onto the land.

The Arboreal Ecosystem

Tropical vegetation with its luxuriance and relative preponderance of large woody plants provides situations in which a host of smaller plants and animals find suitable micro-habitats on the trunks and branches of trees and even on leaves. Some use may be made of the tree irrespective of whether it or its parts are living or dead. The arboreal ecosystem is most complex in the rain forest and becomes less so in the deciduous forests and the savanna woodlands. The typical dependent plants are the epiphytes with which are associated animals having a role in the specialized food chains of plants not directly rooted in the soil.

Epiphytes

Epiphytes are plants which grow attached to the trunks, branches and even leaves of trees, shrubs and lianes (Figure 61). The micro-climate which they occupy would appear to be what is available for plants which combine small size with relatively high light requirement. However, epiphytes appear to have made certain sacrifices in order to satisfy their light requirements, for the micro-habitat lacks soil. In drier types of vegetation or at certain heights they also face shortage of water as well as of mineral nutrients. Epiphytes do not depend on their support plant for their nutritional requirements, and are thus distinguished from parasites. The effect that they exert on their support is the weight that they add to its branches. Although this could in extreme cases lead to the toppling of the tree or the breaking of a branch in strong wind, this does not often happen. Apart from a smothering effect on small trees like cocoa and citrus, it can be said that epiphytes exert little or no ill effect on the support. Epiphytes do, however, provide the principal habitat for certain animals in the ecosystem. For example, the roots of epiphytic flowering plants and ferns often provide the nesting places for arboreal ants.

One of the ways in which tropical epiphytic communities differ from temperate ones is that while temperate epiphytes consist almost entirely of cryptogams (algae, fungi and bryophytes) tropical epiphytes also include vascular plants, largely pteridophytes and flowering plants. It is estimated that there are over 200 genera of

61 Some typical epiphytes
SOURCE: J Y Ewusie (1973)

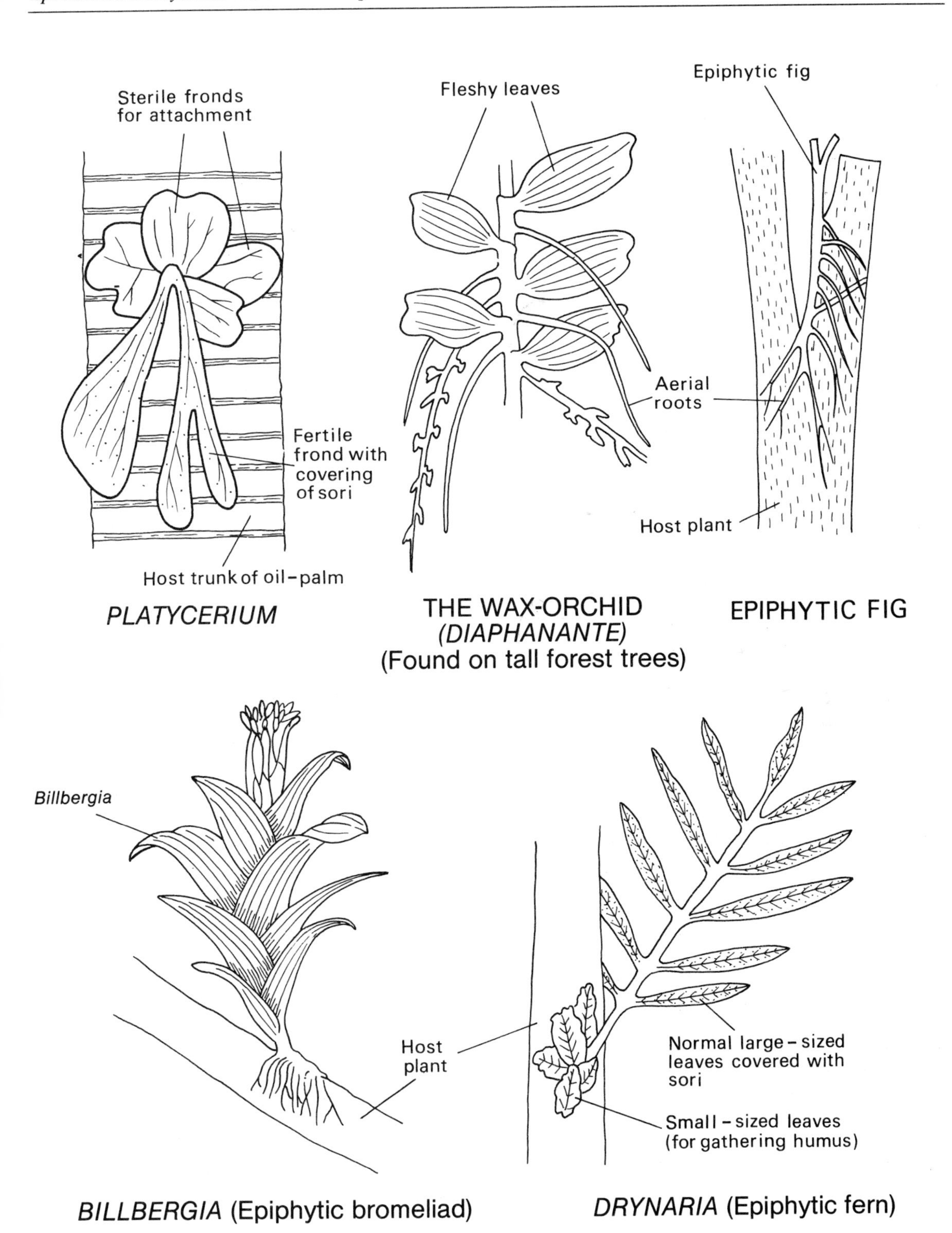

PLATYCERIUM

THE WAX-ORCHID (*DIAPHANANTE*) (Found on tall forest trees)

EPIPHYTIC FIG

BILLBERGIA (Epiphytic bromeliad)

DRYNARIA (Epiphytic fern)

33 families of flowering plants and about 20 genera of pteridophytes which are epiphytes. Among the flowering plant epiphytes are the monocotyledonous families of *Araceae*, *Bromeliaceae* and *Orchidaceae* and the dicotyledonous families of *Asclepiadaceae*, *Cactaceae*, *Ericaceae*, *Rubiaceae* and *Melastomataceae.* Tropical American vegetation is known to be much richer in epiphytes than the African vegetation.

Factors of the Habitat

The factors of the habitat of epiphytes are not quite the same as those of plants living on the ground and some of the special features of the epiphytic habitat are worth highlighting here.

The fact that the substratum is raised above the ground level often means higher illumination and lower relative humidity. The host or the part of it to which the epiphyte is attached may be inclined at any angle ranging from a horizontal to a strictly vertical position. This creates the problem of seed dispersal and the successful attachment and maintenance at that angle. The bark of the host tree may be rather smooth, as is typical of tropical rain forest trees, and when this is the case the problem of anchorage becomes acute. Where the host tree trunk has burrows and fissures or is covered with large and persistent leaf bases as in the case of the oil palm tree, then anchorage presents no problem and the growth of epiphytes becomes luxuriant.

The shortage of soil in the habitat means that epiphytes must necessarily depend on the small quantities of debris in cracks and hollows on the supporting tree, or on what debris they can collect among their own roots and leaves. A number of epiphytes are constructed in such a form as to collect soil. Also the ants which inhabit the root systems of most epiphytes gather dead leaves, seeds and debris which are broken down into humus. The humus improves the water-holding capacity of the soil and also provides the epiphytes with nutrients.

Water shortage can become severe for an epiphyte owing to the high rates of evaporation that are attendant upon low relative humidity. In order to overcome this, most epiphytes are either drought resistant or conserve their water supply. Epiphytes are also able to absorb water quickly when it is available and conserve it for use later. The velamen tissue of the roots of orchids and some aroids is used in this way (Figure 62). The cryptogamic epiphytes, with their small water requirement and their ability to survive in a dry-air condition, do not have the same difficulties with water as flowering plants. This may help to explain the apparent absence of flowering plant epiphytes in temperate climates. Plants with specialized water storage organs are moreover ill adapted to survive freezing conditions.

With this brief introduction to the habitat factors of the epiphytic environment it is now worth examining how some of these factors, and others, determine the distribution of epiphytes.

1. Illumination

The types of epiphytes found are related to three levels of illumination: bright light; dense shade; and a wide range of conditions. These levels affect the distribution of epiphytes at different heights on the trees. In the rain forest, the main classes of epiphytes recognized are: **extreme xerophilous epiphytes** which live on the topmost branches and twigs of the taller trees, examples of which are some bromeliads and, surprisingly, some cacti; **sun epiphytes** which are usually xeromorphic and occur mainly within the crowns and on the larger branches of the upper tree storeys and which as a group form the richest of the epiphytic synusiae in terms of individuals and of species; and **shade epiphytes** which occur mainly on the trunks and lower branches and may also occur on the stems of the larger lianes. Shade epiphytes often show typical mesophytic and not xerophytic features.

The heights at which the different groups are found in relation to light tend to be constant within any one type of forest, but these heights are different in other types of forest. Where the top strata are dense these levels become high, in relatively open types of forest they are fairly low, and on more widely spaced trees they are lower still.

2. Inclination

Differences in inclination and aspect between different parts of the same tree may affect the epiphytic vegetation, not only by influencing the colonization of seeds and spores but also by modifying illumination and evaporation. Epiphytic vegetation on the vertical trunk will be different from that on the horizontal branches. For example, it is noted that large epiphytic ferns like *Asplenium nidus* and *A. africanum* tend to prefer the trunks to the branches.

As the inclination of the surface also affects the rate of accumulation of humus, certain 'humus

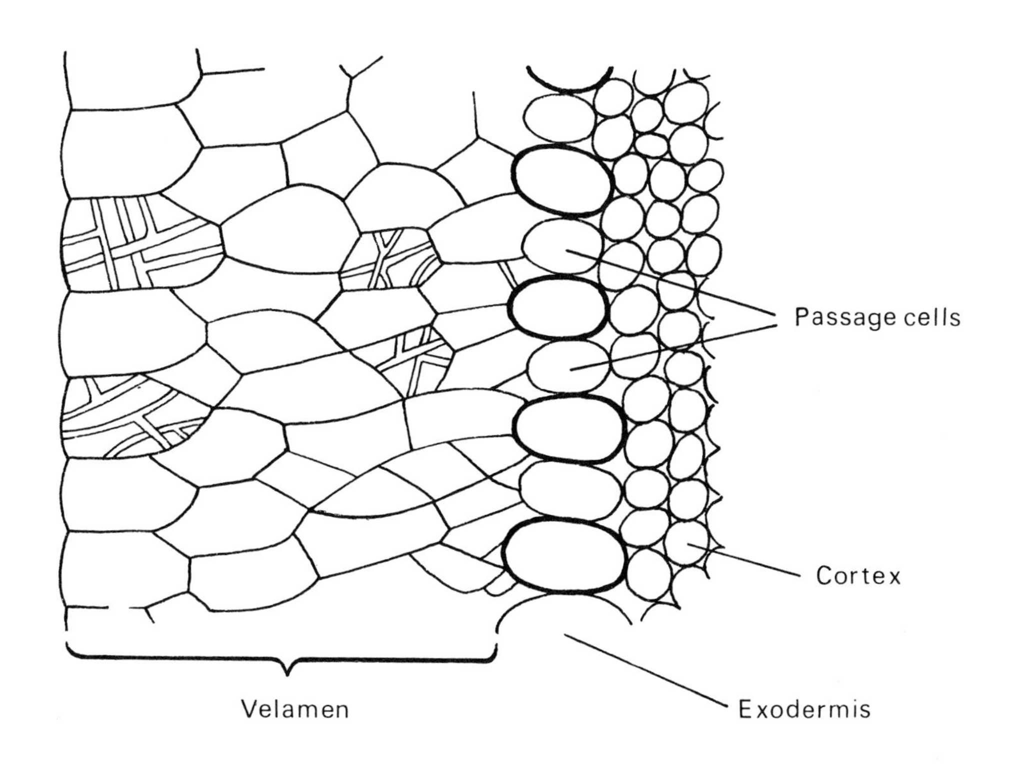

62 Velamen – tissue of epiphytes
SOURCE: J Y Ewusie (1973)

epiphytes' occur only where relatively large amounts of humus can collect, as in the forks of branches.

3. Age

The epiphytic vegetation on a tree may also depend on the age of the tree. For example, the epiphytic vegetation on *Altingia excelsa* when it is a young tree and smooth-barked is different from that on the same plant when it becomes old and scaly-barked.

4. Species of Support Plant

Differences in abundance and floristic composition of epiphytes may also depend on the species of the support tree. The abundance of epiphytes on *Samanea saman*, for example, as compared with that on other tree species commonly planted in tropical towns, is often very marked. It has been shown that the specialization of epiphytes on some species of trees is due not to physical factors but rather to the chemical composition of the bark as well.

Succession

Some epiphytes seem to prepare the way for others. The *Bromeliaceae*, for instance, which have remarkable water-holding powers, often form the starting point for the growth of other less tolerant epiphytes. The first colonizers on trees are usually algae and crustose lichens, followed by mosses and liverworts and then by the resistant ferns like *Platycerium*, which help to improve the conditions for the entry of other ferns like *Nephrolepis*. Finally, flowering plants enter and may dominate.

Adaptations for Nutrition

Epiphytes differ in their adaptations to the collection of water and the accumulation of soil. Four forms of adaptation are found, namely proto-

63 Epiphytic bromeliad
SOURCE: W H Brown (1935)

epiphytes, nest and bracket epiphytes, tank epiphytes and hemi-epiphytes.

Proto-epiphytes are rather simple forms with no special structures for the collection of water or soil. The roots or rhizomes creep on the support plant in order to exploit a large area of substratum. They have xeromorphic structures with various types of organs for water storage. A characteristic anatomical structure is the special non-living tissue at the outside of the aerial roots, the velamen (see Figure 62). When rain falls, the velamen tissue becomes full of water. When this water dries up in dry weather, the velamen then serves as a layer against further loss of water and excessive heating. Examples of such proto-epiphytes include the epiphytic orchids (see Figure 63), ferns and species of *Peperomia* and *Vaccinium*.

Nest and bracket epiphytes accumulate humus and debris from which the roots obtain water and mineral substances. The roots of nest epiphytes form a dense interwoven mass which looks like a bird's nest. Ants nest in these roots and help to make humus available. Ferns, aroids and orchids are among the nest epiphytes. Bracket epiphytes have bracket-like leaves. Good examples are *Platycerium* and *Dischidia*.

Tank epiphytes have long broad stiff leaves which form a rosette with their sheathing bases overlapping to form a reservoir or tank which holds water. These are known to occur only in the *Bromeliaceae* found in the American tropical forests. Insects fall into the water in the tanks and other animals breed in them. These provide humus and minerals which are absorbed by special epidermal structures on the leaves rather than by the roots.

Hemi-epiphytes develop long aerial roots which eventually reach the ground from where they are able to obtain water and nutrients like normal land plants.

Dispersal Mechanisms

Nearly all epiphytes have fruits, seeds or spores with special mechanisms for dispersal. These may be suited for dispersal by animals or by wind. Often the fruits are fleshy with sticky pulp. Thus the seeds easily stick to the bark of the tree or are deposited by birds on telegraph wires where they may germinate. The very small seeds of orchids and the spores of ferns are very light indeed and are easily carried by the wind. Where the seeds are heavier they often have parachute appendages.

Associated Animal Communities

The spaces between the leaves of epiphytic orchids and bromeliads are often inhabited by earthworms, ants, scorpions, spiders, tree frogs, lizards, grasshoppers or snakes. Small animals with flattened shape are found in the bases of banana and palm fronds, and between the leaves of tank bromeliads. The animals in the tree trunks or fallen logs obtain from there their food as well as shelter and protection. Wood boring insects make holes and form their own micro-habitats, but when the holes are filled with water, mosquitoes and other aquatic insects begin to breed in them.

The fauna on the trees, like the flora, change with the varying stages in the development of the epiphytic synusia. Thus the successions of animal and plant communities may be closely correlated.

The Arboreal Habitat on an Oil Palm Tree

The oil palm tree is commonly found in tropical forests and savanna woodlands. Although the trunk has no true branches, it has a number of large compound leaves which are commonly referred to as branches. When they are cut, numerous leaf bases remain which provide a rough surface to the trunk and pockets in which water and humus may accumulate to support various epiphytes. The large compound leaves above cast shade on the trunk and so modify the environment considerably. This affects also the temperature and humidity locally.

When the trunk is examined it will be found in most cases that there are various stages of colonization. First on the lower levels are the crustose lichens and mosses like *Polytrichum*. These are followed on the middle region by *Platycerium* which helps to improve the conditions for the entry of other ferns like *Nephrolepis* and *Microgramma*. These may be followed by such flowering plants as *Coleus, Pantodon* and *Commelina*. Much bigger flowering plants or shrubs, like *Ficus* spp., may enter later and dominate the rest of the community.

Animals are quite common on the oil palm tree. First there are those animals which are buried in the wood and debris; they include boring larvae and adult beetles and termites. Then there are animals which live beneath the bark or in cracks or crevices of the bark. A number of insects, especially ants, feed on the fruits of the palm tree and of epiphytic plants on the tree. The leaves also seem to harbour insects like grasshoppers which may feed on them. In the water which collects in the leaf bases may be found breeding mosquitoes. Predatory spiders, lizards, chameleons, geckos and tree frogs may also be present to feed on insects. Mice and rats may also be found since they feed on the fruits of various epiphytes. Snakes, when present, feed on the frogs, and at times a bird of prey may visit the palm tree to feed on mice and lizards. Occasionally also other larger animals like bats, squirrels and monkeys may visit the tree in order to feed on the fruits.

Chapter 11 Tropical Savanna and Desert Ecosystems

Tropical savanna ecosystems consist of at least three principal types. These are savanna woodlands, thorn woodlands and savanna grasslands. In some tropical countries there are further subdivisions of each of these, but for our purposes here, the three broad types may be enough. The reader would naturally have to consult local works for details as they apply to his or her region. Tropical desert ecosystems, on the other hand, are relatively uniform.

Savanna Woodlands

Savanna woodland vegetation is found under conditions somewhat drier than those of deciduous forest. The vegetation is consequently more open with widely-spaced trees, except along water courses. The spacing of the trees allows enough light to penetrate to the ground for a distinct grass layer to be formed, unlike in the forests. Savanna woodland trees are generally broad-leaved. Grasses and xerophilous shrubs are conspicuous, but lianes and epiphytes are poorly represented.

Savanna woodland vegetation in one form or another is found very widely in tropical and subtropical regions of the world. Among the places of occurrence are East, Central and West Africa, much of Cuba, parts of the Caribbean, the Guianas, Brazil, northern Argentina, much of India and China and northern and eastern Australia.

Structure

The tree species are scattered and open. The trees are often about 12–15 m high and only rarely form a closed canopy with their branches. It is only in situations where the water content of the soil is high, such as near water courses or in depressions, that closed canopies are found. Such local vegetation along the rivers is referred to as **fringing** or **gallery forests** and those in depressions are called **forest outliers**. The trees are often stunted, show resistance to drought and appear leafless during the dry season. They often have thick fissured barks and are mostly fire resistant. The crowns of the trees are sometimes flattened in shape.

Although the tree species vary, they comprise a high proportion of members of the *Leguminosae*. In West Africa, for example, the common trees are *Daniellia* spp., *Isoberlinia* spp., *Lophira* spp., *Butryospermum parkii* (shea butter-tree) and *Uapaca saman*. In the South American and Cuban savannas there are such conspicuous palms as species of *Mauritia* and *Copernicia*. The West Indian and Venezuelan savannas are typified by trees of low gnarled growth like *Curatella* and *Byrsonima*. In southeastern Asia there is not much savanna vegetation in comparison with regions of tropical Africa and America which have a comparable climate.

The grass layer, containing other herbaceous plants as well, may be as much as 3 m high, but the grasses may be much taller, and indeed are sometimes taller than some of the trees during the rainy season.

There are also short grass savannas on shallow or clay soils. The grasses often grow in tussocks or scattered groups (see Figure 64). Among the numerous grass genera in tropical savanna woodlands, *Andropogon* and *Hyparrhenia* are common. The West Indies lowland savanna woodlands contain grasses like *Axonopus*, *Trachypogon* and *Paspalum*. Among the geophytes are ground orchids like *Eulophia* and aroids like *Stylochiton*, as well as a number of plants belonging to the *Zinziberaceae*. A typical aspect of the woodland savanna is shown in Figure 64.

64 Woodland savanna. *Butryospermum–Parkia–Tamarindus* association. *Icacina senegalensis* in the foreground
SOURCE: C J Taylor (1960)

There are, of course, a number of local variations of the woodland savanna such as the parkland type where the trees are in clumps or the orchard type which consists of small trees spaced out evenly.

Seasonal Changes in Vegetation

Savanna woodlands often exhibit changes in the aspect of the species during the year. In general one can recognize three periods of noteworthy changes, which may be described as the cool and dry period, the warm and rainy period, and the hot and dry period.

1. The Cool and Dry Period

During this period of the year fresh leaves and flowers generally develop on the trees and the grasses begin to grow in tussocks. The geophytes appear. This is the time when the dry season has just been broken, and the grasses have not yet developed to overshadow the herbaceous plants which are intolerant of shade.

2. The Warm and Rainy Period

This period is marked by a dense tree canopy and a thick screen of grasses, which make it difficult for an observer to see through for a long distance, if at all.

3. The Hot and Dry Period

During this period the grasses wither and the trees shed their leaves. Bush fires sweep through the dried grasses but do not seriously damage the trees which are mostly fire resistant.

Habitat Factors

1. Rainfall

Rainfall is the most important ecological factor determining the boundaries of the savanna woodland. Annual rainfall is between 900 and 1150 mm and the rains fall during one period of the year and with only one peak. The period of the rainfall in relation to the dry period is responsible for the seasonal changes in vegetation (phenology) already described.

The presence of rivers in savanna country results in a closer development of trees along their banks (fringing or gallery forests) with some of the species being typical of forest vegetation except for their lower height. Also, accumulation of water in large depressions leads to similar development of forest species (forest outliers).

2. Fire

Fire reaches most parts of the savanna woodland almost every year during the dry period. Fire may be started by farmers when clearing the land for farming or when burning the old grass in order to stimulate the growth of new and fresh leaves to be eaten by cattle and other grazing animals. Hunters also use fire to drive out animals while hunting.

Fires seem to stimulate some perennial plants to regenerate from their underground buds. While some trees are damaged most of them are not affected as they have to be fire-resistant.

3. Animals

Because of the grasses, large numbers of hoofed grazing animals abound and the large carnivores, such as the lion, prey on them. These animals exert much influence on the vegetation.

Fire, farming and grazing activities have also combined to reduce the outer fringes of tropical forests into a type of savanna that is described as 'derived' savanna. This is often dominated by the alang-alang grass (*Imperata cylindrica*).

Successional Relationships

It is presumed that the true climax of the savanna woodland consists of a canopy of small trees and shrubs, and that the present biotic climax is the result of the effect of fire and farming. This is supported by the fact that when such activities are excluded from an area of savanna woodland, the density of the trees increases in the course of time towards the putative climax vegetation.

Economic Importance

Cereal farming is one of the common activities of the woodland savanna areas, and cattle rearing is the most frequent occupation except in parts where insect pests like the tsetse fly seriously affect the health of the cattle and make their rearing difficult.

Thorn Woodlands

Thorn woodlands form a major plant community that is characterized by a sparse cover of small leaved and thorny tree species and a low grass layer (Figure 65). Here the dry period is very prolonged, lasting seven months or more, and is more severe than that of the savanna woodlands. Many trees are of the *Acacia* type with compound leaves with very small leaflets. A number of the plants bear thorns, prickles and scales, and some have water conserving devices. The rain falls for a brief period which is marked by intensive growth and flowering. The rainfall totals from 250 to 900 mm per annum. Maximum daytime temperatures are rather high all the year round and range from 43 to 46°C, but night temperatures may be much lower. The thorn woodland exists nowadays as a biotic climax historically brought about by fire and now maintained by grazing. The thorn woodlands show a graduation in dryness towards desert conditions, and in some tropical regions different names are given to the part of this vegetation that borders on the desert, such as Sahel savanna in West Africa. In the driest areas grasses are either lacking or appear as widely separated clumps with bare soil between them. As the grass cover becomes thinner so fire ceases to operate as an important factor.

Tropical thorn woodlands are extensively developed in dry regions. Among such regions are northeastern Brazil and elsewhere in subtropical and tropical South America, the larger and more northerly Caribbean islands and Central America and Mexico. They also occur in the Sudan, the Gulf of Aden, East, Central and southwestern Africa, India and in central, northern and eastern Australia.

Structure

Thorn woodlands are often found on sandy or limestone soils that are very permeable to water. In depressions or ravines where there is a local increase in humidity, savanna woodland vegetation is found. There are various forms of thorn woodland depending on the texture or the dryness of the soil and the amount of tree cover. The dominant trees, which may grow to about 18 m, are remarkably xerophilous; the leaves may be reduced to scales and the stipules and twigs to spines. The tree canopy cover is generally more than 20 per cent.

65 Thorn woodland in Australia
SOURCE: W H Brown (1935)

Some of the common trees and shrubs in the thorn woodland of West and East Africa are *Bligha sapida*, *Sclerocarya* sp., *Adansonia digitata, Balanites aegyptiaea* and *Commiphora*. The branched palm, known as the doum palm (*Hyphaene thebaica*) is also found in the driest thorn woodlands. The date palm (*Phoenix dactylifera*) is also planted in various places. The grasses are much shorter than in the savanna woodlands, and they differ again in not growing in large tussocks. These grasses often have a characteristic feathery appearance due to awned spikelets. Some of the genera represented are *Chloris* and *Aristida*.

With the reduction of leaves, a number of trees are present which have green bark, due to secondary photosynthetic tissue development. Also a number of the woody plants store water in their swollen trunks or roots for the dry season. An example is the Brazilian bottle tree (*Cavanillesia arborea*) which has a swollen barrel-shaped trunk. Giant succulents such as the *Euphorbiae* and the *Cactaceae* are also present. The scattered trees often consist of thorny *Acacia* spp. or other plants belonging to the *Leguminosae*, such as *Tamarindus indica* and *Parkia clappertoniana*. The woody plants often have much branched roots since there is severe competition among them for water. The roots also penetrate very deeply in order to get to a moist horizon or to the water table. The grasses which have rather shallow roots are mostly unable to survive the dry season.

Xeromorphic herbs such as the terrestrial euphorbias, bromeliads, agaves and cacti, all with prickles on leaves or stem, are characteristic of the dry thorn woodlands as desert conditions are approached. It is rare to find epiphytes here. The epiphyte *Tillandsia* flourishes in isolated localities and lichens may be common on twigs. Geophytes shoot up during the brief rainy season and go through their life history within a few weeks. Along water courses the tree density is increased and fringing forests are found.

In Kenya thorn woodland with less than 20 per cent tree cover is referred to as wooded grassland, and when there are only shrubs (and not trees) the vegetation is called bushland. These often occur in areas that are drier.

Seasonal Changes in Vegetation

There is a distinct rhythm of life in the thorn woodlands during the brief rainy period. As soon as the rains begin, the geophytes as well as other ephemerals shoot up quickly and produce flowers and fruits within a few weeks and die before the wet period is over. With the onset of the long dry period that follows, the grasses wither and may be burnt. Their remaining basal pieces remain dormant until the advent of the rains again. The trees also remain leafless and appear lifeless in the dry season. They often flower at the beginning of the rainy season.

Habitat Factors

1. Rainfall

The wetter parts of the thorn woodlands receive about 640 to 900 mm of rain a year, while the drier parts receive as little as 250 mm a year. The dry period is severe and lasts for seven months or more. The relative humidity is often about 25 per cent or less. The plants found here cannot survive without adequate adaptation as described above.

2. Temperature

The high maximum temperatures of 43 to 46°C have also necessitated the physiological adaptations of the plants to the environment.

3. Soil Factors

The catena or soil sequence (discussed in Chapter 8) affects the vegetation in the relatively wetter parts of the thorn woodlands, especially in Africa.

4. Animal Grazing

Wild life exists and animal husbandry thrives in the wetter parts of this vegetation. These animals, of course, exert much influence on the vegetation by uprooting grasses and leaving bare,patches of ground. The species of plants which are more palatable, such as *Themeda triandra* and *Cymbopogon* of East Africa, are naturally more grazed while the less palatable ones like *Anogeissus* in West Africa are avoided. Selective grazing thus contributes to the alteration of the species composition of the vegetation. The animals migrate during the dry period of the year to sources of water like the river courses.

5. Farming activities

Except on rocky areas, farming of maize, millets and other grains, like the guinea corn of West Africa, is widespread. Litter from the cattle and sheep is used to fertilize the land and sometimes to construct homesteads. Without irrigation, farming activities can be undertaken only for very brief periods of the year. Thus most of the farmers here may migrate during the dry season to the forest or other wetter types of vegetation to seek temporary work.

Successional Relationships

It is presumed that the original vegetation of the present thorn woodlands was more dense, and that the present vegetation is a biotic climax brought about by fire which eliminated some trees and preserved the fire resistant species.

Economic Importance

As mentioned above, animal husbandry involving cattle, sheep, goats and camels, and grain farming, make this vegetation economically valuable. In some countries where parts have this type of vegetation, the migration of labour for agriculture takes place to the wetter areas. Human population densities may locally be very high.

Tropical (Savanna) Grasslands

Grasslands in tropical and subtropical regions usually exist in the form of savanna, which consists of widely spaced trees among the grass, and in this way differs from temperate grasslands which are often treeless except along water courses. This evidently is another demonstration of the high incidence of woodiness in tropical vegetation. Usually the dominant grass vegetation belongs to the *Gramineae* or grass family, but at times grass-like plants belonging to the *Cyperaceae* or sedge family may dominate large areas of savannas. Hygrophilous 'meadows', which occur in the temperate climates, are rarely found in the tropics, and this has not been easy to explain.

As has been mentioned under the discussions on the savanna woodlands and the thorn woodlands, it now seems difficult to explain the nature of any savanna type of vegetation in terms that do not involve the action of man. In other words, it is becoming difficult to justify the existence of a tropical grassland climate, even though some grasslands seem to be natural. It appears that many of the grasslands have resulted from the effects of fires or browsing animals. Exceptions to this are undoubtedly edaphic climaxes resulting from local soil conditions especially where, in alluvial areas, heavy clay soils near the surface impede drainage and restrict root development. An example of this is found in the Asipo savannas of Trinidad.

Tropical savanna grasslands are found in many parts of the tropics throughout the world. They are found in the West Indies, Central and South America to the north and south of the Amazon forests and in southwestern North America. They occur also in many parts of Africa such as the Sudan and the Zaïre region, central Madagascar, India and Asia and the central and northern parts of Australia.

Structure

The trees of the savanna grasslands are distributed at varying intervals. They are usually stunted and gnarled but in some places they can be tall. Many of them are deciduous but a number of evergreens are also present. The species are often distinctly different from forest species and often include palms as well as some plants of peculiar habit. They comprise less than 20 per cent of the vegetation. The grasses are often fast growing and can attain such heights that make them exhibit their dominance by interpenetrating the lower branches of the trees. A grass like *Pennisetum purpureum* (elephant grass) may attain a height exceeding five metres. *Acacia* and other *Leguminosae* are common trees, and in Africa there is also the

66 *Adansonia – Tamarindus* association
SOURCE: C J Taylor (1960)

baobab (*Adansonia digitata*) which has a large swollen water-storing trunk (Figure 66).

Savanna grasslands have more rainfall and are therefore less arid than the thorn woodlands. The soil is also less sandy and permeable than is usual in the thorn woodlands. Thus the grasses grow better.

Wild Life

The scarcity of trees and the abundance of grasses, coupled with the seasonal cycle of rain and drought have determined the nature of animal communities in the tropical savannas. Tree animals are not common, and the few that may exist are confined to thickets and isolated woodlands. The abundance and variety of grasses, herbs and browse make savanna an ideal habitat for herbivores. Savannas thus have a high animal biomass. Among the herbivores, the large ones are unable to live continuously in the same place throughout the year and have to travel in the dry season for water or to a new feeding ground. The smaller herbivores are more limited in their movements and have to compensate in other ways for scarcity of food and water in the dry season. Some rodents burrow and feed on the food reserves in bulbs, roots and rhizomes, and others hibernate.

The presence of a large collection of herbivores is naturally accompanied by an equally varied assemblage of predators and scavengers in the tropical savannas. In some savannas, notably those in East Africa, the biomass of wild life is the highest that can be found in any type of vegetation. One reason for the high wild life biomass in East Africa is the presence of a high proportion of palatable plants in great variety, particularly the grasses *Themeda triandra, Hyparrhenia, Cymbopogon* and the introduced *Cynodon dactylon*.

Termites (*Macrotermes*) are common on many savanna types of vegetation in which they play an important part in decomposing wood. Termites also play a leading role in determining the distribution of woody vegetation in the savanna. They carry large amounts of dead vegetation into the underground spaces in their mounds. In so doing they accumulate rich soil in the savanna. As this rich soil is brought to the surface to construct new compartments, the seeds of certain plants are able to germinate in such soil. These seeds are often brought to the termite mounds by the termites themselves or by other animals that visit such sites. For example, *Acacia tortilis*, which often occurs on termite mounds in East Africa, and whose dead wood is decomposed and carried underground by termites, is shown to be brought to the termite mound by impalas (Lamprey, 1963).

Economic Importance

The economic importance of tropical savannas in supporting a high wild life biomass and their use in the grazing of ranch animals cannot be over emphasized. They also provide agricultural lands which easily lend themselves to mechanization at all levels. Other uses include the production of wood for fuel and building materials, recreation and tourism.

The development of the grasslands, however, brings in its wake a number of urgent and apparently intractable problems. These include the dangers of soil erosion, overgrazing, the devastating effects of the misuse of fire, and the spread of pollution.

Tropical Deserts

The hot deserts are those regions in the tropics and subtropics which receive such a small amount of rain that they support no vegetation at all over large parts and only a scanty growth of a few scattered, appropriately adapted, plants in others. Other plants consist of very short-lived ephemerals which show up only during the brief rains if and when these occur. Deserts are evidently climatic climaxes where they occur naturally, but there are various areas within the savannas where man-made deserts have been created through overfarming, fire and grazing and trampling by domesticated animals. These deserts are, of course, biotic deserts.

The hot deserts consist of the Sahara and Arabian deserts which occupy much of northern Africa and southwestern Asia respectively. These extend eastwards into northwestern India and northwards and then eastwards into temperate central Asia. Other extensive hot deserts are found in central Australia and the southwestern portions of North America. Less extensive deserts occur in southwestern Africa and western South America.

Habitat Factors

1. Water

The greatest limiting factor to the growth of plants here is water. In the hot deserts rainfall rarely exceeds 250 mm per annum, and it is less than this amount in many places. The rain falls in a rather unpredictable fashion and so in some places there may be no rain at all for several years and these areas may completely lack plant life. When the rain does come it falls in such a heavy storm that much of it runs off the ground.

The scarcity of water and the dryness of the atmosphere have exerted much influence on the structure and physiology of the plants. The role of water in the desert is demonstrated by the presence of more plants where more water exists.

2. Relative Humidity

The relative humidity in daytime in the hot deserts is generally less than 50 per cent on the average, but it often drops to as low a level as 5 per cent. This is made even worse by the frequent winds which prevail. Under these conditions succulent plants avoid excessive loss of water by closing their stomata during the day and opening them at night when the water loss is less than it would otherwise be if the stomata opened during the day.

3. Temperature

In view of the excessively clear and dry atmosphere, the sun makes the day blazing hot, while the night is fairly cold with dew and mists occurring in the early mornings.

Structure

In hot desert areas the land consists of bald expanses of flat or rolling plains that are sunbaked and which are covered in places with wide tracts of yellowish sand dunes, brownish gravels, rugged rocks or broken scarps of bare hills. The whole area bears the stamp of aridity. In the slightly more favourable situations some dry plants of peculiar shapes are dotted about. It is only in the very few spots with higher soil water content, the oases, that a relatively luxuriant vegetation is found.

Vegetation cover varies on different tracts or areas. In the western Sahara, for example, there are:

(a) pebbly-clayey areas where the vegetation consists of cushion plants and succulents;
(b) sandy or gravelly beds of dry water courses covered with tamarisks (*Tamarix* spp.);
(c) sand dunes with sparse cover of heath-like bushes and grass tussocks;
(d) rocky plateaus which consist of split stones

67 Near-desert vegetation, Arizona, USA
SOURCE: W H Brown (1935)

and broken rocks in the fissures of which one may occasionally find an isolated plant; and
(e) saline depressions in which may grow some low-lying halophytic shrubs in small patches.

The near-deserts of America bear the characteristic giant cacti such as *Carnegiea gigantea*, the small pin-cushion cacti (*Mammillaria* spp.), and the creosote bushes (*Larrea* spp.) (Figure 67). The Australian deserts are also remarkable for their highly peculiar plant forms. The South African desert bears the remarkable gymnosperm (*Welwitschia mirabilis*) and the desert melon (*Acanthosicyos horrida*).

As mentioned above, desert plants show various adaptations which enable them to withstand the harsh conditions under which they live. Many of the shrubs have long roots which penetrate the soil to considerable depths for water or damp layers. They generally penetrate below 10 m, and many of them, such as the tamarisks, can even reach a depth of 50 m. Other plants, particularly cryptogams, dry up almost entirely during the peak of the drought, but are able to absorb water from the atmosphere when the humidity improves. Other ways of enduring the drought include a densely tufted life form, investment with hairs and spines, and storage of water in the massive stems or swollen organs of succulents as in the cacti and cactus-like *Euphorbiae*. *Welwitschia* depends on obtaining more water from dew on its leaves than through its roots.

The ephemerals or drought escapers spring up with the advent of the brief rains and quickly pass through their whole cycle of development, bear flowers, fruit and scatter the seeds before the wet period is over, this often taking a matter of days. In the Sahara desert *Boerhavia repens* is found to germinate and set seed within ten days only. During the brief rains, geophytes too send up aerial shoots, flower, fruit and dry up before the wet spell is over.

The shrubs have various adaptations with respect to foliage. Some have very small evergreen xerophilous leaves, or leaves reduced to scales so that photosynthetic function is transferred to the green shoots or the leaf-like or succulent stems. Those shrubs that have larger leaves bear them only during the brief rainy spell but shed them after that period. In general desert plants show a number of xerophytic characteristics such as excessive development of fibrous tissues, thickened epidermis, sunken stomata, reduction in the transpiring surface, and covering of the epidermis by a thick cuticle of wax. The cell sap often has a high osmotic value. The seeds are often able to remain dormant for years without losing their viability, and so germinate whenever they receive enough water. The camel is the well-known desert animal, being able to do without water for long periods.

Oases develop in the few places where a lasting supply of water exists. These are often found along the banks of rivers or in areas where ground water rises to or near to the ground surface. Here date palms (*Phoenix dactylifera*) and a variety of tropical and subtropical agricultural crops are cultivated. The wild trees along dried-up river courses can be quite large, but they have very small leaves or thorns. On the sandy or gravelly beds are grasses. Camels and the desert type of gazelle are more common here.

Successional Relationships

Although the hot deserts presently lie in a climatic zone that would inevitably account for their ecological status, yet, at least in the Sahara desert, there is evidence that they supported a relatively higher level of plant and animal life in the recent past than they do now. This evidence comes from the discovery of the bones of amphibious animals such as the hippopotamus, suggesting a climate that fluctuated between very wet and very dry conditions which now does not obtain.

Economic Importance

Deserts are very difficult areas for human habitation in view of the harsh environmental factors. As has been demonstrated by the oases, deserts are not really poor habitats, and can support agricultural crops if only water is supplied. Many desert areas are known to be rich in minerals as well as in petroleum oil.

It seems that man's ability to utilize the deserts depends on the development of his technology. Already in Nevada in America, through the application of irrigation, air-conditioning and modern building technology, desert areas have been converted into thriving and brisk cities with luxuriant farms.

Chapter 12 Fire in Tropical Ecology

Long before the advent of man, periodic fires started by lightning or volcanic action swept the vegetation of various areas, except in the very wet rain forest. With the coming of man accidental fires have been kindled, and in modern times broken bottles and cigarettes ends are a common cause. Man as well brings about deliberate firing of vegetation to aid his farming and the hunting of game, and now uses fire in grassland management (Figure 68).

68 Effect of fire on forest vegetation
SOURCE: C J Taylor (1960)

In view of these sources of fire, and the fact that accidental firing has gone on for millions of years, most ecologists believe that the existence of a substantial portion of the tropical savannas can be attributed to fire. When these fires are experimentally excluded from woody savannas secondary succession proceeds towards the original forest-type climax vegetation. The vegetation of practically the whole of the continent of Australia and the tropical savannas has evolved in response to annual fires. Some ecologists, on the other hand, have contended that climate is the deciding factor with fire playing a minor role in the establishment of the savannas, but this view receives little current support (Figure 69).

69 Effect of fire on savanna vegetation
SOURCE: J Brian Wills (1962)

It is, of course, conceded that some areas are too dry for most woody plants, and that grasses would persist even if fire were excluded. The coastal savannas in some parts of West Africa with rather low annual rainfall can be cited as examples. There are areas too where the soil is unsuited to the growth of shrubs, even though the climate appears favourable for tree growth. Examples may be given of the patches of grassland which occur on oxysols (see Chapter 8) in tropical rain forests in some tropical areas. In spite of all this, it is evident that periodic fires have in general tended to convert much woody vegetation to grassland, leading to biotic climaxes including derived savannas at forest boundaries. West (1965) has made a special study of this in eastern Central Africa. Where succession is proceeding towards the original forest further periodic fires arrest the succession.

Heat and Temperature

Fire has many consequences on the individual plants in the vegetation and also on the soil. In order to understand these and to be able to make an intelligent appreciation of the results of experiments in this field, it is necessary to understand the physics of vegetational fires. The amount and rate of heat released when a given vegetation burns depends on such factors as weather conditions, topography and the nature of the fuel.

The weather conditions at the time of the burn, including wind velocity, as well as those of the immediate past, naturally affect the dryness of the fuel and the moisture content of the soil. Fires move faster upslope than on level ground, and much slower still downslope. Thus up a 10° slope fire spreads twice as fast, and up a 20° slope four times as fast, as on the level. The kind, amount and nature of the fuel that is available at the time of the burn would also affect the amount and rate of heat generated in a vegetational burn. As far as wind is concerned it is observed that backfires, that is, those moving against the wind, are hotter and produce their maximum temperatures nearer the ground, while headfires, those moving in the same direction as the wind, are less hot and produce their maximum temperatures well above the ground. It appears that headfires are the more common.

A number of researchers have found that the height of maximum heat generation in vegetational fires is often well above the ground, sometimes as high as 20 cm. Some of these works will be considered here. Pitot and Mason (1951) used thermocouples to measure temperatures at elevations of 0.0, 0.5 and 1.4 m in a savanna burn in Senegal. They recorded temperature ranges of 90–140, 285–560 and 140–375°C respectively. Earlier, Masson (1949) found that when the grass cover was over one metre tall the ground level temperature rose to 715°C. In derived savanna in West Africa where the fuel cover was much heavier than in Senegal, Hopkins (1965) found that when the burning took place in the early part of the dry season temperatures at the soil surface exceeded 538°C but did not rise much above the elevation of one metre, whereas when the burn took place later in the dry season this temperature extended up to an elevation of three metres.

In Venezuela, when the savannas burn, (Bentley and Fenner, 1958) soil surface temperatures do not often rise above 90°C, but the same authors found that in California the soil surface temperatures in the burning of a stand of an annual grass were up to 121°C. Cook (1939) observed that temperatures at the base of a burning grass tussock in southern Africa as fire swept over the veld exceeded 600°C. He found, however, that at a depth of 5 mm in the soil the temperature rose very little and that this higher temperature lasted only six minutes. Masson (1954), who worked on a heavy stand of grass in West Africa, found that with burning the soil surface temperature rose as high as 720°C, while the soil temperature at a depth of 2 cm rose to only 14°C. From these and other studies made outside the tropics it would appear that when African grasslands burn, soil surface temperatures are considerably higher (up to 720°C) than elsewhere, where they have seldom risen above 100°C.

Fire in Grass and Woody Vegetation

Whereas most trees are killed by fire, most grasses are adapted to survive. There are a number of reasons for this. First of all, less fuel is available when grasses burn than when trees burn, and so less heat is generated with the burning of herbs. It is observed that grass fires attain much lower temperatures than do forest fires, where temperatures may reach 1150°C. Also, these high temperatures persist much longer in forests than is the case in grasses. Secondly, the perennating buds of grasses and geophytes or seeds lie just at or below the ground surface. As shown above, the highest temperatures in vegetational fires are well above the ground. Thus shallowly placed buds and seeds are able to escape damage. Trees, on the other hand, have their buds within the level of the highest fire temperatures. Thirdly, in view of the relatively short time required for herbaceous plants to redevelop normal shoots, the soil is not left bare long enough to permit new micro-climatic conditions to prevail. In the case of trees, the fact that they take a much longer time to regenerate means that the soil surface micro-climate may become so altered that regrowth is greatly hampered. Fourthly, perennial grasses produce abundant seeds one or two years after germination, while most woody plants reach seed-bearing age only after several years. Thus, fire inhibits seed production in woody plants, so that in course of time new trees and shrubs diminish while grasses increase.

What seems to be an exception to this trend is reported in the Venezuelan savanna. Here the fire tends to kill the grass in circular patches centred on isolated trees; but, somehow, before the grass can re-invade the burnt area, the seedlings of new woody plants join the sprouts from the old ones in initiating a small grove. In this way woody plants tend to expand rather than to be restricted in these Venezuelan grasslands as a result of the action of fire.

It is obvious that in general there is much loss of energy when the massive perennial stem of the tree burns, whereas this is much less so with herbaceous plants in view of the very little supporting tissues involved.

It should be mentioned here that fire commonly favours forbs over grasses in both annual (Biswell, 1956) and perennial grassland (Cushwa and Redd, 1966) but there are apparent exceptions in which grasses are favoured at the expense of forbs, as reported by Cook (1965) in the Transvaal.

Time of Burning

Recent analysis of various studies of the effects of fire on vegetation (Daubenmire, 1968) has shown that many of the apparently contradictory results can be resolved when the time of application of the fire is taken into account, all other factors being equal. The time of year, or even the time of day, when the fire is applied has been shown to be important in nearly every effect of burning.

Examples illustrating this are given under the different effects discussed below. Komarek (1965) has drawn attention to the fact that one of the sources of difficulties in trying to compare results of the effects of fire on vegetation, arising from intentional and accidental fires, is that whereas accidental fires often occur in the dry season, intentional fires are not so restricted but are applied at any time of the year in order to maximize the desirable effects of fire in grassland management.

Burning Versus Mowing

It seems that when grassland is mowed and the cut material is removed, the effects produced are very much like those brought about by fire. Thus in Natal, southeastern Africa, Scott (1951) reports that annual mowing of grassland maintained it in much the same condition as annual burning. On the other hand, when *Trachypogon* savanna in Venezuela was burned during the dry season in November there was a greater increase in shoot production and plant height than was produced by clipping on the same date. However, when the same treatments were repeated during the rainy season rather opposite effects were produced.

Effects of Fire on Soil and Vegetation

Erosion Effects

One of the common assumptions is that, as a result of burning, erosion is accelerated. The basis for this assumption is that when much burning occurs in the dry season in the relatively drier savanna areas, the soil is exposed in places for some months. When torrential rains arrive later they beat directly on the bare ground which has little vegetation to take the shock of the impact. This would appear to be quite a reasonable way of explaining erosion in certain areas.

However, reports of the effect of fire on erosion have not been as consistent as one would have thought. Du Plessis and Mostert (1965) and West (1965) reported that in the drier parts of southern Africa annual burning accelerates erosion and runoff even on rather flat land. Shaw (1957) reported similar results from subtropical Queensland following the annual burning of *Heteropogon contortus* grassland. On the other hand, Edwards (1942) found that regular burning of grass did not cause erosion in Kenya. Nye and Greenland (1960) made similar reports on tropical savanna in parts of West Africa. It is not easy to resolve these apparently contradictory results. What would seem to be of greatest importance here is the time in the dry season when fire is applied. As already stated, if the vegetation is burnt early, the land is bare for a much longer time before the rains and erosion is likely to occur, whereas if the burning occurs late in the dry season, the soil will not be bare for long and erosion may not occur. In view of this Cook (1965) has recommended that burning as a management practice should be carried out fairly late in the dry season if erosion is a serious problem in the area concerned.

Other Fire Effects on Soil

The effects of burning on soil humus have been of much interest. In southern Africa, Cook (1939) observed no differences in the humus of the upper 38 mm of the soil between plots burned annually for six years and unburned ones. Later Edwards (1942) showed that the humus in a grassland in Kenya, protected from burning for ten years or more, increased; this naturally gave the impression that burning might reduce the humus content of the top soil. The work of Moore (1960) helped to appreciate the problem. He found that 30 years of annual burning in Nigeria led to an increase in the humus content of the upper 20 cm of the soil by 17 per cent where the burning was light and was applied early in the dry season. On the other hand, the humus was reduced by 12 per cent where the fires were kindled late in the dry season and were thus much hotter. Moore's findings evidently show that early burning must stimulate productivity resulting in the higher humus levels in the top soil, whereas late burning does not.

Soil moisture is also affected by some of the changes that are induced by the burning of grasslands. In Central and southern Africa various workers have shown that fire reduces the water-retaining capacity of the soil, so that the grass becomes more susceptible to drought injury (Cook, 1939; Phillips, 1919; Scott, 1934; and West, 1965).

The burning of plant tissue is known to result in the volatilization of nitrogen and sulphur. Other nutrients are changed into simple salts which are more readily available to the soil and plants because of their water solubility. In view of the great mass of woody material in trees, a considerable quantity of ash is produced following burning, and this serves as a fertilizer to the soil, while the release of nutrients by burned grass is negligible by comparison. This has been supported by the work of Nye (1959) in Ghana as shown in Table 19.

With the exception of nitrogen and sulphur which volatilize, no other loss of nutrients is brought about by fire. The main effect of fire, then, is bulk release of nutrients at one time to the soil surface instead of the gradual release that normally takes place as a result of decay by decomposers. As a result of the rapid release, some of the nutrients leave the habitat through the action of wind or water without benefiting the plants or animals. Some may also be leached through the soil horizons so fast that they cannot be absorbed by soil colloids and utilized.

Moore (1960) has reported that in the Nigerian savanna, mild fires at the beginning of the dry period resulted in increases in cation exchange capacity, in available phosphorus, in exchangeable calcium, magnesium and potassium and in percentage base saturation. However, hot fires coming late in the dry season were found to reduce the cation exchange capacity and exchangeable calcium and potassium, while available phosphorus and the exchange capacity remained unchanged and the percentage base saturation only slightly increased. Edwards (1942) also reported that in Kenya the base exchange capacity of regularly burned grassland decreased, probably as a result of a reduction in the humus content of the soil.

Table 19. Estimated quantities of nutrients released by burning tropical vegetation, in kilogrammes per hectare

	PHOSPHATE	POTASSIUM	CALCIUM	MAGNESIUM
Savanna	8	46	35	26
Tropical rain forest	127	830	2560	351
	(16 times as much)	(18 times as much)	(73 times as much)	(13 times as much)

As far as soil nitrogen is concerned, Cook (1939) has reported a small reduction in the total nitrogen of the upper 38 mm of soil resulting from regular burning of *Themeda* grassland in southern Africa. The results here appear quite contradictory although in many cases it seems that there is a definite increase in nitrogen following burning. Moore (1960) studied the situation in derived savanna in Nigeria. In plots protected from burning he found nitrogen to be the lowest. He also found that when derived savanna in Nigeria was burned early in the dry season when the heat was moderate, nitrification was much increased; but when the savanna was burned later in the dry season the hot fires produced an opposite effect. Nevertheless both times of firing appeared to raise the carbon/nitrogen ratio almost equally when compared with unburned savanna. The nutrient (and nitrogen) content of the soil is often increased where burning increases the proportion of legumes in the community. In Tanzania, however, legumes have gradually disappeared with annual firing so that the soil has lost much of its nitrogen in the course of time.

Soil biota are also affected by burning. Meiklejohn (1955) conducted burning experiments in Kenya and studied the microflora of the top 25 mm of soil for five months. She noticed a fall in the total number of organisms in the microflora for the first three months before any recoveries took place. Although fungi appeared to have been completely destroyed at first, they were back to normal two months later, with *Penicillium* dominating as before the burn. Aerobic nitrogen fixers, such as *Clostridium*, even though poorly represented, persisted through the burning. After five months, sampling showed that total nitrogen had increased. This was taken to have resulted from the destruction of nitrifiers so that the conversion of much nitrogen into soluble forms that might be leached had not taken place.

Some Factors Modifying the Effects of Fire on Plant Populations

We have already observed that burning late in the dry period does more damage to the vegetation itself than burning at the beginning of the dry season. Allied to this is the observation that the drier the climate the greater the damage by fire. Bossman (1932) observed in southern Africa that fire is detrimental to grassland only in areas with less than 750 mm rainfall per annum. In Australia, it was reported by Burbidge (1943) that *Triodia pungens* is usually killed by fire in inland areas but not along the coast.

A number of characteristic features of certain species also determine the degree of damage that fire can cause to them. One of these is the phenologic condition of the species. Aldous (1943) stated that generally as new foliage of perennials reaches full size, the major part of the food reserves will have been withdrawn from the underground organs, so that fire destroying leaves at this stage injures the plant most severely: Robocker and Miller (1955) observed that when fire ran through a mixed planting of grass species in Wisconsin, species which had started growth early were damaged whereas the late growers were not. In mid-continental grassland of North Africa it is often observed that *Poa pratensis*, whose genus is of northern derivation and starts growth early, is selectively damaged by fire, whereas the native dominants such as *Andropogon, Sorghastrum* and *Panicum*, which are of southern origin, escape damage because they remain dormant until the warm season is more advanced.

Growth form is another plant characteristic modifying fire damage, and species with different growth forms may be affected differently by the same burn. Blaisdell (1953) observed in eastern Idaho that rhizomatous and annual steppe grasses were stimulated by burning, whereas the same fire damaged the suffrutescent plants of the area. Also, in the savannas of central Brazil, Rachid-Edwards (1956) has observed that many species of grasses and other herbs escape damage by the frequent fires because dead leaves form a tunic about the perennating buds. *Aristida stricta* and *Sporobolus floridanus* have closely packed persistent leaf sheaths which exclude oxygen and so do not burn. This provides a good insulation for their buds (Lemon, 1949).

Effects of Fire on Seed Germination, Vigour and Plant Vitality

There is a well-known notion that fire stimulates the germination of certain pasture species. When hot fires descend to the level of the soil surface buds and seeds may suffer lethal temperatures. Pasture seeds which are not killed by fire may be stimulated to germinate earlier. West (1965) observed that the germination of fresh seed of *Themeda triandra*, an African grass, was not only favoured by fire but was significantly increased when treated with dry heat. In tropical Australia, Shaw (1959) observed that seedlings of *Heteropogon contortus* appeared in great numbers only on burned areas. He took samples of the soil both before and after burning and germinated some of the seeds of the grass in them. He observed that equal numbers of seedlings were produced in the glasshouse environment as in the field, and he concluded that, in this case at least, the stimulation was brought about by the altered environment and not by a change in the condition of the caryopses. Much more experimental work is required on this phenomenon.

Apart from germination, fire shows varied effects on the subsequent size of vegetative organs of herbs. On the coastal plain of Georgia, fire was reported to have reduced the size of the vegetative organs of both *Paspalum notatum* and *Cynodon dactylon* (Burton, 1944) although associated species were not affected. Shaw (1957) also reported that while fire sweeping through *Heteropogon contortus* grassland in Queensland, Australia, killed parts of the bunches, there was a marked increase in the basal areas when compared with those of the unburned areas. In southern Africa, West (1965) observed that where there was relatively more soil moisture, basal area increased after fire, although this declined later as litter accumulated.

Many species of grasses and other herbaceous plants are found to produce scanty seed crops from one year to the other if there is no disturbance, but they are noticed to flower in great profusion during the first or second flowering season after the application of fire (Biswell and Lemon, 1943). Among the species reported to show increased flowering in response to fire are *Aristida* in Florida (Lewis, 1964) and *Themeda* in Africa (Brynard, 1965). Other species show a negative response in flowering to fire; among these is *Trachypogon montufari* in Venezuela.

The low vitality of unburned grass has been explained by O'Connor and Powell (1963) as resulting from self-generated micro-climate of the accumulated litter. Evidence in support of this explanation has been given by Curtis and Partsch (1950) in Florida and Wisconsin by mechanically removing litter and obtaining as much increased vitality as with burning. An increase of 600 per cent in the number of inflorescences and an increase of 60 per cent in their height were obtained by Curtis and Partsch. Not only does the vigour effect of fire show in quantitative terms. There is also the precosity with which the stimulation occurs. A number of writers have stated that the new shoots that come up after a fresh burn do so about one to three weeks earlier. The precosity which also results in the flowering period being advanced may disappear during the course of the season or it may persist in the second post-burn year. It should be mentioned that in some species fire rather delays the sprouting of new leaves.

Since in tropical climates dormancy is correlated with a dry rather than a cold season, grass is able to sprout within a matter of days after the application of the fire. Thus the new leaves of *Aristida stricta* in Florida appear within three days (Lewis, 1964); *Andropogon schirensis* and *Monocymbium corisiiform* in Nigeria sprout within six to ten days (Hopkins, 1963); and *Trachypogon* spp. in Venezuela within nine days.

The general improvement in pasture following the application of fire has been variously interpreted. The improvement in pasture is often expressed in terms of increased vigour or an increased uptake of nutrients. One hypothesis is that this is a direct result of a sudden increase in available nutrients derived from the ash of the fire. In support of this, Heyward (1938) stated that if a fire does not raise soil temperatures above 100°C, soluble salts and ammonia are increased in the soil, and Hart *et al.* (1932) observed that where the soil was high in fertility, burning had less effect on plant chemistry. Laboratory experiments have shown that within the range of 3–11°C at least, the higher the temperature the more rapidly ammonium is oxidized to nitrate (Anderson, 1960), and that raised temperatures almost invariably result from burning. However, different conclusions can be drawn from other evidence, and so there is no general acceptance of this hypothesis yet.

Another possible explanation for the higher nutrient content of foliage on burned areas is that living but senescent parts of previous year's tillers may depress the growth of new tillers by competing with the latter for nutrients. In support of this Mes (1958) found that decomposing material from detached foliage increased nitrogen and ash content of new grass shoots, but intact foliage from the past season, still feebly active, depressed nitrogen and ash content.

Still another possible explanation is that fire stimulation may increase the activity of the roots of the burned plants, this in turn increasing the uptake of nutrients (Mes, 1958).

Fire Effects on Plant Communities

We have already found that different herbaceous species growing together may respond differently to the same fire. The same applies to woody species. Those that are not destroyed by repeated burning are the fire resistant ones (which have rather thick barks) and those that sprout from their roots. Apart from these most trees are very susceptible to fires and the branches may be eliminated or the trees reduced to a thin stand of small individuals. This is one reason why fire is widely used as a means of maintaining grass dominance in environments where woody vegetation would otherwise take over. This is the position in tropical savannas.

In southern Africa, regular burning is used as a means of maintaining the best quality of forage and keeping the desirable *Themeda triandra* dominant (Botha, 1945). This is so desirable that even some loss of productivity is often sacrificed in the process. Burning savanna grasses like *Chrysopogon, Sorghum* and *Themeda australis* in northern Australia, on the other hand, reduced their productivity (Smith, 1960); the later the burn in the wet season the greater the reduction in productivity. In Venezuela, Blydenstein (1963) has reported that January fire in *Trachypogon* savanna reduced shoot production by about a third in the first post-burn season. Burning in March also reduced it slightly, but burning in November or December, which is the beginning of the dry season, resulted in about 50 per cent increase.

From this and other effects of fire that have been described so far it is evident that in an evaluation of the effects of grassland fires on total productivity of the plant community, account should be taken of the other changes in the character of the plants. Thus, a net increase in inflorescence development, for example, may more than offset a reduction in the amount of vegetative growth. Fire may increase basal area without having much effect on production.

Fire Effects on Animals

It is a well-known observation that new grass following recent burns is attractive to hoofed grazing animals probably because of increased palatability, and cattle are found to make great gains in weight when they graze on recently burned grassland as opposed to unburned areas. Many fully grown ungulates are able to escape fire, but their young are often destroyed.

Fire causes much harm to ground nesting birds. Their nests are destroyed and the protective cover which may be used in constructing new ones may also disappear. Their insect food sources may also be eliminated. Mice and other rodents suffer loss of their essential food and grass cover, but they are better protected by virtue of the fact that they rear their young in burrows. Insects are also destroyed by fire which may be said to exert some measure of control over insect pests in the field. *Coleoptera* are found to be abundant on recently burned areas with little or no cover. Grasshoppers appear to follow a little later when the vegetation is recovering from a burn.

Fire and Pasture Management

Fire is now used as a tool in the management of forests and grasslands in a number of tropical and temperate climates. In East Africa fire has been used to keep the grass *Themeda* short enough for easy grazing by cattle and sheep. Farmers in various tropical countries use fire in one way or another in their agricultural practices. Among these is the stimulation of fresh growth of grasses for livestock. In the West Indies and Cuba, fires are used in sugar cane plantations to burn off much of the old leaves in order to facilitate harvesting of the canes.

From our present knowledge of the effects of fire there are a number of lessons which can be drawn for pasture management. For example, the increased palatability of grass on fresh burns raises a problem concerning the use of fire in range management. If animals are given access to the new grass, they crop it heavily and leave insufficient photosynthetic tissue to allow the plants to recover. However, this situation can be exploited to advantage. Grant *et al.* (1963) reported that *Molinia caerulea*, which is mainly unpalatable except just after a burn, can be burned and heavily grazed and thus severely damaged, so that it may in part be replaced by more desirable species.

Also, perennial grasses in the wet tropics are found to have a lower protein content, even when they are young, than equivalent temperate zone species. The increase in nutritive quality after a burn falls rapidly as the foliage matures, and within about five months the foliage becomes poor for the nutrition of livestock. In southern Africa it has been a common practice to burn half a grazing unit every six months in order to maintain a continuous supply of satisfactory forage (Botha, 1945). It is therefore easy to see the benefits of burning grasslands at least once a year in the moist tropics. The fires remove unused grasses that have lost their food value and constitute mechanical hindrance to grazing. The fires also maintain new herbage in satisfactory nutritive condition. In grassland management it would appear advisable that grazing be reduced in advance of burning to allow sufficient litter to accumulate to carry the fire. Also grazing must be withheld for some time after the fire, to prevent the excessive use of the new foliage to the detriment of the plants.

It may be said in conclusion (with Phillips, 1966) that fire is on the one hand a good servant when it is used with discretion and at the right time in order to obtain the right effects, and on the other hand a bad master if allowed to be used indiscriminately. There is no doubt that fire still remains the most convenient and the cheapest method of controlling the selectively utilized pasturage and browse in the heterogeneous terrain and vegetation of most parts of the tropics and subtropics. In tropical vegetation variable palatability and growth form make it difficult to secure uniform grazing. In addition many physical obstacles such as woody growth, termitaria, rocks and topography always make the mowing of natural pastures difficult and expensive. We do not as yet have any satisfactory herbicide or arboricide for large scale use on such habitats, and fire for the present, seems to provide the best answer. It is essential that we intensify our studies on all effects of fire on vegetation so that we shall be better able to utilize fire to our advantage.

Chapter 13 Biological Conservation and Problems of Pollution

The permanent functioning of an ecosystem depends on the maintenance of each part or each trophic level of it in a healthy state. This means that if exploitation, pollution and other human actions of deleterious effect reach a point from which any part of the ecosystem cannot recover, then the whole ecosystem suffers permanent damage with serious impoverishment or death.

Conservation is aimed at keeping actions that are harmful to the ecosystem at such a level that a proper balance is maintained between its various parts so as to enable the system to renew itself and be able to continue to provide man with the natural resources which he requires from the ecosystem. Just how to control the activities of man in order to ensure the survival of any ecosystem, whether living organism or the biosphere at large, is the task of conservationists. Conservation is thus an exercise in the proper maintenance, management and utilization of natural resources and ecosystems for the long term benefit of mankind. All species of plants and animals are potential resources, especially genetic resources, which may be highly localized and totally irreplaceable. When preserved, they also serve as 'controls' in our studies of the effects of man on ecosystems, quite apart from their aesthetic enjoyment by man. The International Council of Nature and Natural Resources defines conservation as 'the rational use of the environment in order to achieve the highest sustainable quality of living for mankind'.

From the above, it should be clear that conservation does not mean the placing of a ban on the use of resources. It can be likened to the use of the interest accruing from a deposit account in the bank. However, in certain situations an apparent ban may be enforced. This may be done primarily to reserve examples of unmodified communities. There are minimum sizes for all ecosystems below which they will continue to deteriorate whether managed or not. Maintaining the survival of a species of animal or plant therefore requires a certain minimum area and number of the species. Moreover, conservation is not simply a matter of leaving natural areas to their own devices. It requires care and is thus expensive. In view of the cost, and of the desire to recover some of the expenses, areas of conservation have often been open to tourists, who are mostly foreign tourists who pay and are regarded as sources of income and foreign exchange. Tourism, it must be appreciated, includes local tourism as well. However tourism, even in its broadest sense, is not by itself a good enough reason for conservation.

Another aspect of conservation is the rehabilitation and restitution of damaged systems, as may occur on abandoned mining sites and overfarmed or overgrazed lands.

In the advanced countries the increases in human population and industrial activities have led to the reservation of areas of natural vegetation. In the developing countries, most of which are tropical, when people look around them and find that they still have so much 'vegetation', they cannot see why they should be asked to set aside any ecological habitats. In most of these countries only forests are reserved, but even here there is pressure on the authorities to make them available for exploitation which is often carried out to the utter destruction of the forests. As for other habitats, hardly any of them are reserved as sources or controls for the study of man's present and future activities. Some habitats are already becoming scarce and soon a number of them will vanish forever. Ecologists in tropical countries have a duty to advise their governments to take steps to reserve adequate samples of all types of ecological habitats in their countries.

Three important factors seem to be making the efforts of ecologists at solving the problems involved appear like chasing a mirage. These are: the unlimited human population increase; an accelerating using-up and impoverishment of the renewable natural resources through unwise land use; and technological growth, which is happening on an increasing scale and either directly or indirectly pollutes the environment, while at the same time speeding up rates of destruction and of population increase. Contrary to what most people think, technological growth does not necessarily increase the standard of living for all, although it may for some.

In most tropical areas, the land use problems are of prime importance. At present in many parts of the tropics the water resources are fading out, permanent rivers are becoming seasonal and many previously perennial rivers are now dry throughout the year. Coral reefs, which serve as spawning sites and nurseries for economically important marine fish, are polluted or dynamited. The rates of soil erosion are frightening; man-made deserts are spreading in various directions; previously forest-clad mountain slopes lie bare to the very rock like the bones of a skeleton. The savanna vegetation deteriorates or is converted to arid lands and the lowland forests become savannas. With the disappearance of the vegetation go the wild animals, which, particularly in Africa, represent a tremendous potential in the form of proteins, hides and tourist revenues. The bad treatment of all these resources creates a vicious circle that affects human life.

Efficient, ecologically based control and management guidance over population, land use and technology are necessary for the future of any human society. If such measures are not taken in the tropical countries within the next twenty years or so, there is little hope. Yet, the present ecological situation of most parts of the tropics would not appear to be too desperate in comparison with many other parts of the world. This gives us time to plan carefully; and such planning must, of course, be based on ecological realities.

It is reassuring to note that as a step in the right direction, African countries have signed a convention on conservation. The preamble to the African Convention on Conservation of Nature and Natural Resources, signed by the African Heads of State at Addis Ababa in 1968, states that they are 'fully conscious that soil, water, flora and faunal resources constitute a capital of vital importance to mankind' and that they are 'desirous of undertaking individual and joint action for the conservation, utilization and development of these assets by establishing and maintaining their rational utilization for the present and future welfare of mankind'.

Article II of the Convention deals with what it calls a 'Fundamental Principle' and states that 'the contracting states shall undertake to adopt the measures necessary to ensure the utilization of resources in accordance with scientific principles and with due regard to the best interests of the people'. Article XII on 'Research' states 'the Contracting States shall encourage and promote research in conservation, utilization and management of natural resources and shall pay particular attention to ecological and sociological factors'. In Article XIV there are three obligations concerning 'Development Plans', which are formulated as follows:

1. The Contracting States shall ensure that conservation and management of natural resources are treated as an integral part of national and/or regional development plans.
2. In the formulation of all development plans, full consideration shall be given to ecological, as well as economic and social factors.
3. Where any development plan is likely to affect the natural resources of another State, the latter shall be consulted.

There are a number of significant ecosystems in tropical countries which ought to be preserved. Among these are samples of tropical rain forests, mangrove swamps, estuaries, coral reefs, inland lakes, tropical montane flora and fauna, peat bogs, specialized forests like the luxuriant Miombo Forest of Shaba (formerly Katanga) and a number of other biomes. There is also a number of areas with endemic species of plants and animals which ought to be preserved.

The management, utilization and exploitation of water, soils, plants and animals must be related to the rate of renewal and maintenance at optimal level of all these resources. In doing so the complex interactions between all these natural resources must be understood and continuously followed by ecological investigations in order to avoid environmental calamities. The ease with which so many natural balances in the form of productive habitats and ecosystems have been upset in various parts of the tropics, and the seriousness of the ecological and social consequences of such environmental deterioration, form the challenge facing ecologists today.

The tropical countries have serious environmental problems, often of quite a different character from those which today are so debated in the industrial countries, where the consequences of the technological revolution are increasingly threatening the environmental quality and human health. Any economic and technological development that ignores ecological factors may cause long-term economic and social disorders which may hamper or even make impossible the purposes it was meant to fulfil. In many cases, the results of development may be entirely negative and constitute an environmental degradation of such a magnitude that it takes generations to repair it if the damage is not irreversible. Such an unnecessary destruction of the ecosystem will continue if there continues to be lack of respect or understanding of ecological realities in development plans.

Thus, there must be a much closer harmonization between development goals and ecological considerations before economic and social decisions are made. In order to ensure this harmonization, some knowledge of ecology and conservation is needed by those involved, especially physical planners and engineers. For example, if an engineer in West Africa were aware of the ten year cycle of fluctuation in rainfall in Africa, he would be able to design his bridge to cope with the peak of such fluctuations. If our agriculturists appreciated the frequency of major droughts they could warn farmers in time. We should not tacitly allow the engineers to look at their assignments, which naturally destroy the ecosystem, as isolated projects without considering their ecological implications. Another false impression in this regard is that any artificial construction is looked upon as if it were a superior creation. The fact that a farm, for example, is good now blinds us to any evils that have been perpetrated in the process of its establishment. Most car owners have come to realize that a small fault in an engine can render the whole unusable, but the same people do not have the same understanding of the soil, the vegetation or natural water.

We can now consider the important parts of the ecosystem with regard to the problems of conservation and pollution.

The Soil

In our discussion of the role of the different parts of the ecosystem (Chapter 6), we noticed that the most indispensable part of the natural ecosystem is the decomposers which inhabit the substrates of ecosystems, namely the soil or water. The first aim of conservation should be to prevent the destruction of the soil and the water.

The easiest way to destroy the soil is to remove its cover and so expose it to heavy rains which erode the deep layer of spongy humus-rich top soil containing the decomposers. As soon as this happens the original vegetation of the area will have gone virtually forever. This is because by removing the tree vegetation, one removes the bulk of the fertility at one disastrous blow. The soil which has thus lost its humus-rich top soil cannot absorb much water which therefore runs off it, and so only very poor vegetation is maintained. The erosion which follows such action is serious in many tropical countries (Figure 70) with the

70 Top: Gully erosion
Bottom: Sheet erosion
SOURCE: J Brian Wills (1962)

peak occurring perhaps in Ethiopia where about 2000 tonnes of soil per square kilometre is lost per year.

The modern practice of agriculture has meant the increasing use of artificial fertilizers and pesticides, leading to pollution of soil and water. No doubt these help considerably in increasing food production. Unfortunately the organochlorine pesticides such as DDT, Aldrin and Dieldrin are persistent for a long time in the soil and in the environment as a whole, including the plants and the animals which are used as food by man. These pesticides are now being used in an increasing number in many tropical countries in the cultivation of agricultural crops. If no precautions are taken the problems of their accumulation in useful and beneficial birds, fish, mammals and man, sometimes to lethal levels, may become unsurmountable. The aim of conservation here should be to find pesticides which do not persist long in the soil and although the search is going on, the solution is not yet in sight.

Another source of damage to soil occurs through oil leakage from storage tanks. In Nigeria, oil leakage in the Rivers State has polluted the soil and destroyed many hectares of mangroves and farms. Mining activities, also, have destroyed soils in many tropical countries. Where erosion or mining has denuded the soil, a programme of hydrological management and revegetation should be carried out. Research should also be carried out on plants that can grow or initiate succession on denuded or poisoned soil resulting from dumping of industrial wastes.

Vegetation and Agriculture

The tropical forests have taken millions of years to reach their present almost stable state as an ecosystem. It will be impossible to replace them once we have destroyed them. In many tropical countries many of the forests have been destroyed within a matter of a few years. In the past many of these areas had as their natural vegetation cover rain forests which were among the most stable and productive in the world. To earlier generations these forests must have seemed inexhaustible. For many centuries their exploitation consisted almost entirely of the primitive slash and burn type of farming. Such a farming regime at the time of fairly sparse population left areas which remained abandoned for many years before the next clearance, and the forest had a chance to renew itself.

In recent years two factors have contributed to the destruction of much of the forest, namely the realization that there was a tremendous timber crop, and the increase in population following the advances in medicine that have taken place in the last thirty years. The harm caused by the timber industry to the forests is now incalculable (Figure 71). This is because the tropical forest is certainly one of the most productive kinds of vegetation, but most of the minerals and the energy in the forest are locked up in the vegetation itself. By removing the trees, either as timber crop or by burning them, one is removing the bulk of the fertility and exposing the soil to erosion.

Many have argued that the timber is a valuable export and as such must be exploited to enable the people to make a living and the country to earn foreign exchange. It is true that it would not be a wise conservation policy that would totally ban forest exploitation. Exploitation of the forests must be allowed, but it should be carried out in such a way that they are preserved for further exploitation. The only way to maintain a forest to produce timber in perpetuity is to manage it on a 'sustained yield' basis; that is, every tree removed must be replaced by another. Felling must therefore be highly selective, and phased in such a way that the forest is maintained in as natural a state as possible. If this can be achieved then the future of the forest resource of the tropical countries is assured. For this to succeed there is need for a planned replacement policy. There is also the need to industrialize the exploitation of timber in order to improve the efficiency of the exploitation. This would, among other things, put an end to the wasteful methods by which much timber is felled and left to rot because the methods used are so inefficient. The purpose of a wise conservation policy here is not only to save the forests themselves and the soil. It is also to protect agricultural productivity. Many of the tropical tree crops such as cocoa, rubber and coffee require a fairly high fertility and a high humidity as provided by forest cover, so that as the forests are lost it becomes more difficult to cultivate these agricultural crops.

As has been pointed out before, the use of DDT and similar chemicals in agriculture is serious for human health in view of the residues that remain in the plants and crops (Figure 72). Although DDT has been banned or its use very carefully controlled in many advanced countries, it is still

71 Destruction of tropical forest for timber
SOURCE: J Brian Wills (1962)

used rather freely in a number of tropical countries, as in malaria control in Ethiopia, in the control of termites in Senegal and in the control of diseases on cocoa in Ghana. While many developing countries have yet to obtain better substitutes for DDT, it is essential now to take precautions about its application in order to minimize its dangers, because in the long run the chemicals enter the body tissues of man through the crops on which he feeds.

In many tropical countries where mangroves exist, they have been much destroyed for firewood and charcoal, as in the Ivory Coast and South Vietnam. It is not generally realized that mangroves are particularly suitable sites for rice production. They also serve as sites for the spawning of some marine animals, and so their destruction would lead to the disappearance in time of certain fish, molluscs and crustaceans.

Wild Life

The expression 'wild life' is used mainly to refer to game vertebrates, and to the plants and lesser animals which interact directly with the game species. Other plants and most invertebrates, for example, grasshoppers and snails, are often

72 A farmer spraying his farm with carbide Bordeaux mixture
SOURCE: *Cocoa Growers' Bulletin* (1972)

forgotten. These also form part of wild life in the strict sense of the term, and are in fact more productive than vertebrates.

The destruction of the forests and other natural habitats is accompanied by the disappearance of the wild life in them. The rapid rate at which wild life is disappearing in some parts of the tropics is alarming. In many parts of West Africa, for example, only 50 years ago large areas were teeming with game such as the elephant, the antelope, the buffalo, the hartebeest, the waterbuck and several other large grazing mammals. Many of these have been thoughtlessly exterminated over much of the area by hunters for 'sport' and by tsetse control workers in a futile attempt to eliminate the pest. Now much poaching goes on for the meat of the remaining animals as food. This is often carried out by hunters who work at night, using lamps on their heads to blind the animals and shooting whatever creatures are transfixed by their lamps, including immature animals and pregnant females. One cannot blame the hungry man for trying to fill his stomach, especially when he is trying to fill it with protein which is in chronically short supply all over the tropical countries, but people must be made to realize that the present rate of destruction will result in the permanent loss of the protein source which they need.

Serious thought should go into the planned cropping of wild life for meat, to supplement and even replace where possible the introduced domestic cattle, sheep and goats. Unsuitable breeds are known to graze destructively; they also compact the soil and hasten its eventual destruction by increasing run-off and erosion. However, suitable breeds, developed on farms with proper modern management and feeding, can be highly productive and not cause environmental deterioration. The wild life on the other hand are in balance with their environment. Research, and the commercially successful ventures in East Africa, through their conservation centres such as national parks, nature reserves, sanctuaries and other gardens, demonstrate what can be done with conservation and planned grazing of wild life for meat. The great herds of grazing mammals of East Africa in particular constitute one of the wonders of the world, being the only really large herds left to man. They contain an extraordinary variety of species, in balance with their environment, each species grazing on a different range of plants, the whole living together very successfully. They certainly are among the most productive in terms of protein in the world. In one area in eastern Zaïre, a recent survey in a savanna area shows that the largest and most conspicuous mammals amount to a biomass of almost 250 kg per hectare, a remarkably high figure. What is more important is that the vegetation can support it without damage. In particular savanna grasslands retain high proportions of calcium and nitrogen which enable them to be higher protein producers than the forests.

What is still more interesting is that many of these large mammals can be domesticated. One of the most successful is the eland, a cow-sized antelope in East Africa. It can be milked, its meat production is high and, most important, it is in balance with its environment. So also is the ibex, a type of goat in Ethiopia of which only about 400 are alive today. Many such animals could be managed on a range basis. Careful cropping of mature animals, at such a rate as to maintain the correct population sizes, would produce a certain and regular supply of protein and at the same time maintain the herd size. If more game reserves or national parks were created in tropical countries and the herds of wild life allowed to build up again, not only would tourist revenue increase but so would the food supplies which man receives from them.

Water Resources

As mentioned earlier, the organo-chlorine pesticides such as DDT, Aldrin and Dieldrin, which are long-persistent in soil and vegetation are eventually drained into lakes, rivers and streams where they can poison fish. Eventually, the pesticides get into and accumulate in man through his food. It is abundantly clear that the accumulation of pesticides in the body tissues of animals and man is coming to be recognized as an urgent problem in a number of tropical countries. In a recent Report on the Human Environment in Kenya, for example, it is stated as follows: 'In a study carried out in Nairobi of the adipose tissue of humans and baboons from various parts of the country, residues of DDT, Dieldrin and BHC were detected and none of the samples were free of such residues.' The total DDT in humans aged 25–40 years averaged a mean value of 4.60 parts per million; in baboons the total DDT averaged 0.07 parts per million.

A number of attempts are being made on a small scale by various existing agricultural and veterinary laboratories to set up analytical facilities for pesticide monitoring but, on the whole, the scope and level of accuracy attainable by these well-meaning attempts are quite inadequate. Alternatively, some field workers have been sending samples overseas to such countries as the United States for analysis at inordinate expense and involving long delays in getting results. Urgent consideration should therefore be given to the setting up of regional laboratories in important centres in tropical countries which would provide large-scale facilities for pesticide monitoring to the highest international standards. Such laboratories should be capable of handling quickly samples from all countries in the region. Their setting up should be preceded by a careful assessment of demand over, say, the next decade, on the assumption that all the countries in the region had a potential pesticide problem.

The use of fertilizers also creates problems in that they dissolve in rainwater, and it is impossible to prevent some of them percolating through to rivers and streams through the drainage systems of the soil. Nitrates and phosphates are two of the main constituents of these fertilizers. When they are added to rivers and lakes they cause increasing eutrophy resulting in changes by which many of the more desirable elements of the flora and fauna die off, and fish, in particular, are unable to survive.

Tropical countries have to face the problem of preventing the occurrences that have taken place in the advanced temperate countries. In the advanced countries we find some of the world's largest lakes suffering from pollution. Lake Erie in the United States, for example, has become almost useless for fishing. Even once remote Lake Baikal in Siberia is suffering from pollution.

The problem is even more serious when it comes to pollution in the sea since pesticides in the sea travel all over the world and may affect fishes anywhere. A major source of pollution here is crude oil. Leaks from tankers are becoming more frequent and off-shore drilling rigs have caused bad spillages, notably off Santa Barbara in California and in the Caribbean. The problem is still real in a number of tropical countries where off-shore oil has been discovered. As mentioned above, this has recently occurred in Nigeria's Rivers State, with considerable destruction of mangrove following the pollution or death of the soil in the area.

There is also the problem of weed control in large bodies of water. A number of developing countries have, within the last 30 years, constructed man-made lakes primarily for hydro-electric power but these are also being used for irrigation agriculture, fisheries and transport. Without keeping the water clean it cannot be used for fisheries and irrigation agriculture. More recent examples are the Aswan High Dam in Egypt and the Volta Dam in Ghana. With increasing use of fertilizers it would be all too easy to wreck these dams as a habitat for fish before their immense potential as a food source is developed. The use of herbicides in these lakes and in rivers to destroy water weeds has to be strongly discouraged since it pollutes the water. Even though *Pistia stratiotes* hampers fishing in rivers, the fishermen should be educated against the use of pesticides to kill these weeds.

Estuaries are known to be very productive of marine and fresh water animals and they need to be conserved. It is reported that the construction of a new harbour at Cotonou in Rep. Benin led to a sudden drop in the catch of fish in the adjoining estuary from 10 000 to only 2000 tonnes a year. In some tropical countries the leaves of certain plants are often ground and put into rivers to aid fishing since their contents are poisonous to fish. They in fact kill other animals and plants. This leads to serious pollution of the rivers which become dangerous for drinking purposes. An example of this is *Milletia ferruginea* as used in the rivers in Ethiopia.

The siting of industries in relation to rivers is another vital matter with regard to water pollution because of the effluents discharged. In Kenya, for example, the dumping of molasses from sugar industries into rivers on which they are sited has led to serious pollution of the rivers. If the molasses are utilized in a secondary industry like rum production this can remove the danger. Many overseas firms which set up industries in developing countries often fail to observe the anti-pollution regulations that are normally demanded of them in their home countries. It is necessary for governments of developing countries to promulgate such laws to regulate this problem.

Air

We often forget the air as one of the essential natural resources of man. Pollution by means of soot and smoke from industries has made the air in some industrialized countries most unclean for man and animals. This has already started in many tropical countries. For example, in Zambia smoke containing 250 tonnes of sulphur dioxide is discharged daily into the air from the copper mines. Also, mining activities have produced dusty environments in a number of mining areas. Exhaust fumes from motor vehicles and aircraft also contribute to air pollution. The burning in the open of urban rubbish, the burning of vegetation in the dry season and dry dust in some countries like the Sudan and Ethiopia contribute to pollution of the air.

Conclusion

The problems of conservation which face tropical countries are immense. The resources involved are intimately inter-related and inter-dependent, so that their solution is to be seen in a total approach and not on a single-project basis. This requires careful consideration and the need to seek advice. But the trouble is often that developments are not thought through to their logical conclusions before they are begun, and the ecologists are only brought in to clear up the mess after the project is completed and things begin to go wrong. In many tropical countries these problems are also beset with a mixture of ignorance, apathy and ruthless exploitation for personal gain.

There are no easy solutions to these problems. To reverse or even halt the present catastrophic race to destruction will involve a major effort in education, law enforcement and re-thinking of traditional ideas.

The biosphere belongs to all mankind, and everyone should raise his voice against the pollution of it. It is the duty of all people on the earth to control this pollution, and the newly developing tropical countries should make constant representations against pollution to their own governments and internationally in order to galvanize all countries into action, so that the environmental deterioration that is occurring may be brought to an end.

References

Abbreviations

AN	*American Naturalist*
BJLS	*Biological Journal of the Linnean Society*
EAWJ	*East Africa Wildlife Journal*
JAE	*Journal of Animal Ecology*
JE	*Journal of Ecology*
JF	*Journal of Forestry*
JWASA	*Journal of the West African Science Association*

Adams, C. D. (1958), 'Autecological studies of some Ghana pteridophyta', PhD thesis, University of London.

Ahn, P. M. (1959), 'The savanna patches of Nzima, south-western Ghana', *JWASA*, vol. 5, no. 1.

—(1970), *West African Agriculture*, vol. 1, *West African Soils* (Oxford: Oxford University Press).

Aldous, A. E. (1943), 'Effect of burning on Kansas bluestem pastures', *Kansas Agricultural Experimental Station Technical Bulletin*, vol. 38, p. 65.

Ambasht, R. S., Maurya, A. N. and Singh, U. N. (1971), 'Primary production and turn-over in certain protected grasslands of Varanasi', paper presented to Symposium on Tropical Ecology, New Delhi.

Anderson, G. C. (1964), 'The seasonal and geographic distribution of primary productivity of the Washington and Oregon coasts', *Limnology and Oceanography*, vol. 9, pp. 284–302.

Anderson, O. E. (1960), 'The effect of low temperatures on nitrification of ammonia in Cecil sandy loams', *Soil Science Society of America, Proceedings*, vol. 24, pp. 286–9.

Baker, H. G. and Harris, B. J. (1959), 'Bat pollination of the silk-cotton tree, *Ceiba pentandra* (L), in Ghana', *JWASA*, vol. 5, pp. 10–25.

Bartholomew, W. V., Meyer, J. and Laudelont, H. (1953), 'Mineral nutrient immobilization under forest and grass fallow in Yangambi (Belgian Congo) region', *Publication de l'Institut National Pour l'Étude Agronomique du Congo Belge* (Brussels), Série Scientifique, vol. 57, p. 27.

Bates, M. (1945), 'Observations on climate and seasonal distribution of mosquitoes in eastern Columbia', *JAE*, vol. 14, pp. 17–25.

Bentley, J. R. and Fenner, R. L. (1958), 'Soil temperatures during burning related to postfire seedbeds on woodland range', *JF*, vol. 56, pp. 737–44.

Bernhard, F. E. and Huttel, C. (1971), 'Some responses of the vegetation to the seasonal changes of the climate in a rain forest of Ivory Coast', paper presented to Symposium on Tropical Ecology, New Delhi.

Berrie, A. D. and Vissler, S. A. (1963), 'Investigations of a growth-inhibiting substance affecting a natural population of fresh-water snails', *Physiology and Zoology*, vol. 36, pp. 167–73.

Bhatnagar, G. P. (1971), 'Primary organic production and chlorophyll concentration in Kille Back waters, Porto Novo (South India)', paper presented to Symposium on Tropical Ecology, New Delhi.

Biswell, H. H. (1956), 'Ecology of California grasslands', *Journal of Range Management*, vol. 9, pp. 19–24.

— and Lemon, P. C. (1943), 'Effect of fire upon seedstalk production of range grasses', *JF*, vol. 41, p. 844.

Blaisdell, J. P. (1953), 'Ecological effects of planned burning of sagebrush-grass range on the upper Snake River Plains', *United States Department of Agriculture Technical Bulletin*, no. 1075, p. 39.

Blydestein, J. (1963), 'Cambios en la vegetation despues de proteccion contra el Fuego', *Boletim de la Sociedad de Ciencias Naturales de Venezolana*, vol. 23, pp. 233–44.

Bossman, A. M. (1932), 'Cattle farming in South Africa', *South Africa Central News Agency Ltd. (de Shantz, H. L., 1947), East Africa Agricultural Service*, vol. 10, p. 458.

Botha, J. P. (1945), *Farming in South Africa* (Pretoria: Govt. Printer).

Boughey, A. S. (1968), *Ecology of Populations* (London: Macmillan).

Bourlière, F. (1963), 'Observations on the ecology of some large African mammals', *African Ecology and Human Evolution*, vol. 26, pp. 43–54.

Brammer, H. (1962), in *Agriculture and Land Use in Ghana*, ed. J. B. Willis (Oxford: Oxford University Press).

Bray, J. R. and Gorham, E. (1964), 'Litter production in forests of the world', in *Advanced Ecological Research*, vol. 2, ed. J. B. Gragg.

Braun-Blanquet, J. (1932), *Plant Sociology* (New York: McGraw-Hill).

Brynard, A. M. (1965), 'The influence of veld burning on the vegetation and game of the Kruger National Park', in *Ecological Studies in Southern Africa*, ed. D. H. Davis, pp. 371–93.

Buechner, H. K. (1961), 'Territorial. behaviour in Uganda kob', *Science*, vol. 133, pp. 698–9.

Burbridge, N. T. (1943), 'Ecological succession observed during regeneration of *Triodia pungens R. Br.* after burning', *Journal of the Royal Society of Western Australia*, vol. 28, pp. 149–56.

Burton, G. W. (1944), 'Seed production of several southern grasses as influenced by burning and fertilization', *Agronomy Journal* (American Society of Agronomy), vol. 36, pp. 523–9.

Chapin, J. P. (1932), 'The birds of the Belgian Congo. Part I', *Bulletin of the American Museum of Natural History*, vol. 65, pp. 1–756.

Chapman, G. P. (1970), *Patterns of Change in Tropical Plants* (London: University of London Press).

Chevalier, A. (1948), 'Biographie et écologie de la forêt dense ombrophile de la Côte d'Ivoire', *Revue Botanique*, appl. 28, pp. 101–15.

Cook, L. (1939), 'A contribution to our information on grass burning', *South African Journal of Science*, vol. 36, pp. 270–82.

— (1965), 'Note upon burning experiments at Frankenwald, Transvaal, South Africa', *Tall Timbers Fire Ecology Conference*, vol. 5, pp. 96–7.

Corbet, P. S. (1958), 'Lunar periodicity of aquatic insects in Lake Victoria', *Nature*, vol. 182, pp. 330–1.

Cornell, J. and Orias, E. (1964), 'The ecological regulation of species diversity', *AN*, vol. 98, pp. 399–414.

Cott, H. B. (1961), 'Scientific results of an inquiry into the ecology and economic status of the Nile crocodile (*Crocodilus niloticus*) in Uganda and Northern Rhodesia', *Transactions of Zoological Science* (London), vol. 29, pp. 211–356.

Coulter, J. K. (1957), in *Malayan Agriculture Journal*, p. 40.

Curtis, J. T. and Partsch, M. L. (1950), 'Some factors affecting flower production in *Andropogon geraridi*', *Ecology*, vol. 31, pp. 488–9.

Cushwa, C. T. and Redd, J. B. (1966), 'One prescribed burn and its effects on habitat of the Powhata Game Management Area', *United States Forest Service Notes*, Series 61, p. 2.

Daniker, A. V. (1929), 'Neu-Caledonien, land and vegetation', *Vierteljahrsschrift der Naturfurschenden Gesellschaft in Zürich*, vol. 74, pp. 170–97.

Dasman, R. F. (1964), *African Game Ranching* (London: Pergamon Press).

Daubenmire, R. (1968), 'Ecology of fire in grasslands', *Recent Advances in Ecology*, vol. 5, pp. 209–66.

Davis, D. E. (1945), 'The annual cycle of plants, mosquitoes, birds and mammals in two Brazilian forests', *Ecological Monographs*, vol. 15, pp. 243–95.

Dobzhansky, T. (1950), 'Evolution in the tropics', *American Scientist*, vol. 38, no. 2, pp. 209–21.

Dougall, H. W. and Glover, P. E. (1964), 'On the chemical composition of *Themeda triandra* and *Cynodon dactylon*', *EAWJ*, vol. 2, pp. 67–70.

Du Plessis, M. C. F. and Mostert, J. W. C. (1965), 'Run-off and soil losses at the Agricultural Research Institute, Glen', *South Africa Journal of Agricultural Science*, vol. 8, pp. 1051–60.

Duthrie, D. W. *et al.* (1937), 'Soil investigations in the Arena Forest Reserve, Trinidad'. *Imp. For. Instit. Paper* no. 6.

Edwards, S. D. C. (1942), 'Grass-burning', *Empire Journal of Experimental Agriculture*, vol. 10, pp. 219–31.

Eigenmann, C. H. (1912), 'The freshwater fishes of British Guiana, including a study of the ecological groupings of species and the relation of the fauna of the plateau to that of the lowlands', *Memoirs of Carnegie Museum*, vol. 5, no. 67, pp. 1–5, 78.

Evans, F. G. (1956), 'Ecosystems as the basic unit in ecology', *Science*, vol. 123, pp. 1127–8.

Evans, G. C. (1939), 'Ecological studies on the rain forest of southern Nigeria. II. The atmospheric environmental conditions', *JE*, vol. 27, pp. 436–82.

Evans, L. T., Wardlaw, I. F. and Williams, C. N. (1964), in *Grasses and Grasslands*, ed. I. C. Bernard (London: Macmillan), p. 102.

Ewusie, J. Y. (1968), 'Preliminary studies on the phenology of some woody species of Ghana', *Ghana Journal of Science*, vol. 8, pp. 126–50.

— (1969), 'Some observations of the annual pattern of flowering of some tropical woody plants', *Ghana Journal of Science*, vol. 9, pp. 74–8.

— (1972), 'Preliminary studies on the floral mechanism in *Sida stipulata*', *JWASA*, vol. 17, pp. 11–18.

— and Quaye, E. C. (1977), 'Studies on daily periodicity in some common flowers', *New Phytologist*, vol. 78, no. 2, pp. 479–86.

Fischer, A. G. (1960), 'Latitudinal gradients in organic diversity', *Evolution*, vol. 14, pp. 64–8.

Frankie, G. W., Baker, H. G. and Opler, P. A. (1974), 'Comparative phenological studies of trees in tropical wet and dry forests in the lowlands of Costa Rica', *JE*, vol. 62, pp. 881–919.

Gates, D. M. (1962), *Energy Exchange in the Biosphere* (New York: Harper & Row).

Gill, L. S. and Hawkesworth, F. G. (1961), 'The mistletoe: a literature review', *United States Department of Agriculture Forest Service Technical Bulletin*, no. 1242.

Glover, J. (1963), 'The elephant problem at Tsava', *EAWJ*, vol. 1, pp. 30–9.

Grant, S. A., Hunter, R. F. and Cross, C. (1963), 'The effects of muirburning Melinia-dominant communities', *Journal of the British Grassland Society*, vol. 18, pp. 249–57.

Greig-Smith, P. (1952), 'Ecological observations on degraded and secondary forest in Trinidad, British West Indies', *JE*, vol. 40, pp. 316–30.

Groat, T. B. (1969), 'Seasonal flowering behaviour in central Panama', *Annals of Missori Botanical Garden*, vol. 56, pp. 295–307.

Haddow, A. J. (1952), 'Field and laboratory studies on an African monkey, *Cercopithecus ascanius Matchie*', *Proceedings of the Zoological Society* (London), vol. 122, pp. 294–7.

Harris, B. J. and Baker, H. G. (1958), 'Pollination in *Kigelia africana Benth*', *JWASA*, vol. 4, no. 1, pp. 25–30.

Hart, G. H., Guilbert, H. R. and Goss, H. (1932), 'Seasonal changes in the chemical composition of range torage and their relation to nutrition of animals', *Colonial Agriculture Experimental Station Bulletin*, vol. 543, p. 62.

Hartland-Rowe, R. (1955), 'Lunar rhythm in the emergence of an ephemeropteron', *Nature*, vol. 176, p. 657.

Hasselo, H. N. and Swarbrick, J. T. (1960), 'The eruption of the Cameroon Mountain in 1959', *JWASA*, vol. 6, no. 2, pp. 96–101.

Heyward, F. (1938), 'Soil temperatures during forest fires in long leaf pine forest soils', *JF*, vol. 35, pp. 23–7.

Holdridge, L. R. (1967), *Life Zone Ecology* (San José, Costa Rica: Tropical Science Centre).

Holdsworth, M. (1961), 'The flowering of rain flowers', *JWASA*, vol. 7, no. 1, pp. 28–36.

Holtum, R. E. (1964), *Plant Life in Malaya* (London: Longman).

Hopkins, B. (1965), 'Observations on savanna burning in the Olokemiji Forest Reserve, Nigeria', *Journal of Applied Ecology*, vol. 2, pp. 367–81.

Jenik, J. and Hall, J. B. (1966), 'The ecological effects of the harmattan wind in the Djobobo massif (Togo Mountains, Ghana)', *JE*, vol. 54, pp. 757–79.

Jochems, S. C. J. (1928), 'De begrocling der tabak slanden in Deli en hare beteeknis voor de rebaks-cultuur', *Mededelingen*, 2nd series, no. 59.

Kaul, V., Zutshi, D. P. and Vyas, K. K. (1971), 'Biomass production in aquatic plants of different life-forms', paper presented to Symposium on Tropical Ecology, New Delhi.

Kellog, C. E. and Duval, F. D. (1949), *Publication de l'Institut National Pour l'Étude Agronomique du Congo Belge* (Brussels), Série Scientifique, vol. 46.

Kennoyer, L. A. (1929), 'General and successional ecology of the lower tropical rain forest at Barro Colorado island, Panama', *Ecology*, vol. 10, pp. 201–22.

Koelmeyer, K. O. (1959), 'The periodicity of leaf change and flowering in the principal forest communities of Ceylon. Part I', *Ceylon Forester*, new series, vol. 4, pp. 157–89.

Komarek, E. V., Snr (1965), 'Fire ecology – grassland and man', *Tall Timbers Fire Ecology Conference*, vol. 4, pp. 169–220.

Lamphrey, H. F. (1963), 'Ecological separation of large mammal species in the Tarangire Game Reserve, Tanganyika', *EAWJ*, vol. 1, pp. 63–92.

Laudelont, H. and Meyer, J. (1954), 'Les cycles d'éléments minéraux et de matière organique en forêt equatoriale congolaise', *Transactions of the 5th International Congress of Soil Science*, vol. 2, pp. 267–72.

Laws, R. M. (1966), 'Age criteria for African elephant', *EAWJ*, vol. 4, pp. 1–37.

Lawson, G. W. (1955), 'Rocky shore zonation in the British Cameroons', *JWASA*, vol. 1, no. 2, pp. 78–88.

Lebrun, J. (1936), *Repartition de la forêt equatoriale et des formations végétales limitrophes* (Brussels).

Lemon, P. C. (1949), 'Successional response of herbs in the longleaf-slash pine forest after fire', *Ecology*, vol. 30, pp. 135–45.

Lewis, C. E. (1964), 'Forage response to month of burning', *United States Forest Service Notes, Southeast*, SE-35, 4.

Lowe-McConnell, R. H. (1956), 'The breeding behaviour of Tilapia species (*Disces cichildane*) in natural waters: observations on *T. takoromo poll* and *T. variabilis bonlonger*', *Behaviour*, vol. 9, pp. 140–63.

— (1969*a*), *Speciation in Tropical Environments* (London: Academic Press).

— (1969*b*), 'Speciation in tropical freshwater fishes', *BJLS*, vol. 1, pp. 51–75.

MacArthur, R. H. (1965), 'Patterns of species diversity', *Biology Review*, vol. 40, no. 4, pp. 510–33.

— and Levins, R. (1967), 'The limiting similarity, convergence and divergence of co-existing species', *AN*, vol. 101, pp. 377–85.

MacDonald, W. W. (1956), 'Observations on the biology of chaorobids and chironomids in Lake Victoria and on the feeding habits of the "elephant snout fish" (*Momyrus kannune Forsk.*)', *JAE*, vol. 25, pp. 36–53.

Mall, L. P. and Singh, U. P. (1971), 'Seasonal variations in the standing biomass, total annual production and calorific energy of *Iseitema indigofera* community of grassland – Ujjain, India', paper presented to Symposium on Tropical Ecology, New Delhi.

Martin, F. J. and Doyne, H. C. (1932), *Soil Survey of Sierra Leone* (Freetown: Department of Agriculture).

Masson, H. (1949), 'La température du sol au cours d'une brousse au Sénégal', *Bulletin agricole du Congo Belge* (now *Bulletin agricole du Congo*), vol. 40, pp. 1933–40.

— (1954), 'Température du sol au cours d'un feu de brousse au Sénégal', mimeographed report to 2nd Inter-African Soils Conference, Leopoldville, 9–14 August 1954.

Mead, A. P. (1928), 'The forests of the Fiji islands', *Empire Forestry Journal*, vol. 7, pp. 47–54.

Medway, Lord (1972), 'Phenology of a tropical rain forest in Malaya', *BJLS*, vol. 4, pp. 47–146.
Meiklejohn, Jane (1955), 'The effect of brush burning on the microflora of a Kenya upland soil', *Journal of Soil Science*, vol. 6, pp. 111–18.
Mitchell, B. L. (1965), 'Breeding, growth and ageing criteria of Lichten Stern's hartebeest', *The Puku*, vol. 3, pp. 97–104.
Moore, A. W. (1960), 'The influence of annual burning on a soil in the derived savanna zone of Nigeria', *Transactions of the 7th International Congress of Soil Science*, vol. 4, pp. 257–64.
Mosby, H. S. (1963), *Wildlife Investigational Techniques*, 2nd ed. (Washington, DC: Wildlife Society).
Njoku, E. (1958), 'The photoperiodic response of some Nigerian plants', *JWASA*, vol. 4, no. 2, pp. 99–111.
Nye, P. H. (1959*a*), 'The relative importance of fallows and soils in storing plant nutrients in Ghana', *JWASA*, vol. 4, pp. 21–9.
— (1959*b*), 'Some effects of natural vegetation on the soils of West Africa and their development under cultivation', *UNESCO Abidjan Symposium. Tropical Soils and Vegetation*, pp. 59–63.
— (1961), 'Organic matter and nutrient cycles under moist tropical forest', *Plant and Soil*, vol. 8, pp. 333–46.
— and Greenland, D. J. (1960), 'The soil under shifting cultivation', Commu. Bur, Soils. Tech. Commun. 51–156.
O'Connor, K. F. and Powell, A. J. (1963), 'Studies in the management of snow-tussock grassland', *Journal of Agricultural Research*, vol. 6, pp. 354–67.
Odum, E. P. (1959), *Fundamentals of Ecology*, 2nd ed. (Philadelphia and London: W. B. Saunders).
Odum, H. T., Cantlon, J. and Kornicker, L. S. (1960), 'An organizational hierarchy postulate', *Ecology*, vol. 41, pp. 395–9.
Olaniyan, C. I. O. (1968), *An Introduction to West African Animal Ecology* (London: Heinemann).
Overland, L. (1960), 'Endogenous rhythm in opening and odor of flowers of *Cestrum nocturnum*', *American Journal of Botany*, vol. 47, p. 378.
Owen, D. F. (1966), *Animal Ecology in Tropical Africa* (Edinburgh and London: Oliver & Boyd).
Owen, G. (1951), *Journal of Soil Sciences*, vol. 2, p. 20.
Papadakis, J. (1965), *Crop Ecology Survey in West Africa* (Food and Agriculture Organization).
Patrick, R. *et al.* (1966), *The Catherwood Foundation Peruvian Amazon Expedition Limnological Studies*, Monograph of the Academy of Natural Science of Philadelphia, no. 14, pp. 1–495.
Pearse, A. S. (1939), *Animal Ecology*, 2nd ed. (New York: McGraw-Hill).
Pereira, H. E. (1957), *Journal of Agricultural Science*, vol. 49, p. 459.
Petrides, G. A. and Pienaar, U. de V. (1971), 'Calculation and ecological interpretation of prey preferences of large predators in Kruger National Park', paper presented to Symposium on Tropical Ecology, New Delhi.
Petrides, G. A. and Swank, W. G. (1965), 'Estimating the productivity and energy relations of an African elephant population', *Proceedings of the 9th International Grasslands Congress* (São Paulo).
Phillips, E. P. (1919), 'A preliminary report on the veld-burning experiments at Groenkloof, Pretoria', *South African Journal of Science*, vol. 16, pp. 285–99.
Phillips, J. (1966), *The Development of Agriculture and Forestry in the Tropics: Patterns, Problems and Promise* (London: Faber).
Polunin, N. (1960), *Introduction to Plant Geography* (London: Longman).
Pitot, A. and Masson, H. (1951), 'Quelques données sur la température au cours des feux des brousses aux environs de Dakar', *Institut Africain Noire Bulletin*, vol. 13, pp. 711–32.
Rachid-Edwards, M. (1956), *Universidade de São Paulo Faculdade de Filosofia, Ciencias e Letras Boletim 209*, Bot. 13, pp. 35–68.
Richards, P. W. (1952), *The Tropical Rain Forest* (Cambridge: Cambridge University Press).
Robocker, C. W. and Miller, B. J. (1955), 'Effects of clipping, burning and competition on establishment and survival of some native grasses in Wisconsin', *Journal of Range Management*, vol. 8, pp. 117–21.
Room, P. M. (1972), in *Cocoa Growers' Bulletin*, no. 18 (Bournville: Cadbury Ltd), pp. 14–18.
Rosenweig, M. (1968), 'Net primary productivity of terrestrial communities: prediction from climatological data', *AN*, vol. 102, pp. 67–74.
Russell, E. W. (1961), *Soil Conditions and Plant Growth*, 9th ed. (London: Longman).
Ryther, J. H. (1963), 'Geographic variations in productivity', in *The Sea*, ed. M. N. Hilled, vol. 2 (New York: Interscience).
Scott, J. D. (1934), 'Ecology of certain plant communities of the Central Province, Tanganyika Territory', *JE*, vol. 22, pp. 177–229.
— (1951), 'Conservation of vegetation in South Africa', pp. 9–12 in 'Management and Conservation of Vegetation in Africa', *Commonwealth Bureau of Pastures and Field Crops Bulletin*, vol. 41, 96.
Shaw, R. H. (1959), 'Bunch spear grass dominance in burnt pastures in southeastern Queensland', *Australian Journal of Agricultural Research*, vol. 8, pp. 325–34.
Shelford, V. E. (1929), *Laboratory and Field Ecology* (Baltimore: Williams & Wilkins).
Shidel, T. (1971), 'A preliminary survey on the soil properties of dry and moist forests of Thailand', paper presented to Symposium on Tropical Ecology, New Delhi.
Slatyer, R. O. (1967), *Plant–Water Relationship* (London and New York: Academic Press).
Smith, E. L. (1960), 'Effects of burning and clipping at various times during the wet season on tropical tall grass range in northern Australia', *Journal of Range Management*, vol. 13, pp. 97–203.

Strickland, A. H. (1947), 'The soil fauna of two contrasted plots of land in Trinidad, British West Indies', *JAE*, vol. 16, pp. 1–10.

Symington, C. F. (1933), The study of secondary growth on rain forest sites', *Malay Forester*, vol. 2, pp. 107–17.

Talbort, L. M. *et al.* (1965), 'The meat production potential of wild animals in Africa', *Commonwealth Agricultural Bureau Technical Communication* no. 16.

Tansley, A. G. (1935), 'The use and abuse of vegetation concepts and terms', *Ecology*, vol. 16, pp. 284–307.

Teixeira, C. (1963), 'Relative rates of photosynthesis and standing stock of the net phytoplankton and nannoplankton', *Boletim do Instituto Oceanografico da Universidade de São Paulo*, vol. 13, pp. 53–60.

Thompson, W. F. and Julia B. (1919), 'The spawning of the grunion (*Leuresthes tenuis*)', Calif. Fish and Game Comm. Fish Bull., vol. 3, pp. 1–29.

Thomson, J. A. (1911), *The Biology of the Seasons* (New York: Holt).

Thorp, J. and Smith, G. D. (1947), *Soil Science*, vol. 67, p. 117.

Vaughan, R. E. and Wiene, P. O. (1937), 'Studies on the vegetation of Mauritius. I. A preliminary survey of the plant communities', *JE*, vol. 25, pp. 289–342.

Vesey-Fitzgerald, D. F. (1960), 'Grazing succession among East African game animals', *Journal of Mammology*, vol. 41, pp. 161–72.

Vine, H. H. (1953), *Nigerian Agriculture Department Special Bulletin*, no. 5.

Vyas, L. N., Agarwal, S. K. and Garg, R. K. (1971), 'Biomass production by deciduous trees in Rajasthan. *Erynthia suborosa Roxb.*', paper presented to Symposium on Tropical Ecology, New Delhi.

Walter, H. (1973), *Vegetation of the Earth (in Relation to Climate and the Ecophysiological Condition)*, Heidelberg Science Library, vol. 15, trans. Joy Weiser (London: English University Press).

Wardlaw, C. W. (1931), 'Observations on the dominance of pteridophytes on some St. Lucia soils', *JE*, vol. 19, pp. 60–3.

West, O. (1965), *Fire in Vegetation and Its Use in Pasture Management with Special Reference to Tropical and Subtropical Africa* (Farnham, Bucks: Commonwealth Bureau for Pastures and Crops).

Williams, C. N. (1960), '*Sopuba ramosa*, a perennating parasite on the roots of *Imperata cylindrica*', *JWASA*, vol. 6, pp. 137–41.

Williams, R. C. and Webb, B. C. (1958), 'Seed moisture relationships and germination behaviour of acid-scarified Bahia grass seed', *Agronomy Journal*, vol. 50, p. 235.

Williams, C. B. (1951), 'The migrations of Libytheine butterflies in Africa', *Nigerian Field*, vol. 16, pp. 152–9.

General Index

Please note that specific geographical locations and genera have not been included.